# Diagnosing Plant Diseases Caused by Nematodes

**Malcolm C. Shurtleff**
Professor Emeritus
University of Illinois, Urbana-Champaign

Adjunct Professor
Texas A & M University, College Station

**Charles W. Averre III**
Professor Emeritus
North Carolina State University, Raleigh

**APS PRESS**

The American Phytopathological Society
St. Paul, Minnesota

Reference in this publication to a trademark, proprietary product, or company name is intended for explicit description only and does not imply approval or recommendation to the exclusion of others that may be suitable.

Library of Congress Catalog Card Number: 00-104287
International Standard Book Number: 0-89054-254-6



Printed in the United States of America on acid-free paper

The American Phytopathological Society
3340 Pilot Knob Road
St. Paul, Minnesota 55121-2097, USA

# Preface

This book focuses on the diagnosis of plant diseases caused by parasitic nematodes commonly encountered by specialists in agronomy, forestry, and horticulture. It is written for plant professionals, including farm advisers (county agents), plant scientists, diagnosticians, students of nematology and plant disease diagnosis, agribusiness personnel, agricultural inspectors, consultants, and others who diagnose diseases caused by plant-parasitic nematodes in the field, in clinics, and in laboratories. Readers should have considerable training in botanical and related sciences and experience in growing plants. This book should serve as a useful reference in diagnosing nematode-induced diseases and identifying these minute roundworms at least to the level of genus and, in two important genera (*Heterodera* and *Meloidogyne*), to the species or subspecies level.

Traditional field and laboratory methods are presented, based on direct observation of nematodes isolated from soil or plant tissue. Indirect methods, such as techniques using serology, polymerase chain reaction (PCR), or molecular probes, although valuable in repetitive assays for specific nematodes, are still in development and are not generally used in diagnostic laboratories. These molecular methods are beyond the scope of this book; readers interested in them should consult a research plant nematologist.

The descriptions of nematode genera and the accompanying illustrations in this book should enable readers to differentiate plant-parasitic nematodes from free-living (microbe-feeding) nematodes commonly found in the soil around plant roots and in symptomatic plant tissue. The information presented is intended to allow the identification of plant-parasitic nematodes to the level of genus. In most cases, to identify species within a genus, readers should consult the publications listed in the references at the end of Chapter 3.

All of the methods described in this book have been evaluated by one of the authors or by respected colleagues; all have potential use for diagnosticians. The nomenclature of nematodes and techniques for identifying them frequently change; the nomenclature used and the methods reviewed herein are not intended to be all-inclusive.

Chapter 1 gives a general description of physical characteristics of plant-parasitic nematodes (their size, morphology, and anatomy) and their means of survival, feeding, and dissemination. This chapter surveys the damage and yield losses due to plant-parasitic nematodes worldwide and describes symptoms of injury they cause in plants, and it outlines steps in diagnosing plant diseases caused by these pathogens.

Chapter 2 describes the collection and submission of soil and plant tissue samples for nematode analysis; methods of extracting nematodes from sample materials; laboratory procedures, including direct observation of nematodes in or on plant tissues, staining of plant tissues, isolation of nematode specimens, preparation of temporary and permanent mounts, and preservation and storage of specimens; and methods of inoculating plants to demonstrate pathogenicity.

Chapter 3 presents a classification of plant-parasitic nematodes, followed by descriptions of 31 genera, arranged alphabetically (from *Anguina* to *Xiphinema*). For each genus, a list of principal hosts, information about parasitism and habitat, body dimensions and descriptions of major morphological characteristics (the stylet, esophagus, vulva, reproductive structures, and tail), a list of similar genera, and general comments are presented. Detailed drawings of nematodes in each genus are included.

The appendix provides additional information about the five genera of nematodes implicated in transmitting plant-pathogenic viruses. It also discusses procedures for virus transmission tests.

An extensive glossary of nematological and related terms is included. The index makes it possible to locate drawings of one or more representatives of a particular genus and a description of its general characteristics.

Permission to reprint illustrations that have been previously published is gratefully acknowledged. Specific contributions are acknowledged in the captions.

The authors are indebted to many staff members for their advice, encouragement, and support in the preparation of this book. We especially thank Joseph N. Sasser and Donald P. Schmitt for kindly reviewing the entire manuscript, Dale I. Edwards for assistance in preparing the section on the collection and submission of samples for nematode analysis, and Kenneth R. Barker for assistance in preparing the section on laboratory methods. Donna Ellington and Zafar A. Handoo, curators of the Nematode Collection Database of the U.S. Department of Agriculture, were kind enough to supply us with some key references and the database for the plant-parasitic genera of nematodes described in this book. Special thanks go to Lenore Gray, who drew many of the illustrations. Joanne B. Averre and Freda L. Shurtleff helped with word processing. We would like to thank all of those involved at APS Press who made the publishing of this book possible.

Any suggestions that readers may have for improving this volume and making it more useful will always be welcomed.

Malcolm C. Shurtleff
Charles W. Averre, III

# Contents

CHAPTER 1

# Introduction to Plant-Parasitic Nematodes

Nematodes (also called nemas, eelworms, or roundworms), the most abundant multicellular animals on earth, are very active animals that belong in the phylum Nematoda (or Nemata) of the kingdom Animalia. They constitute a large and diverse group of unsegmented roundworms and occur virtually everywhere. Nematodes are found in large numbers in films of water in all natural soils, in fresh and marine waters from the polar regions to the equator, in humus and decaying plant tissues, and as parasites in moist tissues of most groups of animals, ranging from other nematodes, earthworms, insects, and molluscs to all types of livestock, household pets, and humans.

Many species that infect vertebrate animals and humans can be more than 5 cm long and are visible without magnification. Free-living microbe-feeding nematodes and plant-infecting species, most of which are in the range of 0.5 to 2.0 mm long, generally cannot be observed in detail without a microscope; some species (e.g., cyst nematodes) can be seen with a magnifying lens. Nematodes possess a high degree of similarity in structure, but with major specific morphologic variation, which is valuable for taxonomic identification. The chapters in this book focus on the plant-parasitic forms.

These organisms are mostly microscopic and generally translucent and thus nearly impossible to see in the soil, except for the females of some species, which swell and become pyriform, reniform, saccate, or lemon-shaped in the adult stage. Most plant-parasitic species are elongate and threadlike (vermiform) throughout their life cycles, and all are of this morphology when they hatch from the egg. Only about 10% of all described nematode species are plant parasites. About 15% of the approximately 20,000 known species are animal parasites, about 50% are marine nematodes, and 25% are free-living.

Different species of plant-parasitic nematodes can inhabit and infect most living parts of plants, including flower buds, leaves, stems, and roots. Over 3 billion nematodes can be present in 0.4 ha (1 acre) of land. Many species occur in the top 30 cm (12 in.) of soil, but some are more common at deeper levels, such as reniform nematodes in pineapple (up to 30 cm from the surface), root-knot nematodes in coffee (more than 30 cm from the surface), and dagger nematodes in grape (up to 1 m or more from the surface). A pound of soil in a cultivated fertile field with a heavy plant cover generally contains 10 to 20 species belonging to diverse genera, including ecto- and endoparasites, predators, and microbe-feeding or saprophytic types. The genera and common names of plant-parasitic nematodes described and illustrated in this manual are given in Table 1.1. Adult females of 24 of the common plant-parasitic genera are illustrated in Figure 1.1.

**Table 1.1.** Genera and common names of plant-parasitic nematodes described and illustrated in this manual (Chapter 3)

| Genus | Common name or names | Page |
|---|---|---|
| *Anguina* | Seed and leaf gall nematodes | 55 |
| *Aphelenchoides* | Foliar nematodes, leaf nematodes, bud nematodes | 57 |
| *Aphelenchus* | ... | 61 |
| *Belonolaimus* | Sting nematodes | 63 |
| *Bursaphelenchus* | Pinewood nematode, pine wilt nematode (*B. xylophilus*) | 65 |
| *Criconemella* (*Mesocriconema*), *Criconemoides* | Ring nematodes | 68 |
| *Ditylenchus* | Stem and bulb nematodes, potato rot nematode (*D. destructor*) | 70 |
| *Dolichodorus* | Awl nematodes | 74 |
| *Globodera* | Round cyst nematodes | 76 |
| *Helicotylenchus* | Spiral nematodes | 78 |
| *Hemicycliophora* | Sheath nematodes | 81 |
| *Heterodera* | Lemon-shaped cyst nematodes | 83 |
| *Hirschmanniella* | Rice root nematodes | 88 |
| *Hoplolaimus* | Lance nematodes | 91 |
| *Longidorus* | Needle nematodes | 93 |
| *Macrotrophurus* | ... | 96 |
| *Meloidodera* | Cystoid nematodes | 96 |
| *Meloidogyne* | Root-knot nematodes | 99 |
| *Nacobbus* | False root-knot nematodes | 111 |
| *Paratrichodorus* | Stubby-root nematodes | 113 |
| *Paratylenchus* | Pin nematodes | 116 |
| *Pratylenchus* | Lesion nematodes, root-lesion nematodes, meadow nematodes | 118 |
| *Radopholus* | Burrowing nematodes | 122 |
| *Rhadinaphelenchus* | Coconut red ring nematode (*R. cocophilus*) | 125 |
| *Rotylenchulus* | Reniform nematodes | 127 |
| *Rotylenchus* | Spiral nematodes, boxwood nematode (*R. buxophilus*) | 130 |
| *Scutellonema* | Spiral nematodes | 131 |
| *Sphaeronema* | ... | 134 |
| *Trichodorus* | Stubby-root nematodes | 113 |
| *Tylenchorhynchus* | Stunt nematodes, stylet nematodes | 134 |
| *Tylenchulus* | Citrus root nematode (*T. semipenetrans*) | 138 |
| *Xiphinema* | Dagger nematodes | 140 |

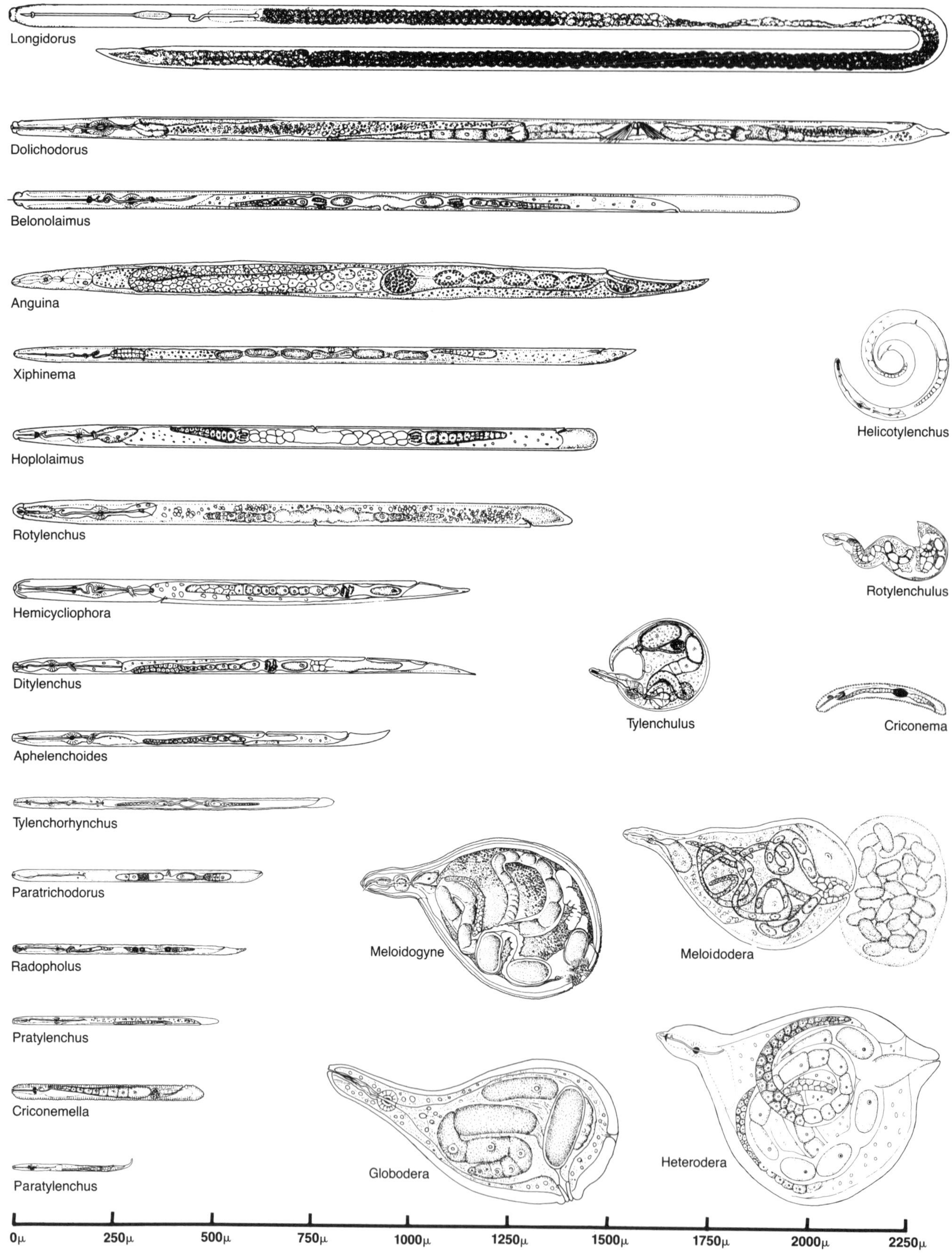


**Figure 1.1.** Morphology and relative size of adult female nematodes in 24 economically important plant-parasitic genera. The scale bar represents 2,250 μm (2.25 mm).

Problems caused by plant-parasitic nematodes are often encountered in specimens received in plant diagnostic clinics. Nematodes occur on plants growing in fields, prairies, forests, nurseries, orchards, turf, greenhouses, and containers. Their worldwide importance as major causes of diseases of a wide range of economic plant species and as constraints on production was not generally recognized until the mid-1940s. The general feeling had been that nematode problems were largely restricted to the tropics and subtropics, with root-knot nematodes being the primary culprit. Crop damage by nematodes is often attributed to other limiting factors (e.g., the air and soil environments, diseases, insects, mites, and large animals), especially in subtropical and tropical agriculture. Nematodes are often not considered or recognized as major limiting factors in growing crops until all other constraints on production have been removed.

Some species feed only on the outermost tissues of plants; some live and feed within tissues; a few stimulate the development of a specialized cell or group of cells adapted for feeding, referred to as nurse cells, giant cells, and syncytia. Damage to annual crops is related to the nematode population density in the soil at planting time. The relationship is somewhat different in perennial crops; damage from only a few nematodes is usually inconsequential but may be serious for future crops, whereas a large population may reduce the yield or aesthetic value of the host or even kill it. Estimates of annual crop losses due to plant-parasitic nematodes range from 3.3% for rye to 20.6% for tomato (Table 1.2). The annual worldwide loss caused by nematodes in crops consumed by humans is estimated at 11 to 14%, for a total of about $80 billion (Sasser, 1989).

Nematodes in the soil move slowly under their own power, perhaps only a few centimeters to about a meter per season. However, they are readily spread by any agency that moves or carries soil particles, including garden and farm equipment, as well as by irrigation, flood water, and drainage water. All types of animals and blowing dust may also spread nematodes. Long-distance spread occurs primarily by means of infected planting stock, nursery plants, soil or media mixes, contaminated farm tools and equipment, and farm produce. A few nematode species that feed on aboveground parts of plants are splashed from one plant to another by rain or overhead watering. Some species can move up wet shoot surfaces under their own power. Nematodes can also spread when infected plants contact adjacent healthy plants.

**Table 1.2.** Estimated annual yield losses of crop plants from plant-parasitic nematodes worldwide[a]

| Common name | Scientific name or names | Loss (%) |
|---|---|---|
| Banana | *Musa* spp. | 19.7 |
| Barley, common | *Hordeum vulgare* | 6.3 |
| Cacao, cocoa | *Theobroma cacao* | 10.5 |
| Cassava, manioc | *Manihot esculenta* | 8.4 |
| Chickpea, garbanzo bean | *Cicer arietinum* | 13.7 |
| Citrus | *Citrus* spp. | 14.2 |
| Coconut palm | *Cocos nucifera* | 17.1 |
| Coffee | *Coffea* spp. | 15.0 |
| Corn, maize | *Zea mays* | 10.2 |
| Cotton (lint only) | *Gossypium* spp. | 10.7 |
| Cowpea, southern pea | *Vigna unguiculata* | 15.1 |
| Eggplant, aubergine | *Solanum melongena* | 16.9 |
| Faba bean, broad bean | *Vicia faba* | 10.9 |
| Forage grasses and legumes | Many genera | 8.2 |
| Grapes | *Vitis* spp. | 12.5 |
| Guava | *Psidium guajava* | 10.8 |
| Melons | *Cucumis melo* | 13.8 |
| Millets | *Eleusine*<br>*Pennisetum*<br>*Setaria* | 11.8 |
| Oat | *Avena sativa* | 4.2 |
| Okra, gumbo | *Abelmoschus esculentus* | 20.4 |
| Ornamentals | Many genera | 11.1 |
| Papaya, pawpaw | *Carica papaya* | 15.1 |
| Peanut, groundnut | *Arachis hypogaea* | 12.0 |
| Pepper (bell or garden) | *Capsicum annuum* | 12.2 |
| Pigeon pea, red gram | *Cajanus cajan* | 13.2 |
| Pineapple | *Ananas comosus* | 14.9 |
| Potato | *Solanum tuberosum* | 12.2 |
| Rice | *Oryza sativa* | 10.0 |
| Rye | *Secale cereale* | 3.3 |
| Sorghum | *Sorghum bicolor* | 6.9 |
| Soybean | *Glycine max* | 10.6 |
| Sugar beet | *Beta vulgaris* | 10.9 |
| Sugarcane | *Saccharum officinarum* | 15.3 |
| Sweet potato | *Ipomoea batatas* | 10.2 |
| Tea, tea-plant | *Camellia sinensis* | 8.2 |
| Tobacco | *Nicotiana tabacum* | 14.7 |
| Tomato | *Lycopersicon esculentum* | 20.6 |
| Wheat | *Triticum* spp. | 7.0 |
| Yam, cinnamon-vine | *Dioscorea batatas* | 17.7 |

[a] This information (adapted from Sasser, 1989) is based on a survey of 371 nematologists worldwide.

## Survival of Nematodes

Soil populations of both microbe-feeding and parasitic nematodes are influenced by many factors. Soil moisture is a major one, especially when rapid and extreme changes occur in the soil environment. Since nematodes are aquatic animals, they require a film of free water coating soil particles or plant parts for motility and survival. Low soil moisture restricts their movement. Heavy rainfall or irrigation may flood the soil and cause oxygen depletion in soil water, causing high mortality in these aerobic organisms. Freezing and high temperatures generally cause a high death rate among soil-inhabiting nematodes. Plants with short life cycles typically overwinter as seeds and will not support nematode feeding and reproduction. Where such plants are present, nematode populations decline, except for species that enter dormancy as a survival strategy.

### NATURAL ENEMIES

Like all living organisms, nematodes have a myriad of natural enemies, including

protozoa, acting as external or internal parasites
predatory nematodes, of three distinct types, according to their various means of feeding:

- by swallowing the prey whole
- by puncturing or slitting the body wall of the prey and removing the internal organs
- by piercing the prey with a stylet (Figure 1.2) and withdrawing the body contents

other minute forms of animal life, including microscopic tardigrades (related to arthropods), oligochaetes (enchytraeids, a type of annelid worms), amoebae, mites, and collembolans (arthropods)

parasitic and predaceous fungi, which capture or destroy nematodes in various ways:

- by producing spores that become lodged in the stoma or adhere to the cuticle, germinate, and form an assimilative hyphal system, which ramifies throughout the body of the nematode
- by producing constricting hyphal loops or excreting adhesive substances to capture a victim and, in either case, perforating the cuticle, so that infective hyphae grow into the nematode and kill it
- by parasitizing nematode eggs or cysts

viruses

bacteria

Nematodes are unable to regenerate body parts destroyed by injury or disease.

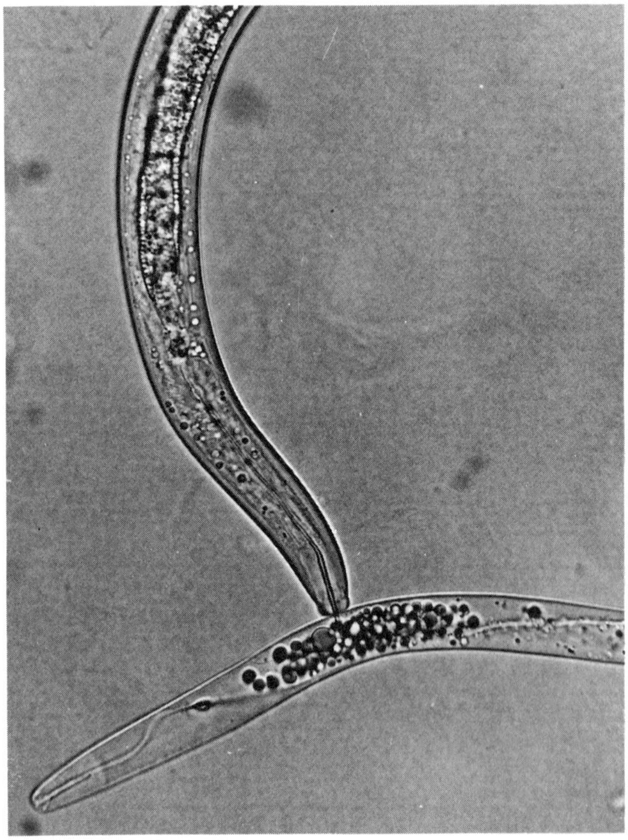

**Figure 1.2.** A carnivorous nematode of the genus *Seinura* feeding on a nematode of the genus *Aphelenchus.*

## Nematodes and Soil Ecology

Most nematodes observed in soils and rotting tissues are not plant-parasitic but feed on microorganisms, including bacteria, fungi, other nematodes (Figure 1.2), algae, protozoa, and minute animals such as tardigrades, rotifers, and small earthworms. Microbe-feeding nematodes (Figure 1.13H–L) obtain their nourishment from organisms associated with decaying organic matter or hyphae. Because they consume microorganisms, some of which cause plant disease, they may have a beneficial effect on the environment. They also influence nutrient cycling. In the soil, nematodes are part of the natural food web of organisms (protozoa, oligochaetes, mites, insects, fungi, bacteria, etc.) that reduce the organic remains of animals and plants to their primary constituents, which are eventually recycled into other organisms. Nematodes are now recognized as playing an important role in the ecology of soils and can be used as indicators of "soil health."

In agricultural soils, many nematode species are antagonistic to and help control organisms that cause important plant diseases. Nematodes have also shown promise as a means of managing certain difficult-to-control insects, such as black flies. A nonparasitic, bacteria-feeding nematode, *Caenorhabditis elegans,* has played an important role in biological systems such as embryological development.

To fully understand nematodes and the plant diseases they induce, a basic background in biology is desirable, including soil ecology, plant physiology, biochemistry, neurophysiology, genetics, and related scientific fields. Molecular biology is becoming increasingly important. All interactions between these animals and plants need to be studied to fully appreciate how they affect the production of agronomic and horticultural plants.

## Symptoms of Plant Injury Caused by Nematodes

Plant-parasitic nematodes are obligate parasites and must feed on living plant cells to complete their life cycles. Most species feed on root cells. They cause disease directly in various ways:

- by mechanically injuring or destroying plant cells and tissues
- by secreting or injecting enzymes and other toxic substances that interfere with the normal physiology of the plant
- by forming specialized feeding sites that physically disrupt the normal functioning of host vascular tissue

Some plant-parasitic nematodes cause tissues at feeding sites to swell by cell enlargement and proliferation, resulting in the formation of galls; some cause foliage to senesce prematurely; and some inhibit apical meristem development by suppressing cell division, resulting in "stubby roots." Plant-parasitic nematodes also cause diseases indirectly in several ways:

- by creating wounds at feeding sites, which provide entry for fungi and bacteria, rendering plants more susceptible to diseases (e.g., root and crown rots, wilts, cankers, and some foliar diseases)
- by creating galls, which provide a nutrient-rich environment for other pathogens
- by vectoring viruses that cause diseases (see Appendix, Table A.1)

The degree of damage depends on many factors, including nematode population density; stress due to drought, inadequate nutrition, or attacks by other pathogens and pests; and the sensitivity of the host.

Plant-parasitic nematodes interacting with other pathogens (bacteria, fungi, viruses, or other nematodes) often cause greater injury and crop loss than they would inflict by acting alone. Two or more pathogens working together induce a *disease complex.*

The symptoms of plant injury induced by nematodes vary with the kind of nematode and its population density, the plant species, the age and vigor of the plant attacked, and the location and extent of the injury. Nematodes cause mechanical damage when penetrating, feeding on, and moving within host tissues. Much of the injury is caused by the reactions of plant tissues to secretions injected by nematodes as they feed.

## Aboveground Symptoms

Symptoms in the shoots of plants infected with root-feeding nematodes include stunting, unthrifty growth, yellowing (chlorosis), early decline, wilting, and symptoms usually associated with nutritional and water deficiencies or excesses, infection by root-rotting fungi or viruses, or insect damage (Plates 1F, I–L, and N and 2K, L, P, and Q and Figure 1.3A–F). Severely infected plants will usually wilt sooner than healthy ones when moisture is limiting. In row crops, such as potatoes and soybeans, nematode-infected plants may fail to form a full canopy, often allowing weeds to grow profusely in affected areas. The leaves of some woody plants may be small and off-color, especially in the upper crown. Early defoliation and twig and branch dieback are other common symptoms. A few nematodes (e.g., *Heterodera glycines,* the soybean cyst nematode) can cause substantial yield losses in the absence of any aboveground symptoms.

Typically, symptomatic plants in the field are scattered in small areas (e.g., 4 by 8 m) to large areas (e.g., 20 by 30 m) in patches or streaks. The pattern of symptomatic plants may be associated with the distribution of nematodes in the soil (they are frequently clustered around growing feeder roots), the soil type, elevation, and cropping history. Usually there is a zone of gradual transition from severely affected to mildly affected to apparently healthy plants (Plate 1F–H). Other examples of field patterns have been described by Shurtleff and Averre (1997).

#### LEAF AND STEM SYMPTOMS

Nematodes that feed within leaf and stem tissues induce a variety of symptoms, including angular leaf spots, lesions, or rots caused by *Aphelenchoides* spp. and *Ditylenchus dipsaci* (Plate 1B, C, and G); internal browning of stems, petioles, and roots of coconut palms infected by *Rhadinaphelenchus cocophilus* (Plate 2Q) and pine infected by *Bursaphelenchus xylophilus* (Plate 1F); and root lesions caused by *Radopholus, Pratylenchus,* and *Criconemella* spp. (Plate 2N and O).

#### NECROSIS, DISCOLORATION, AND TWISTING

Nematodes feeding within leaf and stem tissues (*D. dipsaci* and *Aphelenchoides* spp.) may induce discoloration and severe stunting, twisting, and distortion.

#### BUD DEATH OR DEVITALIZATION

Nematode feeding often induces a proliferation of lateral buds, so that affected plants become bushy. A few species of nematodes feed on or in the terminal bud and kill it as the shoot emerges from the soil.

#### SEED OR LEAF GALLS

*Anguina* spp. produce galls in seed and leaves (Plate 1A), in which many juvenile nematodes can survive for years.

## Belowground Symptoms

The most common symptom of high populations of nematodes feeding on roots is a nonspecific decline of plants (Figure 1.3). The destruction of root cells results in less absorption of water and nutrients by the plants.

#### EXCESSIVE ROOT BRANCHING

Certain nematodes (*Meloidogyne hapla* and *Nacobbus* spp.), when feeding on young roots, stimulate the development of branch rootlets just above the feeding site. The result is excessive root branching (Plate 2A).

#### SUPPRESSION OF ROOT GROWTH

Some nematodes, including *Longidorus, Paratrichodorus* (Plate 2H–J), *Trichodorus,* and *Xiphinema* spp., feed on or close to root tips and inhibit their growth by suppressing cell division in the root tips. This commonly causes stubby roots and a reduction in the size of the entire root system. Plant growth is severely suppressed even though moisture and nutrients may be available. *M. hapla* can cause a stubby-root appearance. In some plant species, the typical response to root-knot nematodes is the formation of a terminal gall, causing the root to be stubby and club-shaped.

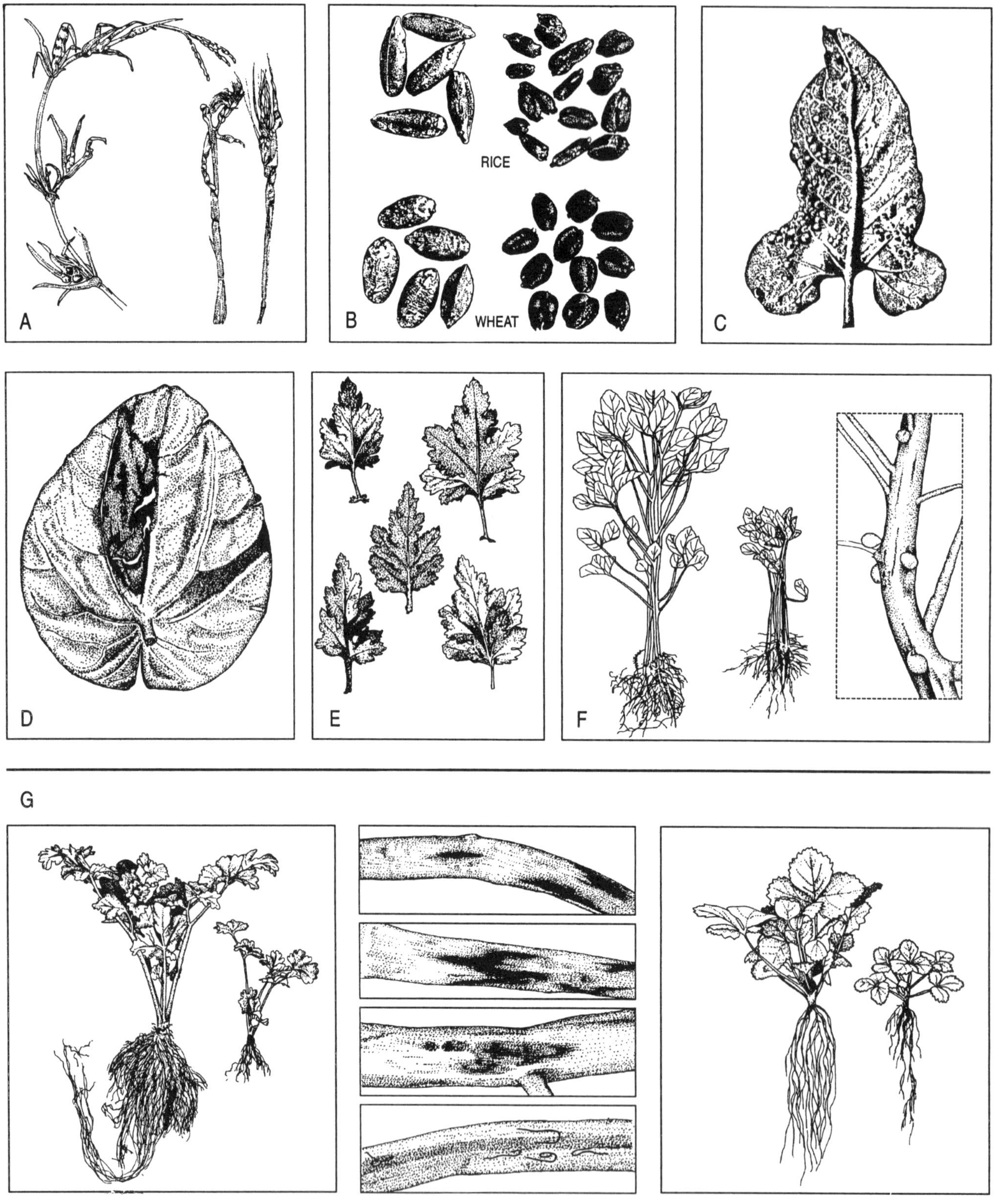


**Figure 1.3.** Examples of plant damage caused by plant-parasitic nematodes. **A,** *Cynodon transvaalensis* with galls caused by *Anguina tumefaciens* (left) and two growing wheat plants parasitized by *A. tritici* (right). **B,** Healthy kernels of rice and wheat (left) and kernels (cockles) infected by *A tritici* (right). **C,** Galls produced by *A. balsamophila* on a leaf of Oregon sunflower (*Balsamorhiza sagittata*). **D,** The lower surface of a peperomia leaf invaded by *Aphelenchoides ritzemabosi.* **E,** A healthy chrysanthemum leaf (center), surrounded by four leaves with sectors of necrotic tissue caused by *A. ritzemabosi.* **F,** A healthy soybean plant (left) and a plant infected by *Heterodera glycines* (right); cysts of *H. glycines* females on a soybean root (inset). **G,** Damage caused by lesion nematodes (*Pratylenchus* spp.): healthy and infected celery plants (left); lesions caused by *Pratylenchus* in roots (center), with nematodes feeding in the root at the bottom; and healthy and diseased strawberry plants (right).

#### ROOT PRUNING

Nematodes feeding near the root tip sometimes kill secondary roots or cause them to be abnormally short. If the roots attain some length before their growth is stopped, numerous short, stubby branches are formed, often in clusters. Affected plants may have few or no symptoms recognized as disease, e.g., lesions, discoloration, or necrosis. Nematodes in many genera cause this injury (Plates 1E and M and 2I, J, and O).

#### ROOT LESIONS

Injury due to feeding by *Criconemoides, Criconemella, Meloidogyne, Pratylenchus* (Plate 2M–O), and *Radopholus* spp. usually appears as small to moderate-sized, brown or black lesions in a restricted necrotic area. The result may be excessive root rotting and dieback of aboveground parts of the plant.

#### ROOT ROT (ROOT BROWNING OR BLACKENING)

Nematodes in many genera that feed on fleshy roots and tubers cause extensive damage resulting in decay. Further damage results from secondary bacteria and fungi of numerous genera that enter the wounded area.

#### ROOT GALLS

Root-knot nematodes (*Meloidogyne* spp.) (Plates 1N–P and 2A–F), *Xiphinema diversicaudatum*, sheath nematodes (*Hemicycliophora* spp.), and others can induce small to large galls on roots. The symptoms vary with the nematode species and its population, the plant species and the age of the plant, and the location of the injury on the plant. Roots infected by cyst nematodes (*Globodera, Heterodera,* and *Punctodera* spp.) may become necrotic within two to four weeks after infection.

Nematode galls on legume roots should not be confused with the nodules formed by nitrogen-fixing bacteria. Bacterial nodulation of legume roots can be reduced where cyst nematode feeding is moderate to severe, depending on the species and race of the nematode.

## Feeding Habits of Plant-Parasitic Nematodes

Plant-parasitic nematodes have several feeding habits, and juveniles, adult males, and adult females of the same species may differ in their habits. They have a hollow stylet, which is usually long and slender (Figures 1.2, 1.4, 1.7, and 1.8) and functions like a hypodermic syringe. Once a nematode has located a suitable host site, the stylet is thrust into a plant cell. Most plant-parasitic nematodes eject enzymes through the stylet to partially digest the cell contents. The partially digested contents are then withdrawn by the nematode through its stylet.

Many ectoparasitic nematode species move from one plant cell to another to feed. These ectoparasites are often called "browsers" or "grazers."

Other species, e.g., root-knot, cyst, and ring nematodes, are sedentary feeders. They establish specialized feeding sites, called nurse cells, giant cells, or syncytia, and a nematode feeds at a single site for the duration of its life. Juveniles of root-knot and cyst nematodes enter root tissue and move to the stele. Ring nematodes modify a cortical cell. Female root-knot and cyst nematodes become immobile once they have begun feeding, and they enlarge as they feed. Root-knot nematodes cause the formation of abnormally large cells (giant cells) at feeding sites. Cyst nematodes induce the formation of enlarged syncytia.

### Migration and Feeding of Parasitic Nematodes

1. **Ectoparasites** remain in the soil and penetrate host tissue with only the stylet or part of the head. They feed on cells near the surface of roots, including cortical cells near the vascular tissue. Ectoparasites may be
   a. **sedentary,** feeding at the same location, sometimes the same cell, for several days (e.g., *Criconemoides* or *Criconemella* [*Mesocriconema*], *Hemicycliophora,* and *Paratylenchus*), or
   b. **migratory,** feeding at many locations on surface or subsurface tissues of roots without becoming attached to the root surface (e.g., *Belonolaimus, Paratrichodorus,* and *Trichodorus*)
2. **Semiendoparasites** usually feed with the head and anterior end of the body embedded in host tissue and may be
   a. **sedentary** (e.g., *Rotylenchulus, Sphaeronema, Trophotylenchulus,* and *Tylenchulus*) or
   b. **migratory** (e.g., *Helicotylenchus, Hoplolaimus, Rotylenchus, Scutellonema,* and *Tylenchorhynchus*)
3. **Endoparasites** enter plant tissues completely or embed a large portion of the body in the host and may be
   a. **sedentary** (e.g., *Globodera, Heterodera, Meloidodera, Meloidogyne, Nacobbus,* and *Punctodera*) or
   b. **migratory** in plant tissues (e.g., *Anguina, Aphelenchoides, Bursaphelenchus, Ditylenchus,* some species of *Helicotylenchus, Hirschmanniella, Pratylenchoides, Pratylenchus, Radopholus,* and *Rotylenchoides*)

## General Description of Plant-Parasitic Nematodes

Nematode species exhibit considerable variation in size and in external and internal structures (Figures 1.4–1.15), which enables them to adapt to a wide range of environments. Most plant-parasitic species are threadlike, vermiform, or eel-shaped and round in cross section (Figures 1.1, 1.4, and 1.5). Nematodes are mostly translucent, except for the contents of the intestine (Figures 1.4 and 1.8), which are generally whitish to gray and semitranslucent. Despite their small size, nematodes have a number of highly specialized systems: digestive, secretory-excretory, reproductive, muscular, and nervous. They lack circulatory and respiratory organs.

Nematodes are covered externally by the cuticle, a more or less translucent, semipermeable, protective, noncellular

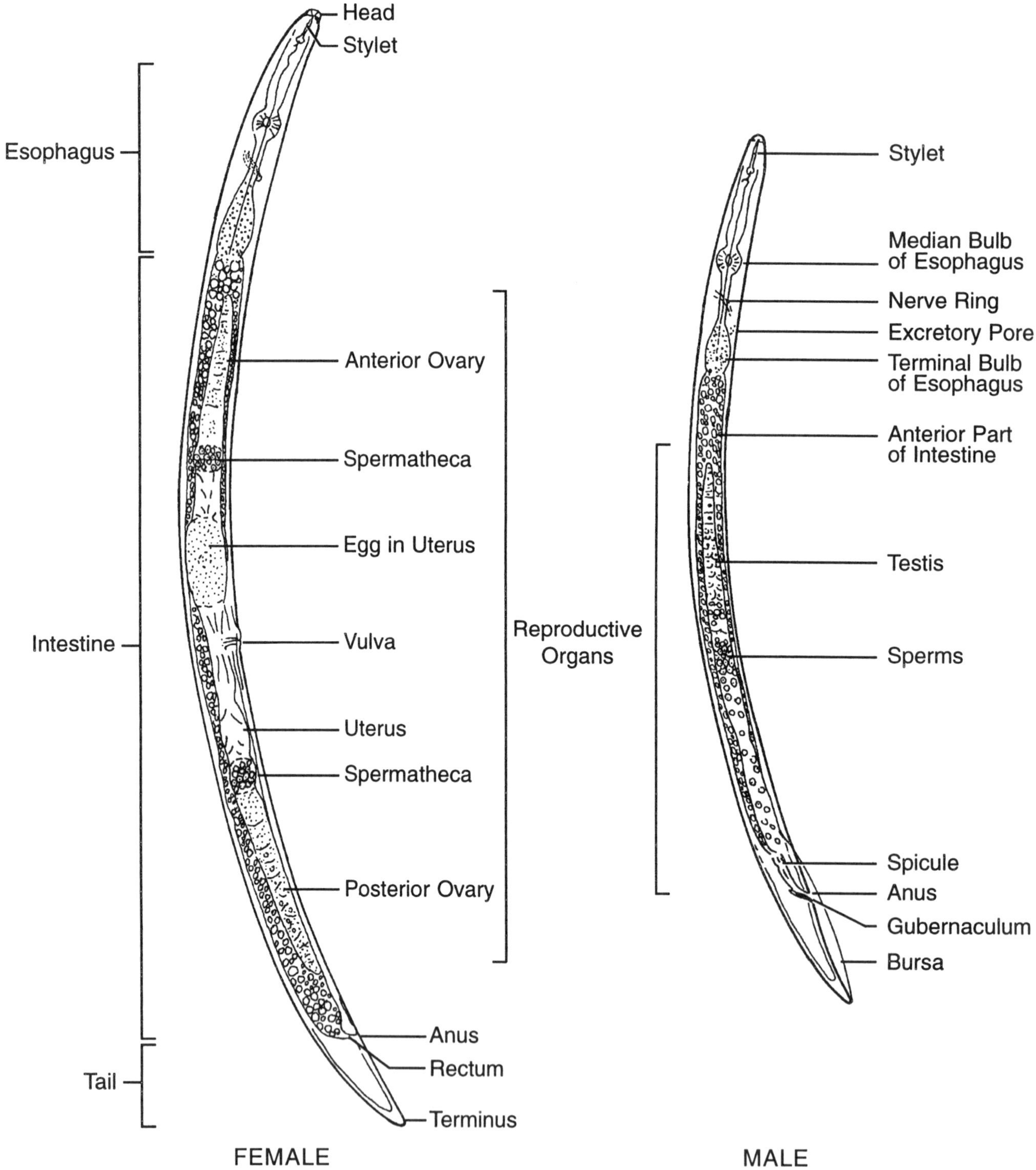


**Figure 1.4.** General anatomical characteristics of typical adult female (left) and male (right) plant-parasitic nematodes (order Tylenchida). Adult males are vermiform, usually smaller than females, never swollen or spheroid, and they have external structures on the tail. Juveniles (or larvae) have undeveloped gonads and are always vermiform when freshly hatched. The terminal gland (basal bulb) overlaps the intestine in some nematodes, in which it is called a basal gland lobe. (Adapted from Taylor, 1971)

covering, usually composed of three principal layers. The surface of the cuticle may be quite smooth or may be marked with dots, dashes, pores, scales, transverse or longitudinal striations, and spination (Figure 1.12). Transverse striations (annulations) encircle the dorsal and ventral surfaces of the nematode body, but in many species the annulations are interrupted on each side by a lateral field (Figure 1.12), bearing one to six or more longitudinal lines called incisures. The lateral fields in some species are raised in ridges or bands. The cuticle may appear partly inflated or winged, especially on the male tail, where it commonly forms lateral alae, which constitute the bursa (Figures 1.4 and 1.9), a caudal structure used in copulation. In some nematodes the cuticle forms projections on the tail, called mucros (Figure 1.14N), which may be quite elaborate and are taxonomically useful.

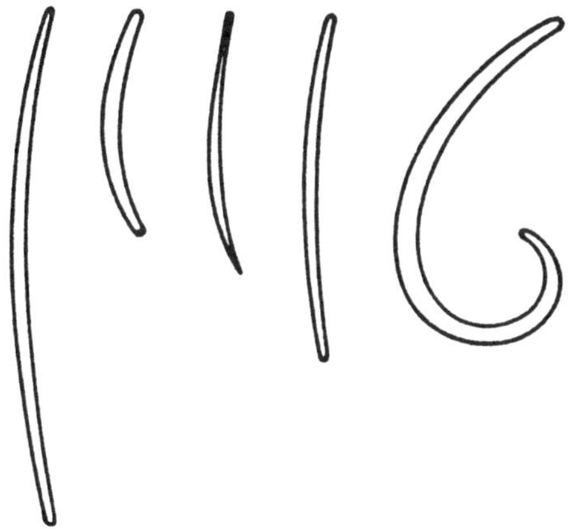

**Figure 1.5.** Various shapes of vermiform plant-parasitic nematodes. (Reprinted, by permission, from Taylor, 1971)

Most nematodes molt four times as they develop from egg to adult. In each molt, the cuticle and the structures formed from it, including the papillae (Figure 1.11), mouth (stoma), rectal linings, and stylet, are shed and replaced by a new cuticle, which grows beneath the old one (Figure 1.13M). The cuticle is composed primarily of collagen, a structural protein. The adult emerges after the fourth molt, grows further, and becomes sexually mature.

Beneath the flexible cuticle is a cellular layer, the hypodermis (or epidermis), which forms a protoplasmic tube extending into the body cavity, with four main longitudinal chords—the middorsal and midventral chords and two lateral chords (Figure 1.10). The hypodermis is a sheath of cells that secrete the cuticle and is apparently largely responsible for the overall architecture of the nematode. The body cavity contains fluid through which circulation and respiration take place.

The general musculature of the body is made up of a single layer of cells attached on their outer sides to the hypodermis and on their protoplasmic inner sides to the body cavity, or pseudocoelom, which is filled with fluid (Figure 1.10). Nematodes also have various specialized muscles, connected to the stylet (Figure 1.8), esophagus (Figure

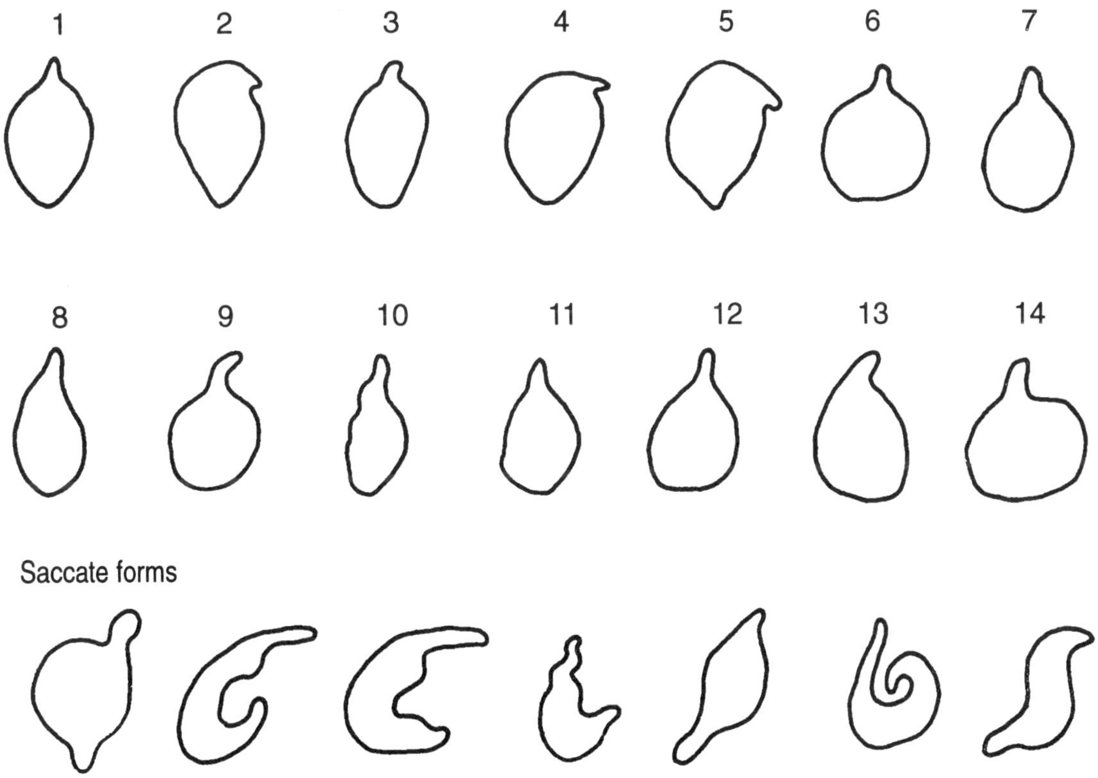


**Figure 1.6.** Variation in the shapes of adult female plant-parasitic nematodes that swell: lemon-shaped (1), globose (2–5, 8–10, and 12), spheroid (6, 7, and 14), pyriform (11 and 13), and saccate (bottom row). (Adapted from Taylor, 1971)

1.10), anus, uterus, vulva, spicules, and cloaca, and the male tail has a specialized musculature that aids in copulation.

The head consists primarily of the slitlike mouth, or stoma, and six lips—a subdorsal pair, a subventral pair, and two lateral lips (Figures 1.8 and 1.11). In many genera, however, the features of the head are greatly modified (Figure 1.13). In some the pairs of lips are fused, so that the nematode has three lips; in some the lips are of different sizes; in still others the head appears almost lipless or caplike. Some non-plant-parasitic nematodes have bristles, or setae, on the lips.

The amphids (Figure 1.11) are located behind the head on each side of the body or, in some species, on the lips. They are the largest and most complex of the cephalic sense organs, probably functioning as chemoreceptors. It is believed that they are important in enabling nematodes to recognize and infect host plants.

Amphids vary greatly in size and shape. They open to the exterior by a prominent pore or aperture. The position and shape of the aperture and the associated pouch of amphids are important in classification. The amphidial openings are in different locations in the two classes of plant-parasitic nematodes (for a classification of plant-parasitic nematodes, see Chapter 3). In the class Secernentea, to which most plant-parasitic nematodes belong, the opening is on the anterior end, usually between the labial disk and the lateral lips. In the class Adenophorea, the amphidial opening is posterior to the lips, down on the body.

The digestive tract (gut or alimentary canal) begins at the mouth, or stoma (oral aperture or buccal capsule or cavity) (Figures 1.4, 1.7, and 1.8), which in most nematodes is surrounded by lips or a single caplike head that bears the sensory structures. In plant-parasitic types, the mouth is armed with a conspicuous, protrusible, usually hollow, spearlike structure called a stylet (Figures 1.2, 1.4, 1.7, 1.8, and 1.13). Examples of stylets are shown in Figure 1.15.

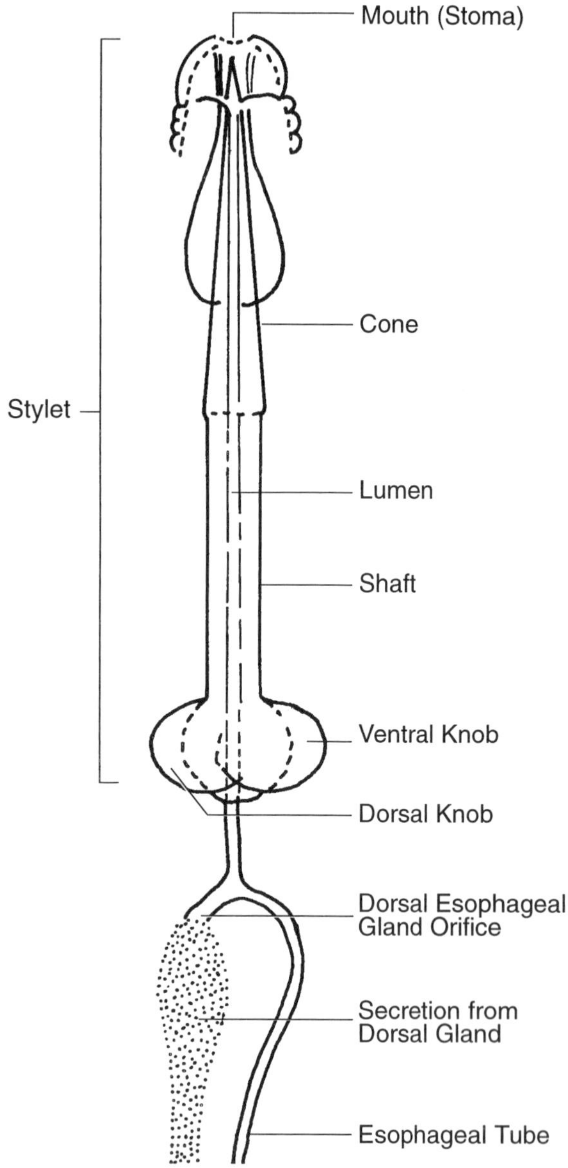


**Figure 1.7.** Stylet and associated structures of a plant-parasitic nematode. (Adapted from Taylor, 1971)

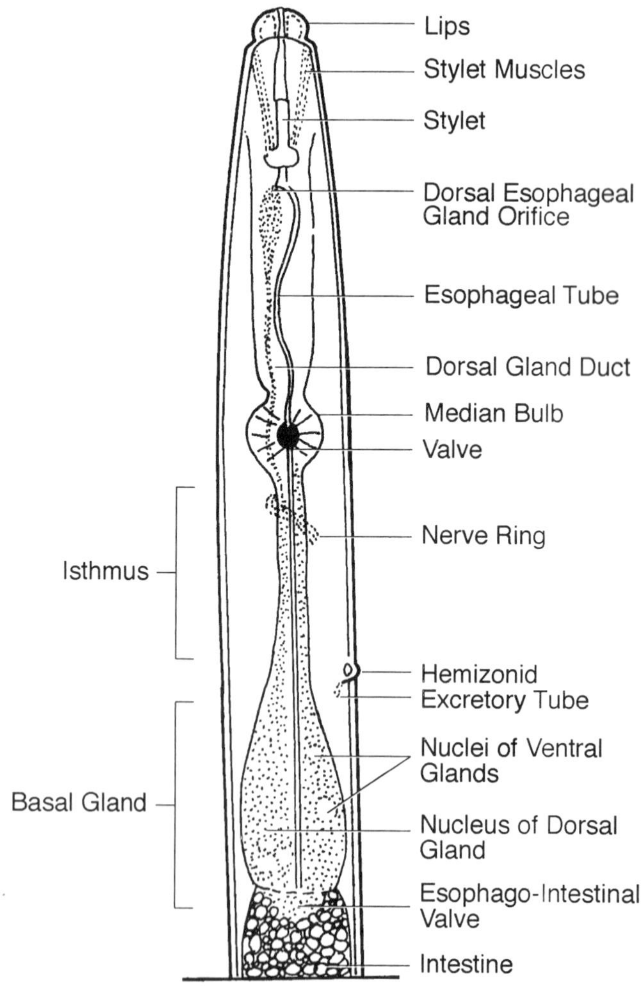


**Figure 1.8.** Anterior portion (head) of a plant-parasitic nematode. (Adapted from Taylor, 1971)

In the Secernentea, the stylet is believed to be derived from fusion of the stomatal lining and is therefore called a stomatostylet. It consists of an anterior cone, a cylindrical shaft, and three more or less rounded basal knobs, to which the stylet muscles are attached.

In the Adenophorea, the stylet is thought to be derived from a tooth and is therefore called an odontostylet. The posterior end of the odontostylet may have an extension and three flanges, which serve as attachment points for the protractor muscles of the stylet. In the plant-parasitic genera *Paratrichodorus* and *Trichodorus,* the odontostylet is curved ventrally, lacks flanges, and is not hollow; these nematodes secrete a hollow feeding tube from the stoma to connect with plants.

The stylet is contained within the cephalic framework, which is lightly to heavily cuticularized. This framework supports the anterior end of the nematode, and within it the stylet protractor muscles are attached. The central cylinder (vestibule) of the cephalic framework serves as a channel for guiding the stylet through the stoma. When a nematode feeds, its stylet is thrust through the oral aperture by contractions of the muscles attached to the stylet knobs (Figure 1.8).

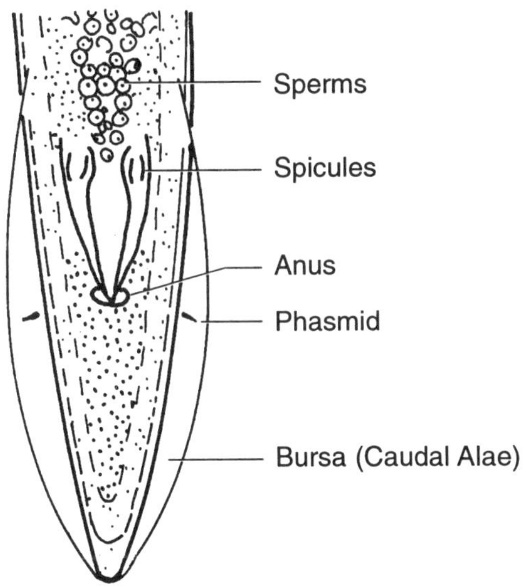


**Figure 1.9.** Tail of a male nematode, ventral view. (Adapted from Taylor, 1971)

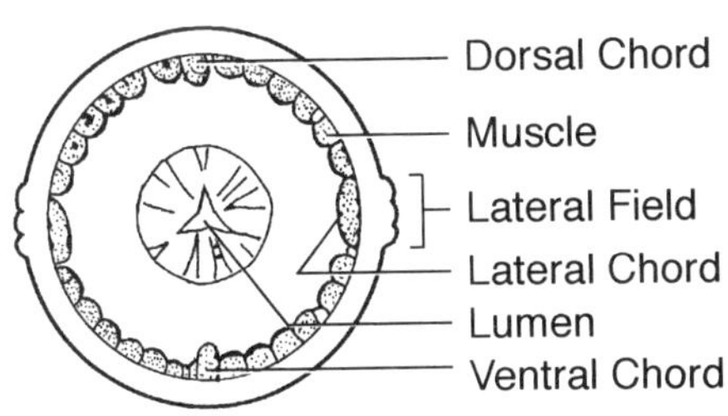


**Figure 1.10.** Cross section of a nematode at the middle of the median bulb in the esophagus, showing dorsal and ventral chords, lateral fields and lateral chords, muscles, the esophagus, and the triradiate structure of the valve. (Adapted from Taylor, 1971)

All plant-parasitic nematodes are equipped with a stylet during some or all of their life stages. Many non-plant-parasitic nematodes also have stylets, which in most cases are short and lack knobs (Figures 1.2 and 1.13H–L); these nematodes feed on fungi, minute animals, or other microorganisms. Predators do not have stylets and are not parasites of vascular plants.

The mouth leads to the cuticle-lined esophagus, called the pharynx in some zoological literature (Figures 1.4 and 1.8). The esophagus is essentially a muscular organ that pumps food into the body from the mouth, passes it backward into the intestine and posterior gut, and pumps saliva into the food source.

In the Adenophorea, the esophagus is divided into a narrow anterior procorpus and a broad posterior corpus. Three

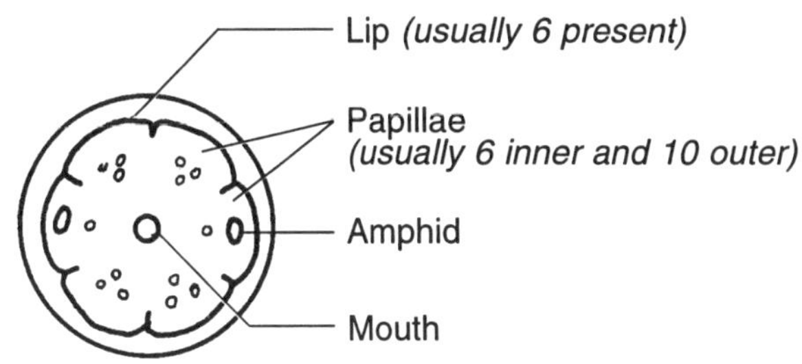


**Figure 1.11.** Head of a nematode as seen from the front (*en face* view), showing the mouth (stoma), papillae, and amphids. (Adapted from Taylor, 1971)

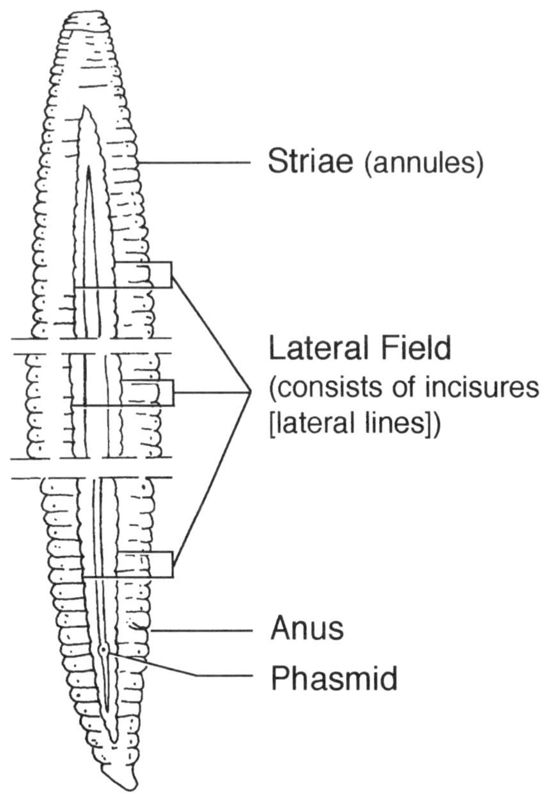


**Figure 1.12.** Cuticular markings at the anterior end, the middle of the body, and the posterior end of a nematode. (Adapted from Taylor, 1971)

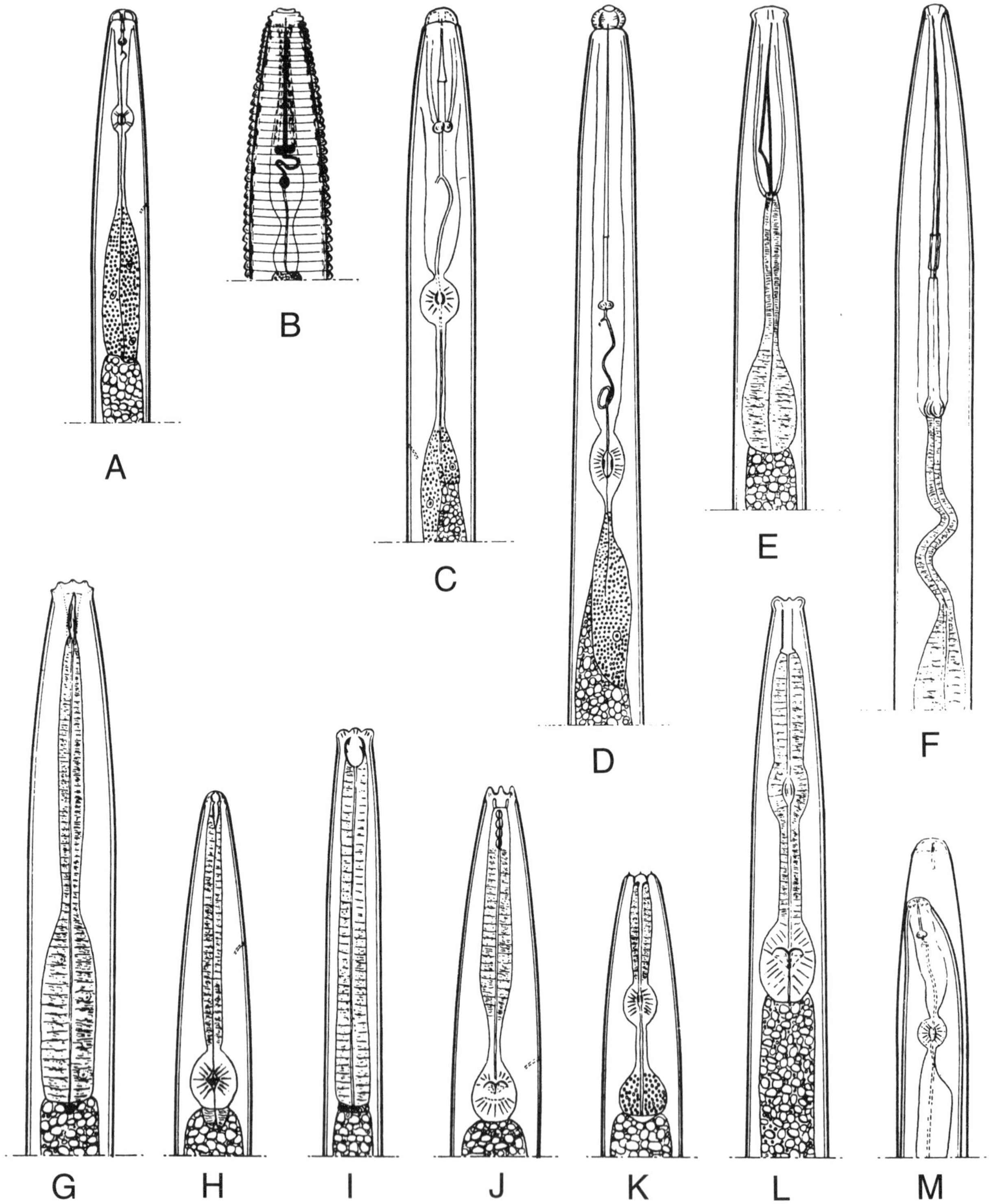


**Figure 1.13.** Variations in the anterior portion (head) of nematodes. Plant-parasitic nematodes are illustrated in **A–F** and **M**; free-living saprophagous or microbe-feeding nematodes in **G**, **H**, and **J–L**; and a free-living predatory nematode in **I**. The sizes and shapes of the stylets and esophagi in **A–F** are representative of many related genera. Nematodes having a typical tylenchoid esophagus, with a distinctive median bulb and stylets with knobs, are shown in **A–D.** Nematodes having a dorylaimoid esophagus, with no median esophageal bulb, are shown in **E–G.** Note the length of the stylet in relation to the width of the lips, the extent of cuticle annulation, and the shape of the lips (offset, flat, or following the contour of the body). **A,** *Ditylenchus.* **B,** *Criconemella.* **C,** *Helicotylenchus.* **D,** *Belonolaimus.* **E,** *Paratrichodorus.* **F,** *Xiphinema.* **G,** *Dorylaimus.* **H,** *Plectus.* **I,** *Mononchus.* **J,** *Cephalobus.* **K,** *Diplogaster.* **L,** *Rhabditis.* **M,** Anterior of a molting plant-parasitic nematode (similar to that shown in **A**) with the stylet attached to the molted cuticle. (Reprinted, by permission, from Taylor, 1971)

to five esophageal gland cells, one dorsal and two to four subventral, empty into the lumen. Each gland is a single large cell connected to the esophageal lumen by a valve. Secretions produced by the esophageal glands may function in egg hatching, penetration of host cells, migration in plant tissues, maintenance of discrete feeding sites, formation of feeding tubes by certain endoparasites, and digestion of cell contents.

In the Secernentea, the esophagus is divided into three distinct parts: a narrow procorpus; a broad, muscular metacorpus, often called the median bulb; and a basal gland lobe, often called the terminal or basal bulb (Figures 1.4 and 1.8). The corpus (procorpus and metacorpus) is usually cylindrical, but its lower end may be more or less swollen. Between the metacorpus and the terminal bulb is a narrow region, the isthmus (Figures 1.4 and 1.8). Many nematodes have various valves and three gland cells (one dorsal and two subventral), which produce digestive juices, located within these structures (Figures 1.4 and 1.8). In the order Tylenchida, the dorsal esophageal gland orifice (DEGO or DGO) is near the base of the stylet (Figures 1.7 and 1.8). In the order Aphelenchida, the DEGO empties into the median bulb.

A nerve ring, or commissure, composed largely of nerve fibers, encircles the esophagus (Figures 1.4 and 1.8), usually at the narrow isthmus. Generally, just posterior to the nerve ring is a clear area in the hypodermis on the ventral side, called the hemizonid (Figure 1.8). It extends through about three annules laterally around each side of the body to the lateral fields. The position of the hemizonid, which is

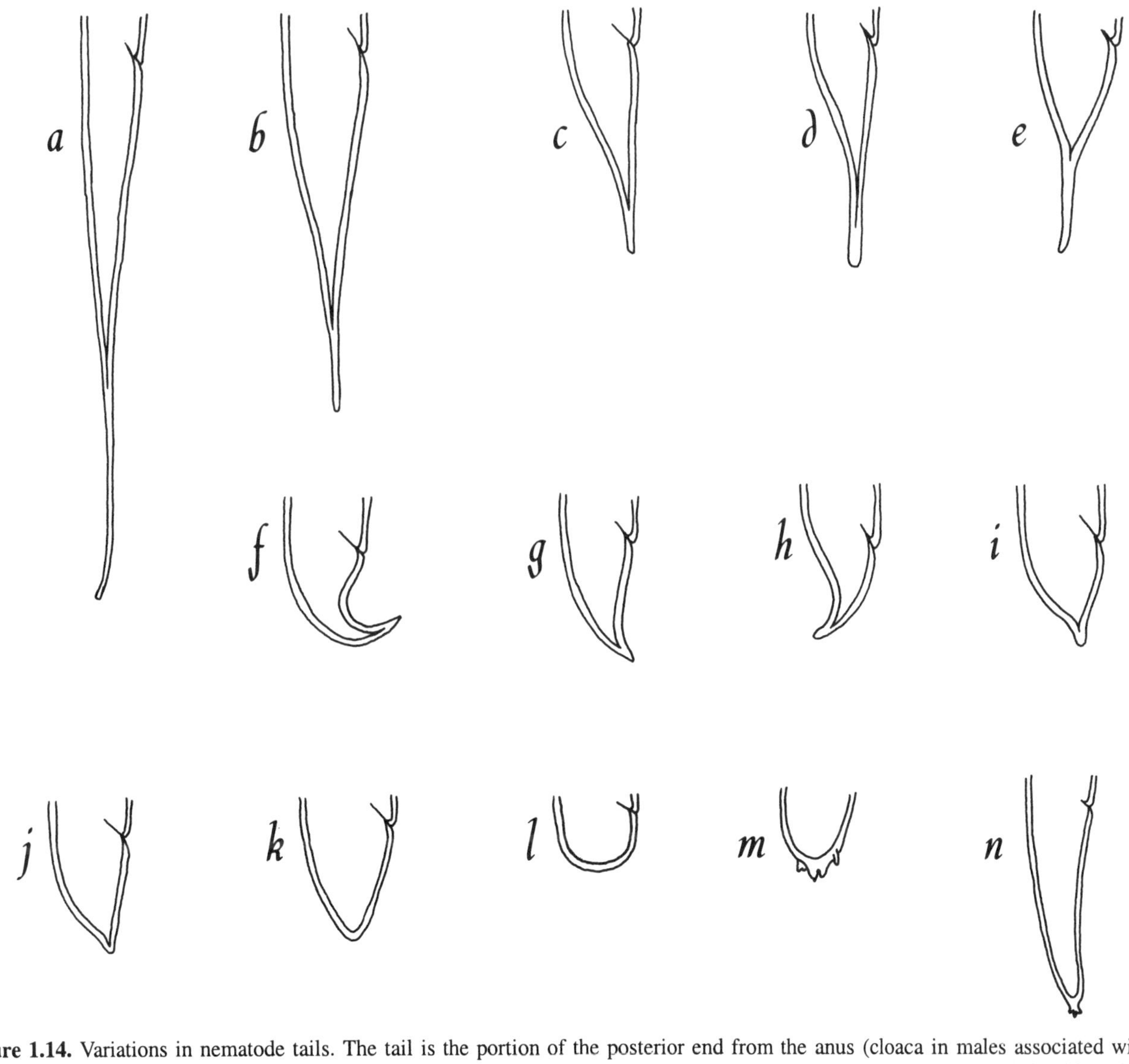


**Figure 1.14.** Variations in nematode tails. The tail is the portion of the posterior end from the anus (cloaca in males associated with the spicules) and the terminus. Some nematodes have terminal ornamentations (mucros), which are cuticular projections and should not be confused with spinnerets associated with the caudal glands of some non-plant-parasitic nematodes. **A,** Filiform. **B,** Attenuated. **C,** Dorsally convex-conoid, then cylindroid. **D,** Uniformly convex-conoid. **E,** Uniformly convex-conoid, spicate. **F,** Conoid, ventrally arcuate. **G,** Conoid. **H,** Conoid, recurved. **I,** Digitate. **J,** Subdigitate. **K,** Bluntly convex-conoid. **L,** Hemispherical. **M,** Multidigitate terminus. **N,** Mucronate (terminus with three mucros).

generally close to the excretory tube and pore, is taxonomically significant, but it is usually a secondary characteristic.

The anterior or cephalic sense organs consist basically of 16 sensillae arranged in three circles and a pair of laterally placed amphids (Figure 1.11). The labial sensillae in the inner circle, like the amphids, commonly open to the exterior and are thought to be chemosensory. The nerve ring receives sensory input from the head through six nerve bundles.

The major chemoreceptor organs in the tail are the usually small, paired phasmids in the class Secernentea (Figures 1.9 and 1.12) and a large sense organ at the terminus, or tip of the tail. The phasmids open to the exterior through a small pore.

Plant-parasitic nematodes in the class Secernentea have a tubular, multicellular excretory gland as part of their sensory-excretory (S-E) system (Figures 1.4 and 1.8). The terminal duct is lined with cuticle, opens through a pore to the exterior, and is commonly visible on the midventral surface of the body. The excretory duct and pore are generally located in the region of the nerve ring (Figures 1.4 and 1.8) and may vary considerably in a single species at different stages. The S-E system is believed to function in secretion and osmoregulation. In some nematodes the S-E system produces a gelatinous matrix, which is secreted around the eggs.

In nematodes in the class Adenophorea, the S-E system is composed of a single gland cell connected to a duct that is not distinctly lined with cuticle. In these nematodes the system functions only in secretion.

The posterior part of the esophagus is joined to the intestine by the esophago-intestinal valve, or cardia (Figure 1.8). The intestine is a simple tube, the walls of which are one cell thick. Digestion occurs in the intestine, and food is stored in its walls. The intestine ends posteriorly in the cuticle-lined rectum and opens at the anus in juveniles and females and at the anus or cloaca in males (Figures 1.4, 1.9, and 1.12).

The reproductive structures of both adult male and female nematodes are usually tubular and well developed within the pseudocoelom. They may be single (monodelphic) or paired (didelphic), straight or doubled back (reflexed), and posterior or anterior, or both when paired.

The female reproductive system (Figure 1.4) is made up of one or two ovaries, an oviduct, a uterus, vagina, and vulva; some species also have a round to elongate spermatheca, an enlarged structure between the oviduct and the uterus, in which sperm is stored. The ovary morphology is taxonomically useful and is diagnostic for many species. The muscular vagina leads to the ventral vulva, which is a transverse slit in many species but is pore-shaped or longitudinal in some species.

Males possess a pair of refractive, curved (arcuate), protrusible spicules, which are used for sperm transfer. The male reproductive system (Figures 1.4 and 1.9) is composed

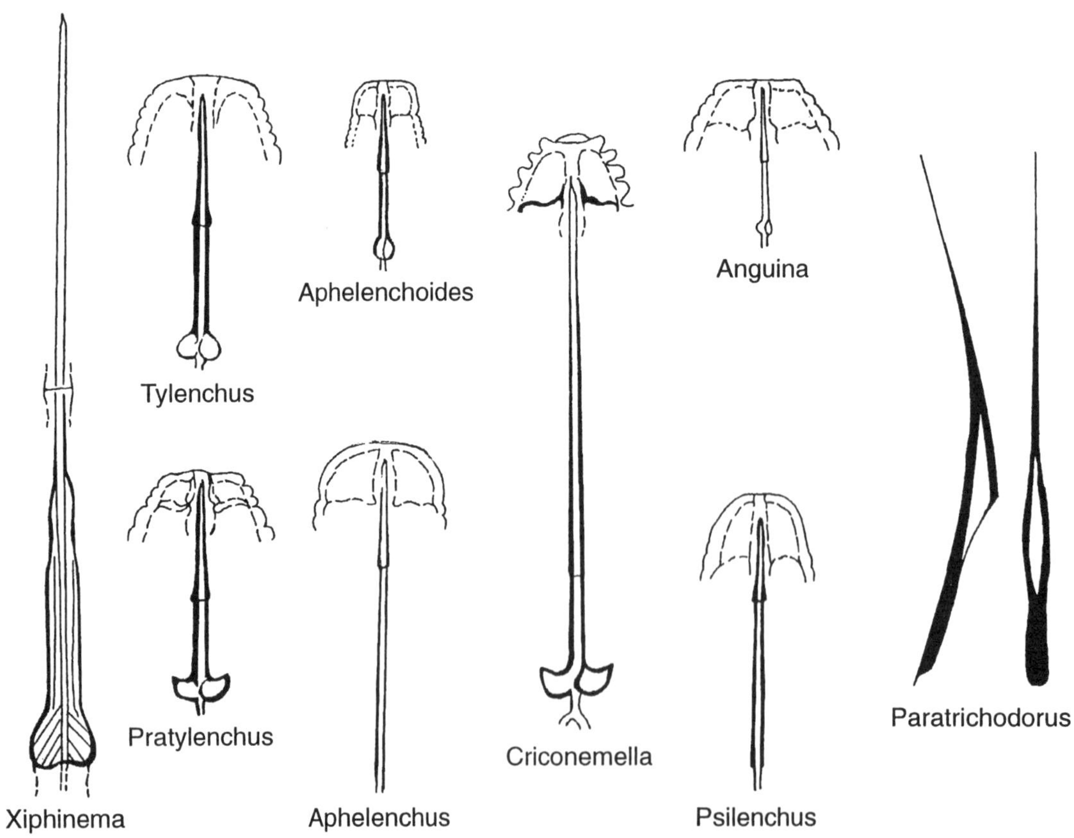


**Figure 1.15.** Various types of nematode stylets. *Psilenchus* and *Tylenchus* are free-living forms and may feed on root hairs.

of one or two testes, the copulatory spicules, lateral accessory pieces, and a gubernaculum; some species also have a bursa, consisting of lateral cuticular flaps called caudal alae. Spicule size and shape is often characteristic of certain groups of nematodes and is used for some identification purposes. Nematode sperm is amoeboid. Males that lack a bursa often have strong copulatory muscles, which cause the tail to curve ventrally when the nematode is killed by gentle heat.

The reproductive structures (or gonads), the ovaries and testes, are fully developed only after a nematode has molted four times and are indicators of sexual maturity. When the gonads develop, mature nematodes may become somewhat wider and longer but generally keep their threadlike shape. In certain genera, however, mature females become obese, pear-shaped, lemon-shaped, fusiform, reniform (kidney-shaped), or irregularly saccate (saclike) (Figures 1.1 and 1.6).

The eggs of most nematodes are generally ellipsoidal and have a translucent shell. They are similar in size regardless of the size of the female (Figure 1.1). The movement of eggs in the female reproductive tract is induced by uterine and body wall muscles. Egg laying in many cases is induced with the aid of a muscular ovejector, which squeezes the eggs out through the vagina. The hatching of eggs seems to be partly chemical and partly mechanical. Development and hatching of eggs may be delayed until they are induced by some environmental stimulus, such as exudates from host roots, or until a dormancy requirement has been met. Nematodes in early juvenile stages can often be observed inside eggs.

Juveniles often predominate in soil samples, but their features are less characteristic than those of adults, and it is generally difficult to identify them below the generic level.

The tail region varies greatly in nematodes of different genera and species (Figures 1.4, 1.9, and 1.14). Juveniles may differ from adults, and males from females, in the shape of the tail. In nematodes in the Secernentea, the tail region contains a pair of organs, called phasmids, one on each side of the body, which probably function as chemoreceptors. They are located in the lateral field (Figures 1.9 and 1.12) and are often difficult to see. Each phasmid consists of a small duct that leads from a single cell and opens to the surface. Nematodes in the Adenophorea do not have phasmids.

## Generalized Life Cycles of Nematodes

Most plant-parasitic nematodes have quite similar life cycles. They develop from eggs through four larval or juvenile stages to become adults. Juveniles in many genera are similar to adults in appearance and structure except for sexual characters. In most species, nematodes hatch from the egg as second-stage juveniles (J2), although in a few genera they hatch as first-stage juveniles (J1). Juveniles commonly move through water films (0.2 to 0.5 mm thick) surrounding soil particles and plant surfaces, searching for a suitable host plant and a site at which to feed.

Between successive stages of development, nematodes molt, shedding the old cuticle and forming a new cuticle, as they increases in size and develop sexually. The process of molting involves three steps: 1) *apolysis,* the separation of the cuticle from the hypodermis or epidermis; 2) the formation of a new cuticle from the outermost surface of the epidermis; and 3) *ecdysis,* the shedding of the old cuticle. Differentiation of tissues and growth of nematodes occur between molts.

Most species feed at every stage after hatching, but there are important exceptions in which nematodes do not feed in some stages. In certain species (e.g., *Rotylenchulus reniformis*) the males and even the juveniles never feed.

In some species, the adults mate, and the females lay fertilized eggs. In other species, males are unknown or very rare and are not necessary for egg fertilization. Those that mate are called amphimictic. Parthenogenetic species do not require mating.

Nematodes of most species usually complete a life cycle in about three to six weeks during the growing season if sufficient moisture is present and if the soil temperature is in the optimum range for the species, usually 20–30°C (68–86°F). However, some species may take up to a year or more to complete the cycle. A few plant-parasitic species (e.g., *Bursaphelenchus xylophilus,* the pinewood nematode) can complete a life cycle in three to seven days.

Juveniles of some species sometimes enter a quiescent stage (e.g., J2 of *Anguina* spp. and J4 of *Ditylenchus dipsaci*) and are thus able to withstand desiccation for long periods, sometimes many years. Other nematodes (e.g., *Bursaphelenchus*) enter a quiescent state in which they retain the cuticle of the previous stage as a loose sheath; in this state they are called dauer larvae. Juveniles of other species (e.g., *Paratylenchus*), lacking a fully formed stylet and esophagus, can survive in soil for long periods without feeding. J2 in eggs of some species (e.g., *Heterodera glycines,* the soybean cyst nematode) enter a dormant state and require long-term chilling before they will hatch. Root exudates from a nearby host plant induce molting to the next feeding stage of the soybean cyst nematode and other species.

## Diagnosing Plant Diseases Caused by Nematodes

As is generally the case in diagnosing plant diseases, the diagnostician confirms a tentative diagnosis based on the plant host involved, the disease signature, root and aboveground symptoms, and species of nematodes known to be in the area. The diagnostician verifies the presence of the pathogen, rather than trying to "key out" nematodes associated with diseased tissues. In some cases, the clinician must be able to *extract* nematodes from plant tissues and soil, *identify* them, and *know* their relative ability to cause damage (Chapters 2 and 3). Information in the following text, the illustrations, and the glossary is presented as a guide to the diagnostician.

Experienced clinicians routinely identify common plant-parasitic nematodes to the level of genus using a dissecting microscope at magnifications of 10 to 30×. A research microscope, with magnifications of up to 1,000× (Figure 1.16), and special staining and mounting procedures are necessary for observing many of the minute anatomical features of nematodes, which are especially important in the identification of species.

The following steps are suggested for diagnosing diseases caused by plant-parasitic nematodes and identifying nematodes to the level of genus. Where disease control is based on crop rotation and planting resistant varieties or cultivars, it may be necessary to identify a nematode to the level of species (e.g., species of *Heterodera* and *Meloidogyne*) or even to subspecies or race (e.g., races of *H. glycines* and *M. incognita*). The key features used to confirm the identity of a nematode are its size, the proportion of length and width of the body, shape and length of the stylet, the type of esophagus, the position of the vulva, the configuration of the gonads, and the shape of the tail.

1. Identify the host and check a host index for a listing of nematodes reported to cause disease in that species. For most diagnosticians, the possibility of discovering a new nematode-incited disease is remote. Printed lists of nematodes on a wide range of plants are readily available. Identification aids can also be found on the Internet.
2. Under a magnification of 20 or 30×, characterize the nematodes associated with symptomatic tissue or extracted from the rhizosphere.
3. The following may be helpful in deciding whether a nematode is a plant parasite:

   a. Plant-parasitic nematodes have stylets (Figures 1.1, 1.4, 1.7, 1.8, 1.13, and 1.15). However, some non-plant-parasitic nematodes seen in diagnostic clinics

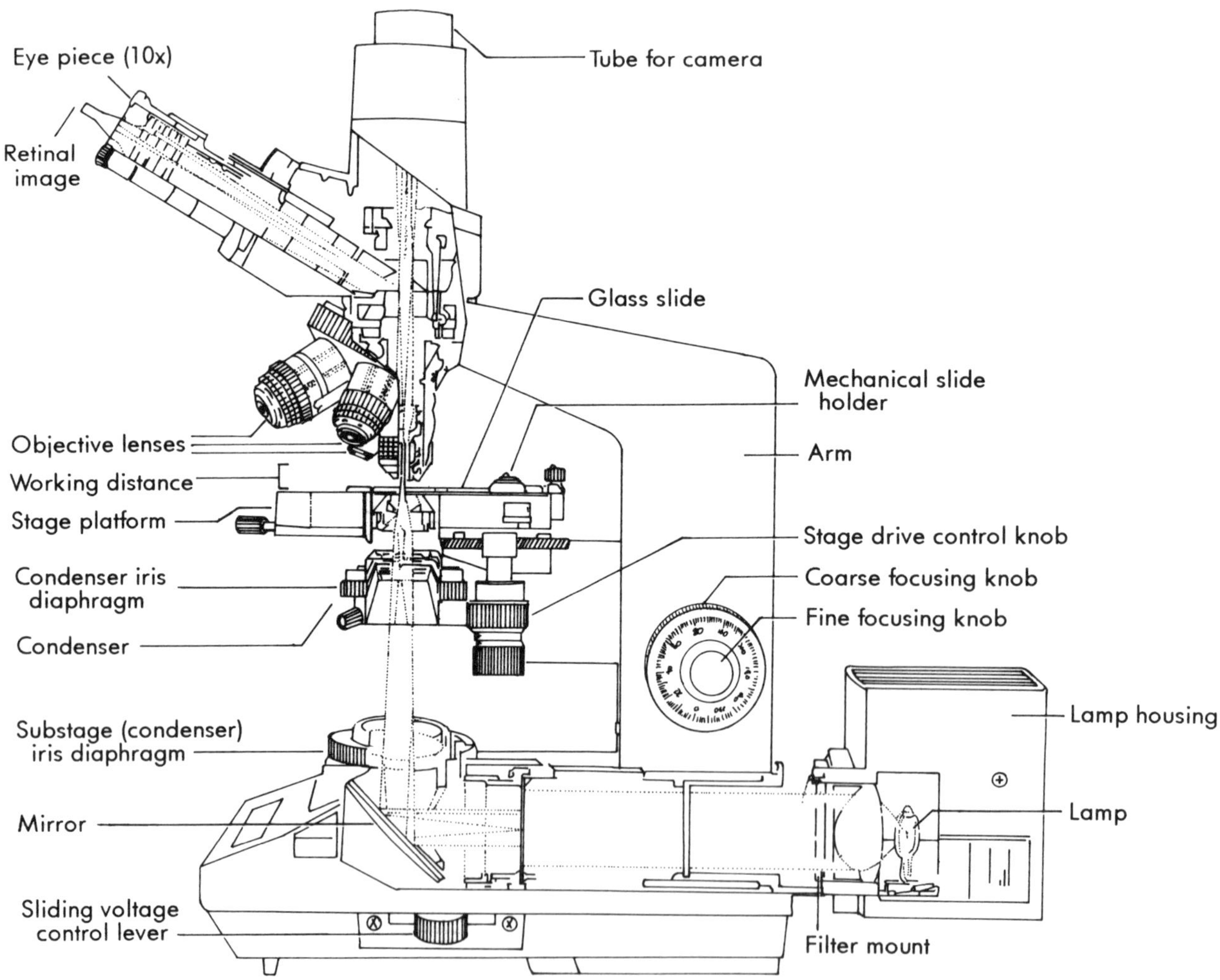


**Figure 1.16.** Basic compound (light) microscope used in many plant diagnostic clinics. Microscopes with phase contrast, dark-field ultraviolet fluorescence, and other capabilities are desirable. Simpler monocular microscopes may be used to see many taxonomically important features of nematodes.

also have stylets (Figures 1.2 and 1.15 and Table 3.6). Nonparasitic nematodes usually have very short stylets without knobs and feed on fungi, algae, insects, other minute soil arthropods, or even plants without causing diseases. Disregard forms without stylets.

b. Most plant-parasitic nematodes have a slow, undulating movement in the dorsoventral (lateral) plane, and one group has an accordion-type movement. Rapidly moving organisms are usually free-living nematodes or other microscopic animals.

c. Plant-parasitic nematodes are not segmented and do not have hairs or appendages (Figures 1.1 and 1.4). However, the cuticle of some nematodes (e.g., *Criconemella* and *Criconema*) is strongly annulated and may appear segmented (Figures 1.1 and 1.13B).

d. Most plant-parasitic nematodes are in the range of 0.5 to 2.0 mm or more in length and are thin and cylindrical (vermiform) with tapering ends (Figure 1.5). However, some sedentary mature females are enlarged and globular or have some other shape (Figures 1.1 and 1.6).

e. Nematodes seen within or partially within living (nonnecrotic) plant tissue are likely to be plant pathogens.

4. Determine whether the nematodes are adults or juveniles. Adults have differentiated gonads. Adult females have a prominent vulva and may have well-developed eggs in the uterus. Mature eggs, both inside and outside the female, may contain a recognizable J1 or J2. Adult males have prominent exterior structures (spicules and sometimes a bursa) near the tail (Figures 1.4 and 1.9). It is important to recognize and differentiate J2 of cyst and root-knot nematodes. Males are rare or even absent in some species of plant-parasitic nematodes.

5. Determine the approximate morphological dimensions. Measurements are usually expressed as a range. If a single measurement is given, assume that the range is on the order of ±10%. Dimensions may be given in relation to some convenient body structure (see the next section, on de Man's measurement code). The following morphological dimensions should be determined:

   a. The length of nematodes of both sexes. Plant-parasitic and free-living nematodes vary greatly in length, from less than 0.3 mm (e.g., the pin nematode *Paratylenchus nanus*) to over 11 mm (e.g., *Paralongidorus maximus*). Most species are in the range of 0.5 to 2.0 mm (500 to 2,000 μm) (Figure 1.1 and Tables 3.2–3.6).

   b. The length of the stylet in relation to the width of the body at the stylet knobs (Figures 1.1 and 1.13). *Short* stylets are less than 2.4 times the body width at the base of the stylet; *long* stylets are more than 2.5 times the body width at the base of the stylet; *intermediate* stylets lie between the two. Also, examine the stylet for the presence of basal knobs or flanges (Figures 1.1 and 1.15). Match the stylet, where possible, to one observed in Figures 1.1 and 1.15. A copy of *Plant-Parasitic Nematodes: A Pictorial Key to Genera* (Mai and Mullin, 1996) should be in the clinic.

6. Determine the shape of the esophagus and the presence or absence of a median bulb containing a sclerotized valve (Figures 1.1, 1.4, 1.8, and 1.10).

7. Determine the approximate distance of the vulva from the lips as a percentage of the length of the body. A vulva near midbody (V = 50; see the next section, on de Man's measurement code) normally indicates the presence of two ovaries; a vulva near the anterior end of the body (e.g., V = 35) or the posterior end (e.g., V = 80) normally indicates the presence of one ovary (Figures 1.1 and 1.4). The ovary may be straight (outstretched) or doubled back (reflexed).

8. Note the shape of the tail: long, pointed, rounded, or blunt (Figures 1.1, 1.4, 1.9, and 1.14). The tail region of the adult female is simple and smooth, with only an anus. The tail area of the male has sclerotized spicules (a copulatory structure) and the associated gubernaculum, and in many plant-parasitic species a winglike structure, the bursa (caudal alae), is present in addition to the anus or cloaca (Figures 1.4 and 1.9). The length of the tail is measured from the anus to the tail tip (terminus) in females and from the anus or cloaca to the terminus in males.

9. Consult Tables 3.2–3.6 to assist with the identification of a specimen. Select five or more nematodes of the proper length. Disregard those that do not fit the characteristics described in the Remarks columns of these tables. Compare the nematodes to the line drawings and descriptions, arranged alphabetically by genus, in Chapter 3.

**CAUTION**

Be sure the nematodes examined are adult females. For species in which males have important distinguishing characteristics, male nematodes can be examined, when present, for clarification.

Examine several nematodes—five or more, if possible.

Nematode samples, especially those collected from crop soils, nearly always contain two or more types of parasitic and free-living nematodes.

## de Man's Measurement Code

The diagnostician should be familiar with de Man's formulas, a widely used measurement code for stating nematode dimensions. Measurements start at the anterior tip. For example, the dimensions of a stunt nematode, *Tylenchorhynchus claytoni,* are stated as follows:

female adult

♀: L = 0.64–0.73 mm; a = 24; b = 5–6; c = 18–19; $V = {}^{29-30}55\text{–}57^{20-25}$

male adult

♂: L = 0.57 mm; a = 26; b = 5.8; c = 14.6; T = 55–68

where

L = total body length (in millimeters or micrometers). The letter L is often omitted in the code.

a = total body length (in millimeters or micrometers) divided by the maximum body width, which is usually at the vulva. For long, slender nematodes the value of a is large. In this example, males are slightly more slender than females.

b = total body length divided by the distance from the lips to the base of the esophageal glands. This ratio is influenced by the fixation of specimens, their nutritional status, and other factors and is therefore of minor value. In the example above, the total body length of males is 5.8 times the distance from the lips to the base of the esophageal glands, and thus this distance is 0.098 mm ([0.57 mm]/5.8). For nematodes with a long esophagus, stylet, or stoma, the value of b is low.

b′ = total body length divided by the distance from the anterior end of the body to the posterior end of the esophageal glands, in nematodes in which the glands overlap the intestine.

c = total body length divided by the length of the tail, measured from the anus or cloaca to the tail tip (terminus). In the example, the total body length of males is 14.6 times the length of the tail, which is thus 0.039 mm long ([0.57 mm]/14.6). For nematodes with a long tail, the value of c is low.

c′ = tail length divided by the body width at the anus or cloaca. This ratio gives a more precise indication of the length of the tail in proportion to the size of the whole body and is less variable than the ratio c.

V = distance of the vulva from the anterior end divided by the body length and then multiplied by 100 to yield a percentage. In the example, the vulva is slightly posterior to the midsection (V = 55–57, or 0.35–0.42 mm from the lips). The raised prefix and suffix numbers ($V = {}^{29\text{–}30}55\text{–}57^{20\text{–}25}$) give the lengths of the anterior and posterior reproductive structures, respectively, as a percentage of body length.

T = length of the testis as a percentage of body length.

The length of the stoma or stylet (and the stomatostylet in tylenchids) is given in micrometers. The dorylaimid spear is composed of an anterior odontostylet and a basal portion, the odontophore, which is measured separately. The lengths of the spicule and gubernaculum are given in micrometers. Other units of measurement are used for convenience and are defined when used. For example, "tail 2 a.b.w." means that the length of the tail is two times the anal body width, or the width (diameter) of the body at the anus.

Living individuals that are anaesthetized provide the clearest view of nematode structures. Permanent microscope slide mounts in glycerin (glycerol) are commonly used for taxonomic descriptions (see Chapter 2 for methods used in the preparation of specimens). The key, tables, and illustrations in Chapter 3 emphasize diagnostic features that distinguish genera and species from closely related nematodes.

### Measuring Nematodes

Accurate measurements are essential for describing and identifying nematodes. Small but consistent differences are often important for distinguishing species. Measurements of short, straight features (e.g., stylet length and body width) are usually made by examining suitably mounted (not flattened) nematodes under a high-power compound microscope with an eyepiece scale, calibrated with a stage micrometer for each objective. A screw micrometer eyepiece, similarly calibrated, is also used for more accurate measurements. A graduated mechanical stage and fine-focusing knob can be used for rough measurements, to within 100 μm or less, and do not need to be calibrated.

For measuring curved whole specimens or structures, first draw them. Most microscopes can be fitted with some form of drawing attachment (e.g., a camera lucida), with which the nematode and the drawing are visible at the same time. The curved drawings can then be measured with a map measurer or a piece of curved wire. Measurements are then obtained by calibrating or comparing with a stage micrometer scale drawn *at the same magnification* as the nematode. Southey (1986) and Thorne (1961) have presented good discussions of making drawings of nematodes for reproduction.

For critical observations of the head and other structures, researchers commonly supplement the compound microscope by using a scanning electron microscope or transmission electron microscope. Because some of the key distinguishing characters of nematodes are at the limits of resolution of the compound microscope, even the best optical instruments are barely adequate.

## References

Agrios, G. N. 1997. Plant Pathology. 4th ed. Academic Press, San Diego.

Ayoub, S. M. 1980. Plant Nematology: An Agricultural Training Aid. Rev. ed. NemaAid Publications, Sacramento, Calif.

Barker, K. R., Pederson, G. A., and Windham, G. L., eds. 1998. Plant and Nematode Interactions. American Society of Agronomy, Madison, Wisc.

Bird, A. F., and Bird, J. 1991. The Structure of Nematodes. 2nd ed. Academic Press, San Diego.

Chitwood, B. G., and Chitwood, M. B., eds. 1950. An Introduction to Nematology. 1974 reprint. University Park Press, Baltimore.

Christie, J. R. 1959. Plant Nematodes: Their Bionomics and Control. University of Florida Agricultural Experiment Station, Gainesville.

Dropkin, V. H. 1989. Introduction to Plant Nematology. 2nd ed. John Wiley & Sons, New York.

Evans, K., Trudgill, D. L., and Webster, J. M., eds. 1993. Plant Parasitic Nematodes in Temperate Agriculture. CAB International, Wallingford, Oxon., U.K.

Goodey, J. B. 1963. Soil and Freshwater Nematodes. 2nd ed. Methuen, London.

Luc, M., Sikora, R. A, and Bridge, J., eds. 1990. Plant Parasitic Nematodes in Subtropical and Tropical Agriculture. CAB International, Wallingford, Oxon., U.K.

Maggenti, A. 1981. General Nematology. Springer-Verlag, New York.

Mai, W. F., and Mullin, P. G. 1996. Plant-Parasitic Nematodes: A Pictorial Key to Genera. With photographs by H. H. Lyon and K. Loeffler. 5th ed. Comstock Publishing, Cornell University Press, Ithaca, N.Y.

Nickle, W. R., ed. 1984. Plant and Insect Nematodes. Marcel Dekker, New York.

Nickle, W. R., ed. 1991. Manual of Agricultural Nematology, Marcel Dekker, New York.

Sasser, J. N. 1989. Plant-Parasitic Nematodes: The Farmer's Hidden Enemy. Department of Plant Pathology and Consortium of International Crop Protection, North Carolina State University, Raleigh.

Shurtleff, M. C., and Averre, C. W., III. 1997. The Plant Disease Clinic and Field Diagnosis of Abiotic Diseases. American Phytopathological Society, St. Paul, Minn.

Southey, J. F., ed. 1978. Plant Nematology. 3rd ed. Ministry of Agriculture, Fisheries and Food, GD1. Her Majesty's Stationery Office, London.

Southey, J. F., ed. 1986. Laboratory Methods for Work with Plant and Soil Nematodes. 6th ed. Ministry of Agriculture, Fisheries and Food, Tech. Ref. Book 402. Her Majesty's Stationery Office, London.

Taylor, A. L. 1971. Introduction to Research on Plant Nematology: An FAO Guide to the Study and Control of Plant-Parasitic Nematodes. Rev. ed. Food and Agriculture Organization of the United Nations, Rome.

Thorne, G. 1961. Principles of Nematology. McGraw-Hill, New York.

Zuckerman, B. M., Mai, W. F., and Rohde, R. A., eds. 1971. Plant Parasitic Nematodes. Vols. 1 and 2. Academic Press, San Diego.

Zuckerman, B. M., and Rohde, R. A., eds. 1981. Plant Parasitic Nematodes. Vol. 3. Academic Press, San Diego.

CHAPTER 2

# Methods

## Collection and Submission of Soil and Plant Samples for Nematode Analysis

The roots of virtually every plant are usually infected by one or more forms of parasitic nematodes. Most nematodes are too minute and translucent to be seen with the unaided eye, but under certain conditions these microscopic roundworms can build up to population levels that cause serious economic or aesthetic damage to plants. Symptoms caused by root-knot nematodes are usually sufficiently distinct to diagnose in the field, and cyst nematodes are relatively easy to diagnose, as the females and cysts attached to infected roots are usually easy to see with the unaided eye. However, most nematode injuries are not readily related to nematodes in the field. Small or even large numbers of one or more species may feed on a plant without causing distinct symptoms. This poses problems for growers, farm advisers (county agents), consultants, and others trying to form a diagnosis.

To correctly diagnose nematode damage, it is usually necessary to extract, observe, identify, and count the nematodes associated with injured plants. Although plant disease clinics usually do not get involved in assaying fields for nematodes, some suggestions for sampling fields and processing samples are given.

It is suggested that readers consult with a plant nematologist or extension (advisory) plant pathologist at the nearest land-grant university for sampling guidelines that work best for the soils and crops grown in their area or region.

### Collecting Samples

A general discussion of nematode sampling is given in *The Plant Disease Clinic and Field Diagnosis of Abiotic Diseases* (Shurtleff and Averre, 1997, pp. 42–44), the first in a series of volumes on plant disease diagnosis. A more complete summary is given here.

#### WHERE TO SAMPLE

Any location where plants do not grow well (when the weather, soil fertility, and other conditions are taken into consideration) could be an area of nematode damage. In a field or an orchard or in turf, such areas may be circular to oval or rectangular in outline; in row crops poor growth may follow the rows. On rare occasions, an entire field may be affected. Trees, ornamental plants, and potted plants may individually exhibit unthriftiness. Soil samples should be taken from the root zone (the rhizosphere) of plants that are not growing properly. Avoid taking soil samples around dead or severely affected plants, because nematode populations on these plants have usually declined.

Aboveground samples suspected of containing stem and foliar plant-parasitic nematodes are taken from plant parts showing a range of symptoms. These include plants with unusual patterns of shoot growth, galls, rots, and abnormal discoloration, such as browning, blackening, or yellowing. To detect the presence of the pinewood nematode (*Bursaphelenchus xylophilus*), large woody evergreens may be sampled by removing, with a pruning saw, sections of branches 15–20 cm (6–8 in.) long, at various locations. Small plants, such as strawberries, should be collected whole, with adhering soil, and handled as described below (see the procedures for sampling herbaceous ornamentals, p. 23).

#### WHEN TO SAMPLE

Familiarity with the life cycles of nematodes in soil, roots, and leaves is important in determining the ideal time for sampling, especially in annual crops, in which nematode population densities fluctuate seasonally. If the identity of nematode species in a sampling area is unknown, however, sampling late in the growing season is recommended, since nematode populations are usually highest then.

Samples should be collected when soil moisture is slightly below field capacity, so that the soil is neither dry nor wet, because excessively wet or dry soil samples are difficult to collect and prepare for analysis.

Soil samples can usually be collected at any time of year, as long as the sampling tool can be inserted into the ground to the proper depth. The best times to sample are usually when damage is first observed or in autumn. For example, an analysis for the soybean cyst nematode or root-knot nematodes in the fall can be used to decide what crop, variety, or cultivar to plant the following spring.

Samples should be protected from direct sunlight at all times and kept cool (not refrigerated) until processed in a qualified nematology laboratory. The ideal storage temperature varies in different regions of the world: 4°C (40°F) in northern and temperate regions, 10–16°C (50–60°F) in warmer regions of the middle latitudes, and 16–18°C (60–65°F) in the tropics and subtropics.

### HOW TO SAMPLE

Nematodes are *never* uniformly distributed in soil. They are normally highly clustered in field soils, being much more numerous close to plant roots and less numerous in other parts of a field because of wetness, soil type, and other physical and biological factors. Thus, it is important to sample a field systematically to obtain a sample that gives an accurate and precise estimate of the overall population.

Sampling is best done with a coring device; many fertilizer dealers will loan these devices, or they can be purchased. Trowels and shovels can be used but are not as good for sampling as a soil probe.

Except in turf and a few other crops, it is usually best to remove and discard the top 3–5 cm (1–2 in.) of soil in which the plants are growing, because extremes of temperature and moisture and the absence of suitable host tissue limit nematode populations. Soil cores are usually taken to a depth of 15–20 cm (6–8 in.). For some nematodes, especially those affecting tree crops and other deep-rooted perennials, cores taken to a depth of 60 cm (24 in.) may be appropriate.

Whenever possible, several subsamples of soil are taken from the fibrous feeder root zones of growing plants, where plant-feeding nematodes are concentrated; around the drip line of trees and other perennials; and in and between rows of annual crops (Figure 2.1). Subsamples can be carefully removed with a small hand trowel, spade, shovel, soil auger (2.5 cm, or 1 in., in diameter), or cylindrical sampling tube. Depending on the size of the plant, four to 20 or more subsamples or cores are taken around the plant, bulked, and *gently* mixed thoroughly. Excessive mechanical mixing grinds nematodes between soil particles and renders the sample useless. Samples should be taken from at least several plants in an area or field, ranging from apparently healthy to moderately affected plants. All soil samples *must contain feeder roots.* At the diagnostic clinic, subsamples will be taken from the total soil sample for extraction of nematodes and counting.

The size of the area sampled and the number of soil cores per sample for a nematode assay are determined by the purpose of the sample. For disease diagnosis, paired samples should be collected: one from healthy plants adjacent to diseased plants (Figure 2.1), comprising 10 to 30 or more soil cores and amounting to about 500 to 1,000 $cm^3$, and a similar sample from unthrifty or stunted plants in one or more planting areas. If differential growth patterns are not evident, one composite sample per sampling site is sufficient. However, a more reliable estimate can be obtained from at least two samples per area.

Samples should not be taken from dead or severely injured plants, since the nematodes affecting them will have died or moved away from their roots. Roots, including feeder roots, should be included in samples, since some nematodes live in them.

**Agronomic and vegetable crops.** Collect soil in the root zone to a depth of 15 to 20 cm (6–8 in.). Collect subsamples from spots around 10 to 20 or more plants showing moderate symptoms or healthy-looking plants at the margin of a severely damaged area (Figure 2.1A, B, and D). In sandy soils in the fall, sample to a depth of 25 cm (10 in.). Light and deep sandy soils allow roots to penetrate several feet deep or more. The greatest nematode populations may be at depths of about 1 m (3 ft) in fields of cotton, alfalfa, and many herbaceous perennial crops and under many tree crops. Use a soil-sampling tube, trowel, spade, or narrow-bladed shovel. Save only about half a cup of soil with roots from each spot that is sampled. Gently but thoroughly mix the collected soil in a clean pail or plastic bag. Collect at least 1 L (1 qt) for processing, and place the sample in a plastic bag.

The number of subsamples (soil cores or borings) needed depends on the size of the area being investigated:

- for a small area, less than 500 $m^2$ (5,000 $ft^2$), take at least 8 to 10 subsamples
- for a medium-sized area, 500 $m^2$ to 0.4 ha (5,000 $ft^2$ to 1 acre), take at least 10 to 15 subsamples
- for a large area, 0.4 to 2 ha (1 to 5 acres), take at least 20 subsamples, and as many as 40, if possible

Each subsample should be from an area of a uniform soil type.

Root symptoms are sometimes diagnostic of a nematode genus or other soilborne problems. If the crop is still in the seedling stage or if the plants are relatively small, carefully dig up three or four plants from the affected area. Gently shake off some of the soil from the roots, and put the soil and the plants in separate bags. Placing plants in a bag with soil can cause them to become dehydrated. If the crop is more advanced, one or two plants will suffice; discard the shoots and gently shake the soil from the roots. Mature plants are usually too bulky for inclusion, and for large plants a representative portion of the root system taken from the soil profile is sufficient.

**Commercial turf and home lawns.** Collect 15 to 30 arbitrarily selected soil cores, 5–15 cm (2–6 in.) deep, from an affected area where the grass is still living and the same number of cores from an adjacent healthy-looking area. A soil sampler is best, but a narrow-bladed trowel may be used instead. Keep only the soil and roots. Gently but thoroughly mix the cores from each site, and save about 500 $cm^3$ (1 pint) of soil from each for processing.

**Woody plants and tree crops.** Remove the upper 5 cm (2 in.) of soil around affected trees and shrubs, and dig down 30–38 cm (12–15 in.) with a shovel or soil auger. Core or auger samples are better than those taken with a shovel. For deep-rooted plants, such as tree crops and ornamental trees, in sandy soils, sampling to a depth of about 1 m (3 ft) may be required to fully determine the associated nematode population. Dig at the outer drip line and in the area between the outer branch tips and the stem or trunk, as shown in Figure 2.1C.

**Herbaceous ornamentals.** Take a small quantity of soil and some feeder roots from several depths in the exposed soil profile around plants where roots are visible. Repeat this process five times around the plant. If possible, avoid taking samples in areas with weeds, and *do not include the roots of weeds.* Mix the soil in the subsamples, and package at least 1 L (1 qt) in a plastic bag for processing.

Use a core sampler for large container-grown plants. Annual ornamentals should be sampled in the same way as agronomic and vegetable crops. Small potted plants may be submitted in the pot together with the soil, or just the entire plant may be submitted.

Many ornamentals, particularly those grown in the home or greenhouse or in a warm region, are susceptible to foliar nematodes, especially with overhead watering. In the moist tropics, lesion and burrowing nematodes commonly move into the lower stems and leaves of some herbaceous plants (e.g., *Aglaonema*). Foliar nematodes can distort or kill buds, stems, leaves, or floral parts of plants and can cause lesions or discolored blotches on the leaves. They are uncommon in many areas, and they rarely occur in the soil. However, if foliar nematodes are suspected, as in greenhouses or in the moist tropics, the entire crown or a half-dozen leaves and

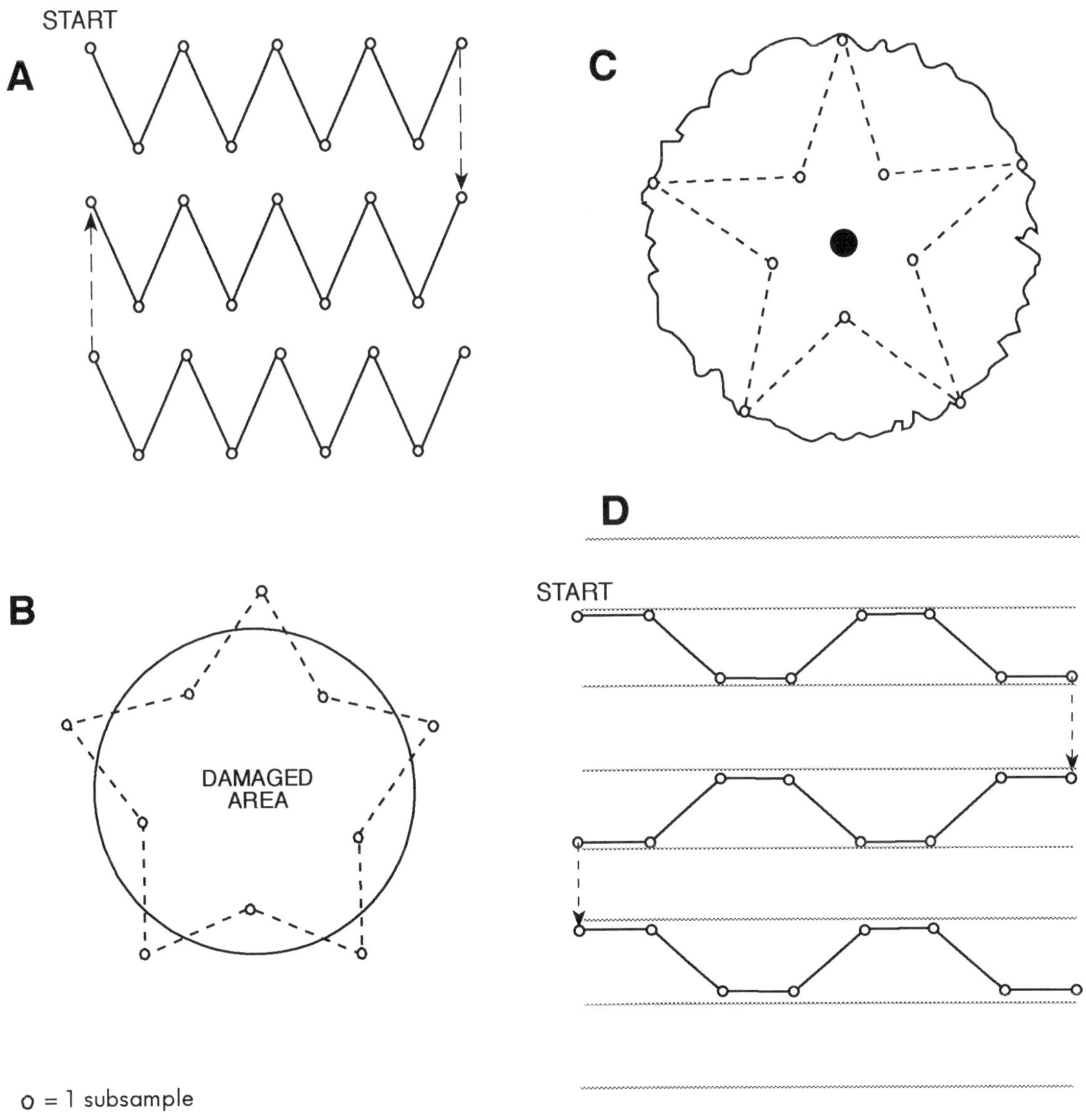


**Figure 2.1.** Sampling patterns for plant-parasitic nematodes in different types of crops. **A,** Sampling a fallow field or a field in which the current crop has no obvious symptoms; the sampling area should not exceed 2 ha (5 acres). **B,** Sampling at the margin of a damaged area. **C,** Sampling within the drip line of a tree or shrub. **D,** Sampling a field planted with a row crop; the sampling area should not exceed 2 ha (5 acres). Sampling every 2 ha is probably not feasible for large farming operations.

the lower portion of the affected plant should be sent. No soil should be included with the sample. Stems and leaves should be packed in moist (*not wet*) cotton, paper, or moss and placed in a plastic bag. Keep samples as close to 4°C (40°F) as possible.

**Bare soils.** In fields with a history of subnormal production, the potential for nematode problems in a subsequent crop can often be determined by sampling prior to planting, either before or after land preparation. Take 20 to 40 subsamples in a systematic zigzag pattern throughout and somewhat beyond the area of subnormal production (Figure 2.1A). Discard the top 1 or 2 in. of soil, and then collect soil to a depth of 20 cm (8 in.), with a soil probe, trowel, or narrow-bladed shovel. Follow the remaining steps outlined above for agronomic and vegetable crops. Limit each composite sample to an area of 2 ha (5 acres).

#### CARE AND LABELING OF SAMPLES

After representative samples have been taken, it is extremely important to properly protect them from drying, exposure to heat (over 27°C, or 80°F) or cold (under about 4°C, or 40°F), rough handling, and anaerobic conditions. In the field, soil and plant samples are usually placed directly in strong plastic bags, quart or half-gallon size (1–2 L), tightly sealed, properly labeled on the outside with waterproof ink, and immediately placed in a cool place, such as an insulated container in the shade. The following information about the sample should be written directly on the bag or on a label securely attached to it:

- name, address, and phone number of the grower and the person submitting the sample
- date of sampling and date of submission
- name of the plant, including the name of the variety or cultivar (if known)
- location of the field or property; field locations can be identified by global positioning system (GPS) coordinates
- portion (percentage) of the planting area represented by the sample
- primary symptoms observed and distribution of symptomatic plants (scattered, clustered, in spots, or uniform)
- extent of damage
- crop history for the past four years, including cultural and chemical management practices
- description of weather conditions for two weeks prior to symptom development
- grower's record or recollections of field history of plant vigor

Some states and countries have nematode soil sample submission forms, on which this information can be recorded. For example, the form used by the University of Illinois at Urbana-Champaign is shown in Figure 2.2. The information requested on the form is essential for diagnosis of nematode problems and making proper recommendations for managing a nematode population.

Samples are best examined within a few hours or days after being taken. However, if this is not possible, they can be stored for one to three weeks in an incubator or refrigerator at 4°C (40°F) in northern and temperate regions, 10–16°C (50–60°F) in warmer regions, and 16–18°C (60–65°F) in the tropics and subtropics.

### Submitting Samples for Analysis

#### PACKAGING SAMPLES

Soil samples for nematode analysis must be handled differently from samples taken for fertilizer recommendations. Nematodes are living animals and require moisture and a moderate temperature to survive. To be recovered and identified, they must reach a nematology laboratory *alive.* All precautions must be taken to prevent rough handling, exposure to temperatures below 4°C (40°F) or above 27°C (80°F), and drying out. Never leave samples in the trunk of a car, in a car with the windows closed, in the back of a truck, or in any other location where the temperature is likely to exceed 32°C (90°F) during transport. Samples can be protected for a few hours in an insulated container.

1. Place a 1-L (1-qt) soil sample in a strong plastic bag. Freezer bags are excellent. Very thin plastic bags are too easily torn. Paper bags and cardboard boxes are unsuitable, because samples dry out in them.
2. Fasten the open end of the bag securely.
3. Label the bag with waterproof ink as outlined above.
4. Place the bag in a *strong* insulated container. Heavy-duty shipping boxes are necessary for shipment of several samples. Seal the container securely to avoid rupture and spillage during shipment.

#### SHIPPING SAMPLES

Do not drop or shake samples. Speed is essential! Deliver or ship the package, preferably by an overnight delivery service, immediately after collection. Send it early in the week, so it will not remain in a post office over a weekend. If a sample must be held for a period of time, store it at 4°C (40°F) in cool latitudes, 10–16°C (50–60°F) in warmer latitudes, and 16–18°C (60–65°F) in the tropics and subtropics.

Address the package to a qualified nematology laboratory (e.g., a plant clinic or an equivalent facility at your land-grant university). Some private laboratories are also qualified to process and identify nematodes. If you do not know of a laboratory or have an address, your local cooperative extension (advisory) office or campus-based offices can help. A complete examination should result in identification of nematodes to the level of genus and, where necessary, to species or race. Before shipping a sample, always make certain that the laboratory receiving it can provide this service.

Clinic No. (s)
N-

Cooperative Extension Service
University of Illinois at Urbana-Champaign

# NEMATODE SOIL SAMPLE FORM

University of Illinois at Urbana-Champaign
PLANT CLINIC, St. Mary's Road, Urbana, IL 61801
(217) 333-0519

Submitted by__________ Date of Sampling__________

Address__________ Date Submitted__________

County__________ Phone__________ Date Received__________

| Clinic Numbers | Sample Number | Soil Type | Present Crop (Summer) | | Previous Crop (Summer) | | Crop 2 yrs ago (Summer) | | Next Crop To Be Grown | Past Nematicide/ Insecticide Treatments | |
|---|---|---|---|---|---|---|---|---|---|---|---|
| | | | Crop | Variety | Crop | Variety | Crop | Variety | | Year | Chemical |
| N- | | | | | | | | | | | |
| N- | | | | | | | | | | | |
| N- | | | | | | | | | | | |
| N- | | | | | | | | | | | |
| N- | | | | | | | | | | | |
| N- | | | | | | | | | | | |
| N- | | | | | | | | | | | |
| N- | | | | | | | | | | | |
| N- | | | | | | | | | | | |
| N- | | | | | | | | | | | |

**Circle Appropriate Information**

Distribution of Symptoms: Scattered Clustered in Spots Uniform
Association with Terrain or Soil Type: Yes No
Weather Conditions Prior to Symptom Development —
Rainfall: Low Medium High Temperature: Low Medium High
Soil Test Information or Fertilizer Application — 2 yrs past_____ 1 yr past_____ Current yr_____
Herbicides Applied This Year__________

COMMENTS:

| For Lab Use | Soil | Roots | Juveniles | Cysts | Comments: |
|---|---|---|---|---|---|
| Date Processed | | | | | |
| Date Read | | | | | |

O-1781

**Figure 2.2.** Form used in analysis of soil samples for nematode assay at the University of Illinois at Urbana-Champaign. Additional information on submitting nematode samples is given on the back of the form.

## Key Points in Collecting and Submitting Soil and Plant Tissue Samples

### COLLECTING SAMPLES FOR DIAGNOSTIC PURPOSES

1. General predictive sampling of fields is done in a systematic pattern. Take samples from healthy-appearing plants at the *margins* of affected areas, to a depth of 15–25 cm (6–10 in.), depending on the type of plant and type of soil. In sandy soils, samples should be taken to a depth of 25–30 cm (10–12 in.) or more.
2. Collect several cores (or shovels) of soil from the affected area, gently mix them, and package 1 L (1 qt) of the mixed soil.
3. Include roots, especially feeder roots.
4. Wash all soil from sampling tools and hands after collecting each sample, to prevent contamination.

### PACKING AND SHIPPING SAMPLES

1. Use nematode soil sample bags or thick, sturdy plastic bags.
2. Securely close the bags with twist ties or light wire, or use zip-lock (reclosable) bags.
3. Clearly label each bag as indicated above, with waterproof ink. Write directly on the bag, or attach a label securely to the exterior of the bag.
4. Place the bags in a strong container, and pack them with insulating material or crumpled newspaper.
5. Complete a nematode soil sample form and include it with the samples.
6. Ship the samples immediately, so they remain fresh. Send them early in the week, to avoid drying out in storage over the weekend.
7. Protect the samples from the sun, heat, freezing, and rough handling. Do not allow them to dry out, but *do not add water* to them.

## Cost

There is usually a service charge for assaying soil samples for nematodes. Check with your cooperative extension (advisory) office or a commercial laboratory for current prices.

## Results and Recommendations

The plant clinic or laboratory will prepare a nematode assay report, stating the results of the analysis and indicating which nematodes may be affecting the crop from which the samples were taken. Some agencies provide recommendations for control. Should the use of a nematicide be suggested, read and follow the manufacturer's directions carefully!

It is important to note that just because some plant-parasitic nematodes were found in a sample does not mean that they are at damaging population densities. Diagnoses and recommendations are dependent on the quality of the soil samples submitted to and received by the plant clinic.

## How To Do Your Own Assessment with Strip Tests

In many areas of the world, diagnosis or prediction of crop damage due to nematodes is reliable. There are places, however, where the desired accuracy of diagnosis and prediction have not been achieved. Since there are infinite combinations of plant species, nematode types, and environmental conditions that may affect damage due to nematodes, it may be desirable to perform a test to evaluate the effects of nematode-induced disease in the field. In such a situation, the grower can conduct a strip test, with a nematicide registered for use on the crop under assessment. The response of plants to nematicide treatment in a test area, after several weeks or months, compared to the growth of adjacent untreated plants, may determine whether a more extensive nematicide application is justified. Another type of test can be conducted in some cases, in which a nematode-resistant variety and a susceptible variety are planted in strips.

A strip test involves chemical treatment of a small amount of land and should be conducted in a problem area. Either a fumigant, applied preplant, or a very effective nonfumigant (granular or liquid formulation), applied at planting time or postplant, can be used, depending on the crop and regulations governing the use of the nematicide. It should be applied to *at least three strips* alternating with adjacent untreated strips. Treated and untreated strips *must be alternated,* because of the wide natural variation in nematode populations across a field, lawn, or orchard.

In annual and herbaceous perennial row crops, each strip should consist of one or more rows. In drilled crops and turf, the treated and untreated areas should be in bands at least 1 × 2 m (3 × 6 ft). In small woody perennials and large woody perennials, a minimum of six and three plants, respectively, should be treated. The size of the treated sample (the number of rows in a strip, the width of the treated and untreated bands, or the number of plants treated) should be sufficient to allow the grower to adequately measure growth and yield response.

Strip tests are most appropriate for annual row crops, small fruits, nursery plants, and established turf. This kind of test does not work well for most established tree crops and established woody ornamentals. It just takes too long for trees to respond and recover, if they do recover.

If possible, a strip test should be augmented by a nematode assay to determine the efficacy of the treatment and to correlate plant response with nematode control. The grower or a farm adviser (county agent) or professional crop consultant should take a soil sample from each treated and untreated strip, six to 10 weeks after application of the nematicide, and the soil should be submitted to a qualified nematology laboratory for analysis. Sampling procedures for various types of crops are outlined above.

## Laboratory Methods

The procedures described in the following sections are routinely used in plant diagnostic clinics for determining the presence and identity of well-known plant-parasitic nematodes. Clinicians should consult a plant nematologist for detailed procedures appropriate for quantifying populations and threshold levels, issuing recommendations for nematode control, or establishing the identity and pathogenicity of a particular nematode, especially a species unknown in the growing area. The following references have excellent sections on methodology: Ayoub (1980), Barker et al. (1985a, 1986), Dropkin (1989), Luc et al. (1990), Southey (1986), Taylor (1971), Thorne (1961), and Zuckerman et al. (1990).

The presence of nematodes in or on diseased plants can be determined by direct observation of symptomatic tissue with a good dissecting (stereoscopic) microscope (an inverted compound microscope may be superior) at magnifications of 10 to 50×, in transmitted or incident light. Methods for extracting and isolating plant-parasitic nematodes from plant tissues and soil for microscopic observation and identification are based on several characteristics of nematodes:

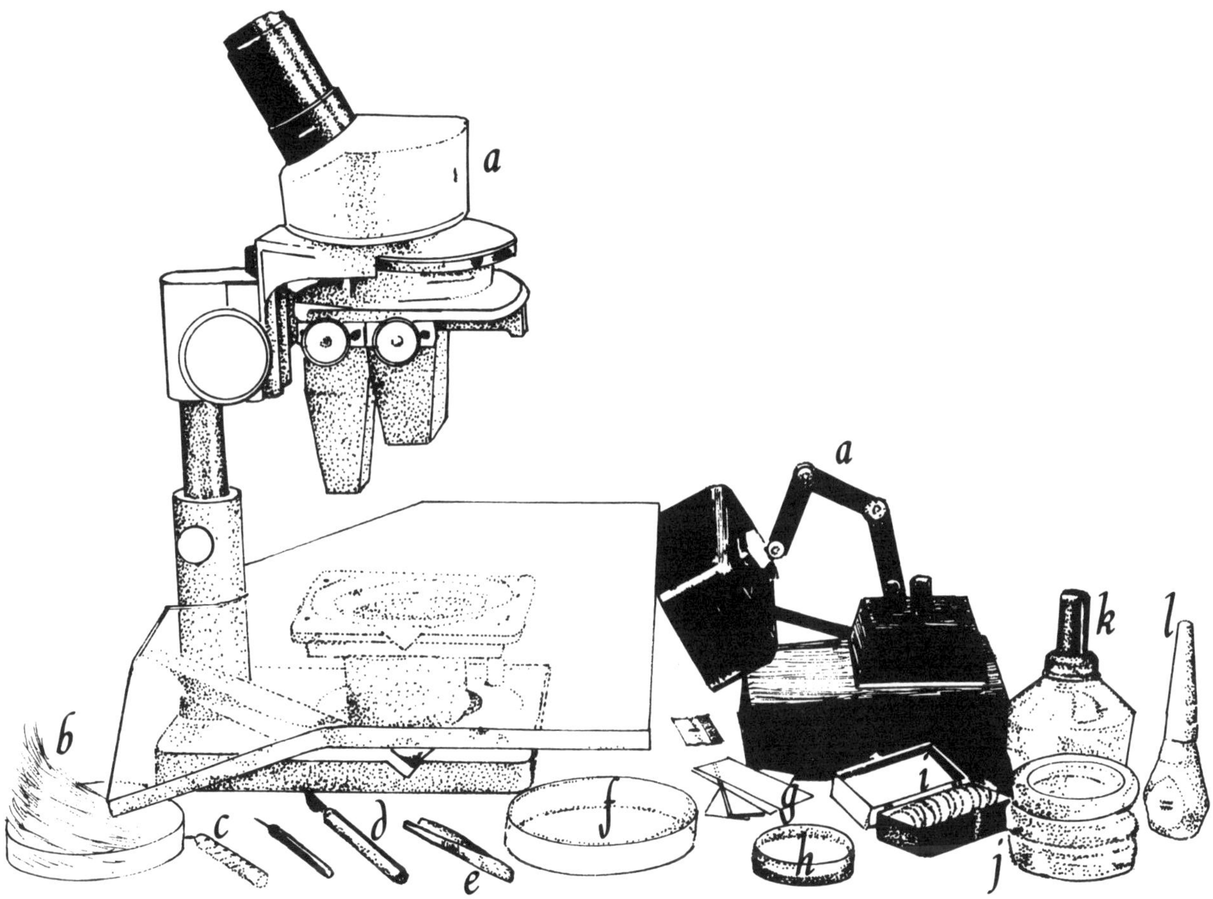


**Figure 2.3.** Equipment commonly used for observing plant-parasitic nematodes associated with diseased plants. **A,** Stereoscopic dissecting microscope with a light source. Magnifications of 10, 20, 30, and 60× are typically used. **B,** Coarse fiberglass wool. **C,** Small candle (birthday cake candle), for applying melted wax to seal coverslips to glass slides. Clear fingernail polish is also commonly used for this purpose. **D,** Sharpened bamboo splinter, for transferring vermiform nematodes, and scalpel. A nylon bristle mounted on a handle, fine insect mounting pin, dental root canal file, porcupine quill, very slender cactus spine, or transfer needle may be used instead of a bamboo splinter. **E,** Forceps, for handling and grasping small roots. Very fine forceps are used to handle cyst nematodes. **F,** Petri dish, used for dissecting small pieces of plant tissue in water under the dissecting microscope. **G,** Microscope slides (25 × 76 mm). **H,** Dish containing class IV microbeads, used as spacers between the coverslip and the glass slide to prevent crushing nematodes. **I,** Coverslips (18-mm, round, no. 1). **J,** Syracuse dishes, used in examining extracted nematodes under the dissecting microscope. Pillboxes and 50-mm petri plates are also commonly used for this purpose. **K,** Alcohol burner, for relaxing nematodes and for cleaning coverslips. **L,** Clear fingernail polish, used to seal coverslips to glass slides. It can also be used to form a spacer between the coverslip and the slide. Also illustrated is a single-edge razor blade. (Adapted from Ayoub, 1980)

motility—active, vermiform endoparasitic and ectoparasitic nematodes will migrate from plant tissues and soil particles suspended over water and settle out

size—samples suspended in water are screened through a series of sieves of known mesh size to separate coarser particles from finer particles

specific gravity—the specific gravity of most nematodes is about 1.05, slightly greater than that of water, and thus they slowly settle to the bottom of a container of water; if the density of the liquid is increased by the addition of sucrose, they will float to the surface

Gel electrophoresis, serology, DNA probes, polymerase chain reaction with random amplified polymorphic DNA (PCR-RAPD), restriction fragment length polymorphism (RFLP), fatty acid profiling, and other molecular procedures have application in rapid and accurate routine assays for specific nematodes. However, they are currently not generally used in diagnostic clinics.

### Direct Examination of Infected Host Tissue

Symptomatic plant tissues are usually examined with the aid of a dissecting (stereoscopic) microscope at magnifications of 10 to 50× to detect the presence of plant-parasitic nematodes. Inverted compound microscopes are becoming more widely used but are more expensive. Some of the instruments, equipment, and supplies commonly used for direct examination and handling of nematodes are shown in Figure 2.3.

1. Very gently wash small roots, stems, leaves, buds, or other plant samples by holding them with a tweezers and dipping them in water to remove loose soil and other debris. Transfer a washed sample to a plastic petri dish or Syracuse dish containing 2 to 4 ml of water, for examination under a dissecting microscope.
2. While observing the operation through a dissecting microscope, cut the sample into smaller pieces (1 to 2 cm long), and dissect galls, if they are present. Carefully slice, shred, or dissect the tissues with strong dissecting needles or a scalpel (Figure 2.4A), and note the presence and position of nematodes in the plant tissues. If no nematodes can be teased out and none are seen on the bottom of the dish, reexamine it an hour or several hours later.

   Most plant-parasitic nematodes found in or on aboveground parts of plants (Table 3.5) emerge in a few minutes and can be found moving about on the bottom of the dish.

   In root samples, large, sedentary endoparasitic nematodes may be seen attached to the surface of roots or in dissected tissue. Semiendoparasites and firmly attached ectoparasites can be seen attached to the surface of roots (Figure 2.4A). Ectoparasitic nematodes may be found on the bottom of the dish if the sample was gently and not excessively washed (Figure 2.4B).
3. In the shallow water near the center of the dish, loop an individual nematode over a thin needle, such as a 00 insect mounting pin, dentist's pulp probe glued to the end of a needle holder, or sharpened bamboo splinter (Figures 2.5 and 2.6), and transfer it to a drop of water on a glass slide (Figure 2.6B). Large, sedentary nematodes and cysts can be very gently grasped with a fine forceps. Dissection and storage in 0.9% sodium chloride (0.9 g of NaCl per liter) helps to reduce the osmotic effect of water, which can cause female nematodes to burst.
4. Prepare a temporary slide mount (Figure 2.7) for viewing under a compound microscope at a magnification of 100× or more to observe morphological and anatomical details necessary to confirm the identity of the nematodes. A detailed description of the preparation of temporary slide mounts is given below (pp. 41–43).

To examine the very small and indistinct structures of nematodes, an observer must be skilled in the use of a compound

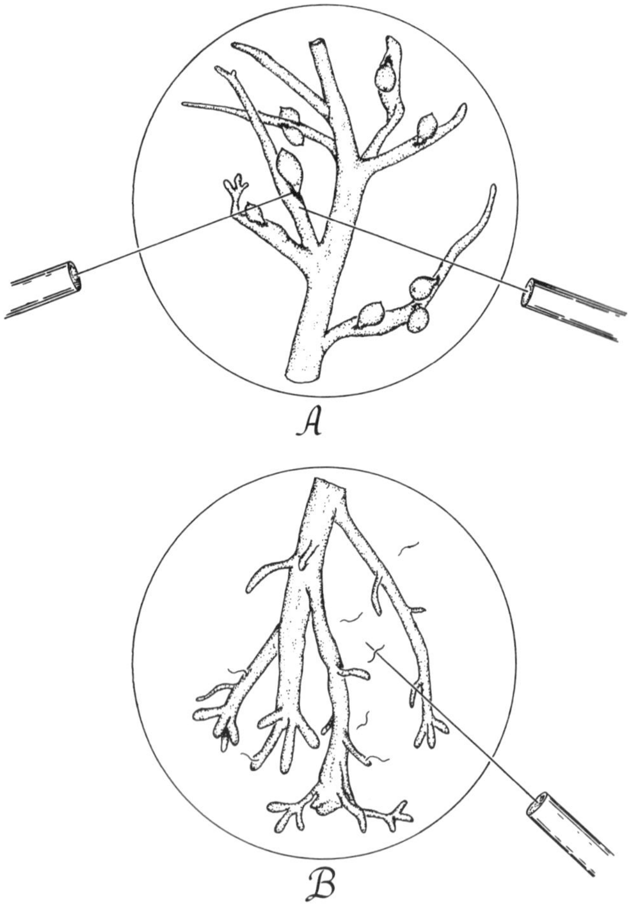


**Figure 2.4.** Washed roots in water in a plastic petri dish, viewed through a dissecting microscope at a magnification of about 15×. Dissecting needles are used to move the roots to better view free and embedded nematodes. Needles, sharp-pointed scalpels, and forceps are used to remove endoparasitic nematodes. **A,** Large, semiendoparasitic nematodes attached to roots (cyst nematodes, in this example). **B,** Ectoparasitic microbe-feeding nematodes on the floor of the petri dish.

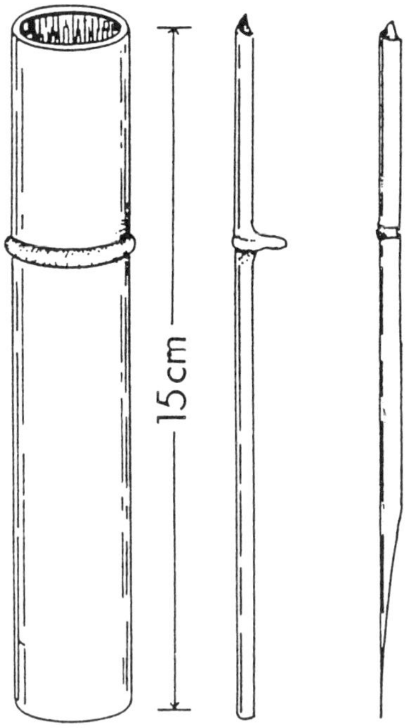


**Figure 2.5.** Nematode transfer pick prepared by splitting and sharpening a small section of bamboo from just beneath the hard outer surface of the cane. The section is placed in a needle holder and brought to a fine point with a sharp knife, scalpel, or razor blade. The final sharpening is done under a stereoscopic dissecting microscope, until the point consists of only two or three fibers.

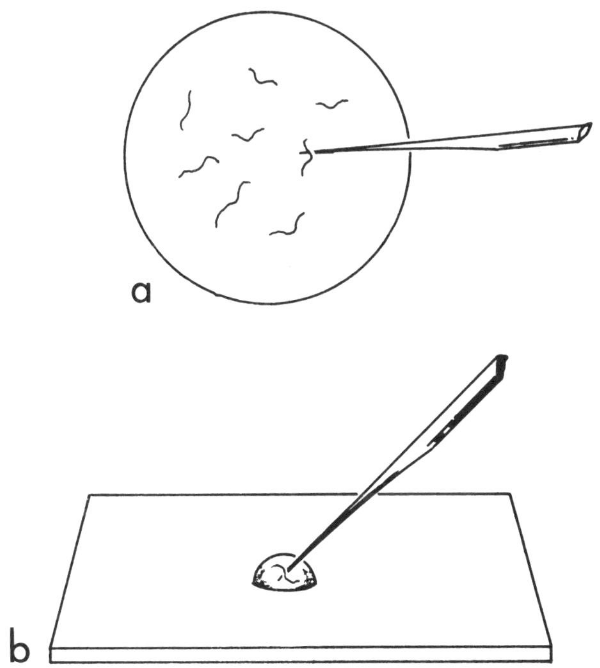


**Figure 2.6.** Bamboo pick used to transfer vermiform nematodes from shallow water in a petri dish to a drop of water on a glass slide. **A,** The tip of the pick is placed under a nematode, slowly raised to the surface, and then quickly removed to the water. A magnification of 20× is optimal. **B,** The tip of the pick, with the nematode, is placed in a drop of water on a glass slide, and the pick is moved in the water to dislodge the nematode.

microscope of good to excellent quality with good illumination potential (Figure 1.16). An oil immersion lens is often needed to identify species. For information on the care and use of microscopes, see Shurtleff and Averre (1997, pp. 103–105).

## Staining Techniques for Observation of Nematodes in Host Tissue

Nematodes are translucent and difficult to see in plant tissues. Staining helps to highlight some of their anatomical features without unduly coloring the surrounding host cells.

### ACID FUCHSIN STAINING METHOD

1. Prepare a stock solution of acid fuchsin: add 25 ml of glacial acetic acid to 75 ml of deionized or distilled water, and dissolve 0.35 g of acid fuchsin in the solution.
2. Wash roots or other host tissue thoroughly under gently running tap water to remove debris and adhering soil particles. Small roots can be stained intact. Larger plant tissues should be cut into pieces 1 to 2 cm long, and thick material should be thinly sliced before staining.
3. Place about 1 g of tissue in a 150-ml beaker. Clear the tissue in diluted sodium hypochlorite bleach (5.25% NaOCl or Clorox) for about 4 min with occasional gentle stirring. Trial and error is needed to determine the right strength of the bleach solution and length of time required for the plant species under examination (for example, tomato clears very quickly, but coffee is extremely difficult to clear). Soak young roots in a mixture of 10 ml of bleach and 50 ml of water, medium-aged roots in about 20 ml of bleach in 50 ml of water, and older or more lignified roots in about 30 ml of bleach in 50 ml of water.
4. Pour out the beaker over a 200-mesh sieve. Rinse the host tissue for at least 45 sec in running tap water, and then soak it in water for 15 min (longer is better), to remove all traces of bleach, which inhibits staining by acid fuchsin.
5. Drain off the water, and place the host tissue in a 600-ml beaker with 30 to 50 ml of tap water and 1 ml of acid fuchsin stock solution, or enough liquid to cover the tissue. Boil the solution for about 30 sec in a microwave oven or on a hot plate in a ventilated area (e.g., under an exhaust hood or while using an exhaust fan) to avoid the vapor of acetic acid. Then remove the host tissue with a forceps, and place it on a 200-mesh sieve.
6. Allow the tissue to cool to room temperature and drain, and wash off excess stain in gently running tap water.
7. If destaining with water is not sufficient, transfer the tissue to acidified glycerin (glycerol) (20 to 30 ml of glycerin plus a few drops of 5N HCl) in a 150-ml beaker, heat the solution to a boil, and allow it to cool. The destaining is complete when the tissue is clear and the nematodes are stained pink to bright red.

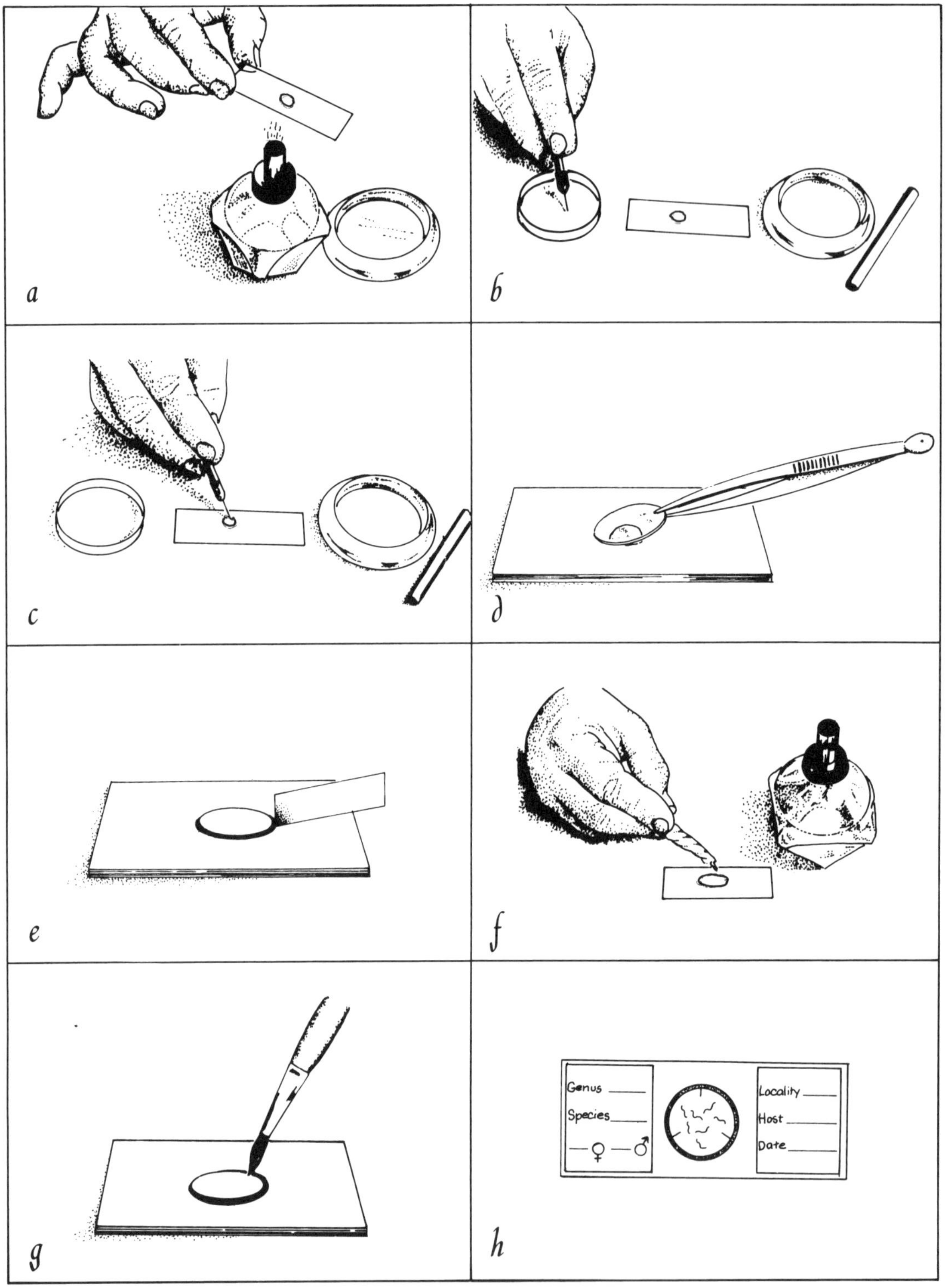


**Figure 2.7.** Preparation of a temporary glass slide mount for observing morphological and anatomical features of nematodes with a compound microscope at a magnification of 100× or higher. **A,** Kill or relax nematodes in a drop of water by passing the bottom of a glass slide several times over the flame of an alcohol burner (this is not necessary if the drop contains fixative or mounting fluid). **B,** Pick up several short sections of glass wool with the moistened tip of a transfer needle. They should be slightly thicker than the diameter of the nematodes. **C,** Place the glass wool spacers on three sides of the drop of water. **D,** Pass a coverslip through the flame of the alcohol burner to eliminate lint and moisture, and then slowly place it over the drop of water. **E,** Remove excess water with the tip of a paper towel or facial tissue while viewing the nematodes under a dissecting microscope, taking care to avoid drawing the nematodes out from under the coverslip. **F** and **G,** Seal the edge of the coverslip to the glass slide with wax (**F**) from the tip of a heated (not burning) birthday candle, using the wick as an applicator, or with clear fingernail polish or some other sealant (**G**), applied with a small brush. **H,** Label the slide. Temporary mounts keep only a few days at room temperature, and somewhat longer if refrigerated.

8. Store the specimens in acidified glycerin.
9. To examine a specimen, mount a small amount of tissue on a microscope slide, cover it with a coverslip, and apply *gentle* pressure. Alternatively, place tissue in a small amount of acidified glycerin on the inner surface of the lid of a glass or plastic petri dish; place the bottom of the dish over the lid, with the outer surface of the bottom touching the specimen, and apply gentle pressure. A grid pattern can be lightly cut with a glass diamond pencil in a petri dish or Syracuse dish, as an said in counting nematodes (Figure 2.8). The sections of the grid should be just wide enough to be seen in the field of view of a dissecting microscope at low power. A flat field and good resolution are needed.
10. Observe and count the nematodes at a magnification of 10 to 50×, under uniform illumination by transmitted light. A hand-tally counter (or a bank of counters) is useful for counting different genera. Stained endoparasitic nematodes in all stages can be counted within tissue sample.

This method of staining can be used for leaves, stems, and other aerial parts of plants, as well as roots. To treat lignified tissues, 1) use stronger bleach solutions, or soak the tissues for a longer time, or 2) soak them in 30% hydrogen peroxide ($H_2O_2$) for an hour, rinse them in water, and stain them as described above. For additional details, see Byrd et al. (1983).

Treating roots with bleach prior to staining with acid fuchsin prevents them from becoming too heavily stained. The time for staining is thus shortened. Enzymatic digestion of plant tissue is helpful for releasing nematodes and makes quantification easier.

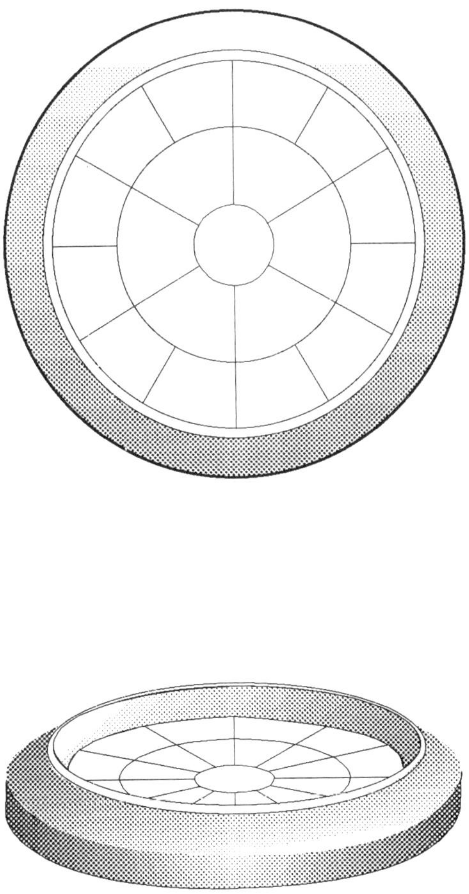

**Figure 2.8.** Syracuse dish for counting nematodes. A grid pattern cut with a glass diamond pencil aids greatly in counting.

#### STAINING EGG MASSES WITH PHLOXINE B

Roots containing egg masses (egg sacs) of root-knot nematodes (*Meloidogyne* species) may be soaked in a solution of phloxine B (15 mg/100 ml of tap water) for 15 to 20 min. and then rinsed in tap water to remove the residual stain. The gelatinous matrices holding the eggs are readily stained pink to red and can be observed and counted in water. For additional details see Holbrook et al. (1983) and O'Bannon and Nyczepir (1982).

#### STAINING PHASMIDS AND SECRETIONS

1. Prepare a 0.1% solution of Coomassie brilliant blue R-250 or G-250 (Sigma Chemical Co.) in 10% acetic acid. To remove insoluble material, filter the stain solution through a Whatman no. 1 filter paper.
2. Incubate nematodes in the stain solution before preparing a microscope slide. Alternatively, stain them directly on the slide by placing about 10 µl (a small drop) of a nematode suspension on a clean glass slide and adding an approximately equal volume of stain solution.
3. Space several pieces of fiberglass wool, about 3 to 5 mm long, around the specimen, or brush on a ring of clear fingernail polish at the margin of the slide to support the coverslip and prevent deformation of the nematodes (Figure 2.7B and C). Slowly lower the coverslip over the specimen (Figure 2.7D), and seal it with clear fingernail polish (Figure 2.7G) or a sealant such as Glyceel or Permount.
4. Immediately observe the specimen under a compound microscope, or incubate it at room temperature and examine it at 30-min intervals over a period of 2 h or longer. Secretions from amphids, phasmids, and the excretory pore stain dark blue. For additional details see Premachandran et al. (1988).

**NOTE** This staining procedure is not suitable for dorylaimid nematodes.

### Extracting Nematodes from Plant Tissues or Soil

Plant-parasitic nematodes can be extracted from soil and plants in several ways, depending on the type of nematodes and plant materials.

To extract nematodes from soil, a funnel extraction method is often used. Small samples of soil (up to 250 cm$^3$, or about a cup) can be placed directly in a funnel and processed by mist extraction or the Baermann funnel method. Samples larger than a cup are normally first processed by Cobb's sieving and gravity method or in an elutriator and then by mist extraction of the residues. The combination of the sieving and gravity method and mist extraction is more effective in processing soil than the Baermann funnel method.

To extract cyst-forming nematodes from soil, the centrifugal flotation method should be used. This technique is also recommended for recovering sluggish and sedentary nematodes, such as *Criconemoides* and *Hemicycliophora,* from soil.

To extract nematodes from roots and other plant parts (e.g., bulbs, tubers, crowns, stems, or leaves), samples are processed by mist extraction, the Baermann funnel method, trituration, soaking in water, or agitation in a shaker.

In addition to the methods described below, numerous other methods of extracting nematodes from soil or plant material have been described by Evans et al. (1993), Luc et al. (1990), Southey (1986), and others.

Regardless of the extraction procedure used, the number of nematodes found should be expressed as a number of individuals per unit of volume (100 or 500 $cm^3$ of soil) rather than as a simple estimate of the number of nematodes present.

The techniques used to extract nematodes from soil and plant tissues to prepare them for observation can greatly distort them, making identification difficult. For example, certain types of nematodes will shrink if immersed for an extended period in the sugar solution used in centrifugal flotation. The appearance of specimens and hence the ability to accurately identify them may also be influenced by the care they receive before, during, and after extraction. Exposure to high temperatures, desiccation, and storage in water for long periods before killing can make nematodes appear less lifelike. Specimens recently killed by gentle heat (65°C for 2 min) and mounted in water appear more lifelike and permit better observation of their taxonomic characters than fixed specimens, especially those mounted in glycerin. Coverslip supports are needed in temporary and permanent mounts to prevent flattening of specimens, which can change the dimensions of taxonomic characters (see Preparation and Mounting of Nematodes for Microscopic Observation, pp. 41–45).

Because of the considerable variation in morphology between individuals of the same genus or even the same species, and the difficulty of observing key taxonomic characters of certain specimens, a minimum of four to six individuals should be observed before a positive identification can be made (Mai and Mullin, 1996).

**Equipment.** Soil processing is relatively simple. It is also relatively inexpensive, except for the cost of fine-mesh metal sieves and an elutriator. The equipment needed is described with the procedures.

**Disposal of residues.** Proper facilities are needed for disposal of soil and plant residues suspected of carrying plant-parasitic nematodes, other pests, or disease agents. Arrangements should be made in advance with state or federal quarantine officials relative to the introduction of such materials and the proper disposal of residues.

#### BAERMANN FUNNEL METHOD

The Baermann funnel method (Figure 2.9) is convenient for extracting active adult and juvenile nematodes from small amounts (30 to 100 $cm^3$) of finely crumbled soil and plant tissues. In this method, eggs will hatch while a sample is incubating, and second-stage juveniles will also be recovered. The method is not effective for extracting sluggish and sedentary nematodes (e.g., *Criconemella, Criconemoides, Hemicycliophora,* and *Xiphinema*) and nonvermiform, sedentary females of *Globodera, Heterodera, Meloidodera, Meloidogyne, Rotylenchulus,* and *Tylenchulus.*

The procedure requires some 12 h, or overnight, to as much as five days (typically three days). Most nematodes are recovered after 24 to 48 h, but the recovery will vary according to the sample size, temperature, time of storage, extraction method, and nematode species, as does recovery by more elaborate techniques (Evans et al., 1993; Southey, 1986). Yields from compacted soils are lower than those from noncompacted soils.

1. Obtain the following equipment and supplies (Figure 2.9A):

   wet-strength facial tissue containing no toxic substances (other suitable materials include double-thickness muslin cheesecloth or nylon gauze, paper towels, cotton wool, and milk filters)—active vermiform nematodes can pass through these materials
   shallow, coarse, molded "basket," 17.5 cm in diameter, with a flat bottom made from stainless steel or plastic screen (*not* copper screen)
   nonfluted glass funnel, 10–15 cm in diameter, with a 10- to 15-cm stem
   rubber or plastic tubing, about 8 cm long, fitting around the stem of the funnel
   funnel support or rack
   spring or screw clamp to close off the tubing
   50- (or 100-) and 250-ml beakers

2. Mount the funnel on the support, attach the tubing to the stem of the funnel, place the clamp near the free end of the tubing, and set the basket on top of the funnel (Figure 2.9B). Fill the funnel with water, open the clamp, tap the funnel stem, and drain off sufficient water to remove air bubbles trapped in the tube. Refill the funnel with water to the bottom of the basket.
3. Place double-thickness tissue (or other "filter" material) on top of the screen, being careful not to tear the tissue, and gently submerge the filter material in the water (Figure 2.9C). It is necessary to select a filter that retains as much debris as possible, with pores large enough for the vermiform nematodes to wiggle through. To extract large

nematodes, such as *Anguina, Belonolaimus, Dolichodorus, Longidorus,* and *Xiphinema,* nylon gauze with apertures about 90 μm in diameter often provides a sufficiently clean extract.

4. Spread about 30 to 100 $cm^3$ of finely crumbled soil or washed plant tissue (e.g., small roots or chopped bulbs, corms, tubers, stems, leaves, or buds), cut into small pieces, uniformly over the filter (Figure 2.9D). Fold the edges of the filter over the sample. Gently pour water along the inner surface of the funnel until it just covers the lower surface of the soil.

**CAUTION**

The edges of the filter must not hang over the sides of the funnel, or water will be wicked out, and the sample will desiccate.

Do not overfill the funnel; the samples must not be completely submersed in water.

Do not pour water over the sample. Add it slowly down the inner surface of the funnel.

Check the apparatus periodically to see that the funnel is level, and replace any water lost by evaporation.

Label the funnel with the sample number and the date, and let it stand at a room temperature of 21–29°C (70–85°F), generally for two to five days. *Do not allow the sample to dry out.* Cover to reduce evaporation. The motion of vermiform nematodes in the sample gradually carries them through the filter material and screen into the water in the funnel, where they settle to the bottom of the clamped tubing.

For soil samples, some workers use water containing methylene blue (2 ppm) and Separan NP10 (2 ppm) (Dow-Elanco Chemical Co.). For plant tissue samples, streptomycin sulfate (1,000 ppm) is added to reduce anaerobic conditions due to bacterial decay of organic matter in the sample.

5. At the end of the extraction period, fully open the clamp to rapidly withdraw 5 to 10 ml of water containing nematodes from the tube and transfer it to a shallow

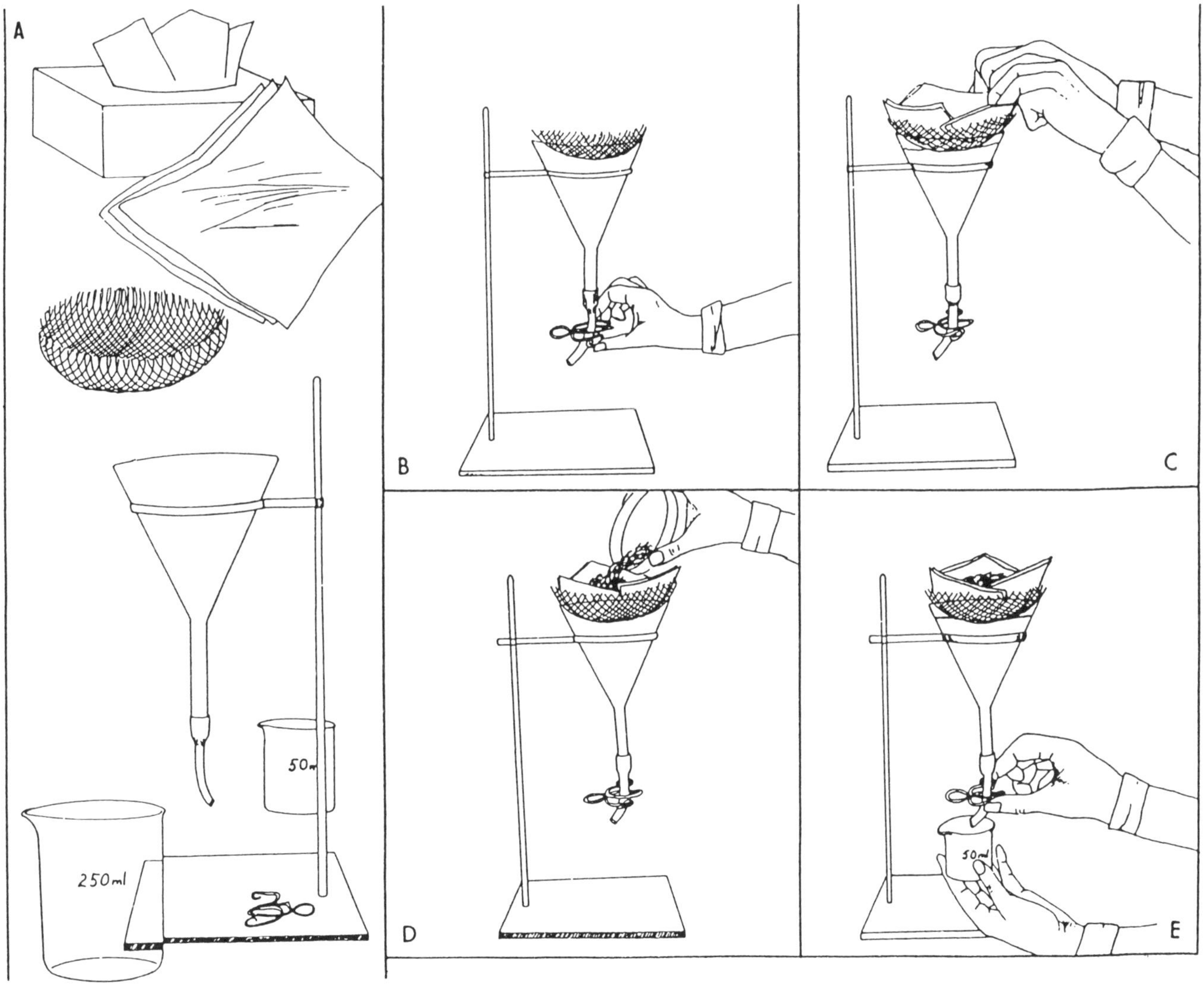


**Figure 2.9.** Extraction by the Baermann funnel method. See the text for a description of the procedure.

viewing dish (such as a Syracuse dish) or a 50-ml beaker (Figure 2.9E). Over 90% of the live nematodes are recovered in the first 5 to 8 ml of water drawn from the tubing. Only a few specimens of inactive species will be recovered. Samples may be stored at about 10–16°C (50–60°F) for examination within a few days. Lack of oxygen at the base of the funnel stem often inactivates nematodes, but they can usually be revived in fresh water.

#### MODIFICATIONS OF THE BAERMANN FUNNEL METHOD

A disadvantage of the Baermann funnel method is poor oxygenation, especially at the base of the funnel, where the nematodes settle, and some nematodes may lodge on the sloping sides of the funnel. Several modifications of this method overcome the problem of lack of oxygenation. Three modifications are described below, and others have been outlined by Southey (1986).

**Dish or tray apparatus.** A shallow evaporating dish, petri dish, or tray can be substituted for the funnel.

1. Spread a thin layer of soil, residue from sieving, or chopped infected tissue on a double layer of cheesecloth, nylon gauze, or paper placed on a flat-bottomed plastic or stainless steel screen about 2 cm (¾ in.) deep.
2. Place the screen, together with the soil or plant material, on supports (glass rods or other nontoxic materials) in an evaporating dish, petri dish, or large tray (e.g., 30 × 35 cm) containing just enough water to wet the lower surface of the sample material. Maintain the water at this level.

CAUTION Do not cover the sample material with water.

3. Place a circle of cheesecloth or paper tissue on top of the material to keep it moist and prevent plant material from floating. Add water as needed, by carefully pouring it between the cheesecloth or paper and the edge of the dish or tray.
4. After 12 to 24 h, gently remove the screen, and examine the water for nematodes. For further extraction, reimmerse the screen in fresh water for another 12 to 24 h.

**Sieve apparatus.** A similar procedure using sieves is suitable for extracting nematodes from large soil samples (100 cm$^3$ or more). See the note on sieves in the section on Cobb's sieving and gravity method (p. 35).

1. Spread a shallow layer of soil, not more than 1 cm deep, uniformly on tissue or cloth covering a 10-mesh sieve set in a pan 15 cm (6 in.) in diameter. Add water to the pan until it just reaches the level of the soil.
2. After extraction overnight or for two to five days, collect the nematodes on a 500-mesh sieve, and rinse them into a viewing dish (Figure 2.8) with a fine stream of water from a wash bottle. For maximum recovery, nematodes should be collected over a four- or five-day period.

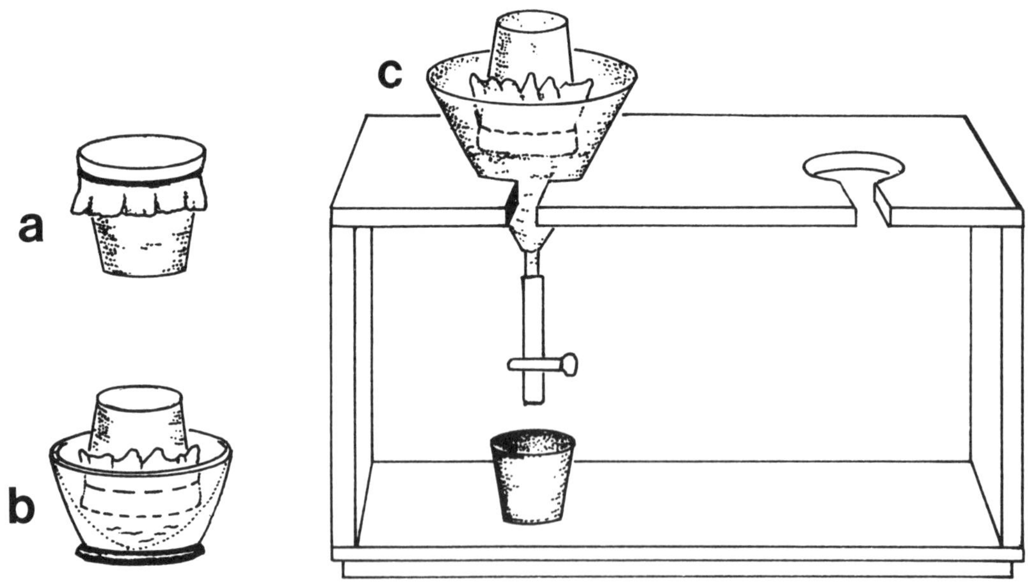


**Figure 2.10.** Simplified Baermann funnel method for extracting active vermiform nematodes from soil and plant samples. **A,** Place a sample in a glass, cover the glass with a piece of cheesecloth, and *tightly* secure the cloth with a rubber band. **B,** Invert the cloth-covered glass in a bowl with a concave bottom, using spacers to support the glass so that the cloth cover rests 5 to 10 mm above the bottom of the bowl. Add water to the level of the cheesecloth. In two to five days, carefully remove the glass. The nematodes will be concentrated in the center of the bottom of the bowl. **C,** Alternatively, place the glass on a funnel setup and collect nematodes as shown in Figure 2.9E. This method is not very efficient.

**Glass and bowl apparatus.** If resources are limited, this simplified method can be used, but it is not very efficient.

1. Place a sample in a small drinking glass, and cover the glass with cheesecloth held over the top with a rubber band (Figure 2.10A).

   CAUTION The cloth covering must be drum-tight.

2. Place wire mesh over the cheesecloth, and invert the glass in a larger glass or bowl with a concave bottom (Figure 2.10B). Add water until it just reaches the level of the wire mesh. Alternatively, the glass can be placed on a funnel setup (Figure 2.10C) in which the nematodes are collected as in the Baermann funnel method (Figure 2.9E).
3. After extraction overnight or for up to five days, carefully remove the inverted glass, still covered with the cheesecloth and wire mesh, and transfer the nematodes to a viewing dish with a pipette.

#### COBB'S SIEVING AND GRAVITY METHOD

Cobb's sieving and gravity method is rapid (taking less than 30 min) and permits the recovery of active, sluggish, and some dead nematodes from large soil samples (1.5 kg, or about 3 lb). Eggs and cysts can also be recovered by this procedure. However, the centrifugal flotation method (pp. 37–39) is better for recovering dead nematodes.

**Sieves.** Sieves of accurately graded sizes are available from scientific supply houses but are expensive. Stainless steel sieves are preferred. Inexpensive substitutes can be made from sections of plastic pipe and nylon mesh cloth of known mesh size. These are fitted over pipelike embroidery hoops. Plastic cups cut across in half nest together for constructing sieves.

The fineness of sieves is designated by a mesh number, which is the number of openings per inch of surface. For example, a 325-mesh sieve has 105,625 openings (44 μm in diameter) per square inch.

Usually only three or four of a set of sieves are used for a particular soil or plant sample. Sieves are selected to match the size of the nematode that is expected to be extracted and the type of soil in the sample. Most adults of large nematodes (e.g., *Anguina, Belonolaimus, Dolichodorus, Longidorus,* and *Xiphinema*) are caught on a 60-mesh sieve; adults of average-sized nematodes (e.g., *Aphelenchoides, Ditylenchus,* and *Hemicycliophora*) on a 150- or 170-mesh screen; adults of small nematodes (e.g., *Criconemella, Paratrichodorus, Paratylenchus, Pratylenchus,* and *Radopholus*) and juveniles of many genera on a 200- or 240-mesh screen; and small juveniles on a 325- or 350-mesh screen. The diameter of the sieve, the quantity of water used, and the amount of debris collected on the sieve affect the number of nematodes retained.

The number and fineness of sieves used in Cobb's method can be varied to better handle different nematode species and soil types. For example, a series of 20-, 100-, 200-, 325-, 500-, and 625-mesh sieves may be appropriate for fine-textured soils. For some soils and nematodes, the suspension is poured through a 60-mesh sieve, and the nematodes are collected on a 325-mesh sieve. To recover cyst nematodes, use a 60- or 80-mesh sieve last in the series, and backwash it into a beaker.

Some soils that contain high levels of organic matter or fine soil particles (e.g., some highly oxidized clayey tropical soils) tend to clog the openings in sieves, making screening difficult. Clogging can be partially avoided by inclining the sieve at an angle of about 30 to 35° to the horizontal while a suspension is poured over it. Gently patting the underside of the sieve into the bucket below and lifting it in and out a few times helps to clear it. In extracting nematodes from such soils, it may be necessary to reduce the size of the sample. In some cases, soil particles must occasionally be dislodged by sonication.

*Generally,* use screens singly. Never stack them and attempt to work a sample through them all simultaneously, as this may reduce the efficiency of recovery.

Sieves are expensive and will frequently need replacement if they are not handled properly. Thoroughly wash all sieves and pans in hot water, preferably using a strong spray from a nozzle. *Do not use soap or detergent,* which tends to plug the openings of fine sieves. Thorough rinsing is essential.

1. Obtain the following equipment (Figure 2.11A):

   several buckets or stainless steel pans (10 L, or 2.5 gal)
   500-ml Erlenmeyer flask
   50-, 250-, and 600-ml beakers
   Syracuse dishes, 50-mm petri plates, or pillboxes
   wash bottle
   series of 8-in. sieves
   3-in. sieve, 500- or 625-mesh

2. Mix soil samples thoroughly by gently tumbling them in plastic bags or on a large sheet of paper or clean plastic. Place a known volume of mixed soil (e.g., 100 to 500 $cm^3$) and about 3.5 L of water in a plastic bucket (bucket A) or a stainless steel pan. Gently mix the soil with tap water, stirring by hand or with a stick to break up all clods, free nematodes from the soil, and suspend them in the water (Figure 2.11B). Work the mass into a thick, muddy mixture. Most of the nematodes will now be suspended in the water. Dry soils should be soaked for several hours. Some workers use water containing a flocculating agent, such as Separan NP10 (12.5 μg/ml), to help break up soil aggregates and release nematodes from the sample.
3. Let the mixture settle for 30 sec to 1 min. Then decant the muddy water through a 10- or 20-mesh sieve into a second bucket (bucket B) or pan, to eliminate trash and heavy soil particles (Figure 2.11C). Avoid pouring the sediment; discard it in a receptacle for disposal. For greater recovery of nematodes, repeat steps 2 and 3 by

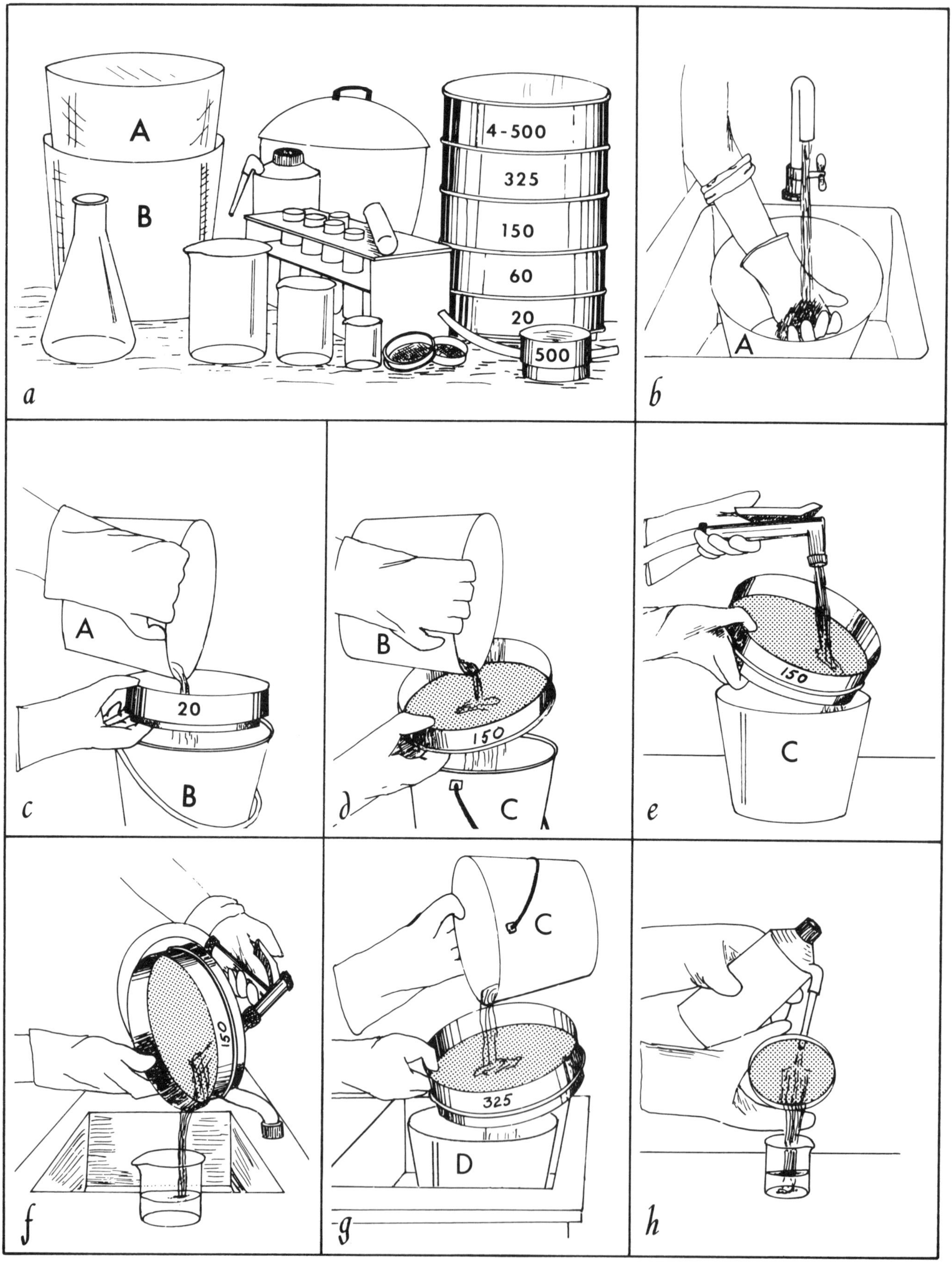


**Figure 2.11.** Extracting nematodes from soil samples by Cobb's sieving and gravity method. See the text for a description of the procedure.

resuspending the sediment in bucket A in an additional 1 L of water and pouring it again through the 10- or 20-mesh sieve into bucket B. Discard the sediment and thoroughly clean the sieve and bucket A. Most of the nematodes in the soil sample will now be in bucket B.

4. Resuspend the nematodes in bucket B by stirring or agitating the water. Let the bucket stand for about 1 min, and then pour its contents through a 150- or 200-mesh sieve into a clean bucket (bucket C or bucket A after it has been cleaned) (Figure 2.11D). Stop pouring *before* the sediment begins to move from the bucket into the sieve.
5. Hold the sieve at a steep angle (35 to 40°), and direct a gentle stream of water onto its upper side, to move the nematodes to the edge of the screen, and collect the water in bucket C (Figure 2.11E). Small nematodes and eggs will be washed through the sieve and will be recovered later.
6. Collect large nematodes on the edge of the upper surface of the sieve by directing a gentle stream of water (about 25 ml) onto the lower surface of the sieve (backwashing), and collect the water in a 250-ml beaker (Figure 2.11F).
7. Repeat steps 5 and 6, using a 325-mesh and then a 500-mesh sieve (Figure 2.11G). Recover the nematodes adhering to the sieves, and backwash them into the 250-ml beaker.
8. Resuspend the small nematodes and eggs collected in bucket C in step 5. Concentrate them on the 325-mesh sieve, and backwash them into the 250-ml beaker.
9. Adult nematodes, juveniles, and eggs in the 250-ml beaker can be concentrated by pouring the contents through a 500-mesh sieve and backwashing with small quantities of water into a viewing dish or 50-ml beaker (Figure 2.11H). A sample can also be concentrated by permitting the nematodes to settle to the bottom of the beaker and then very carefully decanting all but the last 10 to 20 ml of water. An experienced technician can recover a large percentage of the nematodes in a soil sample in about 20 min.

The procedure must be carefully followed. At the end of the process, if the nematodes are mixed with an excessive amount of debris, they can be recovered by centrifugal flotation in a sugar solution (as described in the following section), or by permitting them to migrate through well-shrunk fine-mesh cloth or paper over an inverted glass or a funnel (Figure 2.10).

The material collected from sieves may be examined directly, or it may be placed in a Baermann funnel for additional separation of nematodes from soil particles (as described in the preceding section). Most nematodes pass through the "filter" in a Baermann funnel in 24 h and can be collected almost free of debris.

For direct examination, allow all the material in the 250-ml beaker to settle to the bottom. Slowly pour off the excess water into a sink, leaving about 50 ml in the beaker. Resuspend the material in the beaker, and gently pour 5 to 15 ml into a Syracuse dish, for examination under a dissecting microscope. Finally, isolate nematodes on a glass slide (Figure 2.7), and observe their morphological and anatomical features with a compound microscope for identification of the genus and species.

#### CENTRIFUGAL FLOTATION

In the centrifugal flotation method (Figure 2.12), nematodes are separated from soil and suspended in water, and the suspension is centrifuged to concentrate the nematodes, which are then centrifuged again in a sucrose solution of specific gravity sufficient to cause them to float to the top. The procedure takes only a few minutes and can be used to process large soil samples. However, it is *appropriate only for soil samples.*

Centrifugal flotation recovers most nematodes from soil, including both active and inactive types. It is an excellent method for the recovery of sluggish and sedentary nematodes, such as *Criconemella* and *Hemicycliophora,* and it readily recovers dead, molting, or fixed nematodes and eggs. Recovery of very long nematodes, such as *Longidorus* and *Xiphinema,* or cyst nematodes requires a more dense sucrose solution than that used for standard analysis.

This procedure is generally more efficient for extracting nematodes than the Baermann funnel method or Cobb's sieving and gravity technique. Centrifugal flotation is often used to clean extracts obtained by sieving or elutriation.

1. Obtain the following equipment (Figure 2.11A):

   equipment needed for Cobb's sieving and gravity method (as described in the preceding section)
   centrifuge
   centrifuge tubes and rack

2. Separate the nematodes from a soil sample and suspend them in water as described in steps 1–7 of Cobb's sieving and gravity method (Figure 2.11) or by elutriation. Concentrate the nematodes in about 20 to 30 ml of water in a beaker, and then pour the contents equally into four or six 50-ml round-bottomed plastic centrifuge tubes (Figure 2.12A).
3. Fill each tube to within 0.5 cm (¼ in.) of the top; fresh water may be carefully added to the centrifuge tubes from a wash bottle or pipette to balance them (Figure 2.12B).
4. Load the centrifuge properly, according to the manufacturer's instructions (Figure 2.12C). Tubes on opposite sides of the centrifuge rotor must weigh the same. If an odd number of samples is being processed, prepare a water blank as a counterweight for the rotor. The centrifuge should have a horizontal rotor head with swinging buckets. The *g* force of the rotor is related to its speed (in rpm) as follows: $g = 0.00001118 \times$ (radius of the rotor, in cm) $\times$ (rpm)$^2$.

CAUTION Always follow the directions for use and the safety instructions provided by the centrifuge manufacturer.

5. Centrifuge for 4 to 5 min at 2,900 × *g* to extract vermiform nematodes, 1,150 × *g* to extract cyst nematodes, and 420 × *g* to extract *Criconemella.* Let the rotor come to a rest without braking. The nematodes and eggs will be concentrated in a pellet at the bottom of the tube.
6. Remove the tubes carefully to avoid resuspending the nematodes, and decant the supernatant (Figure 2.12D).
7. Refill the tubes halfway to the top with sucrose solution (Figure 2.12E):

    a. To suspend vermiform nematodes, dissolve 673 g of white sugar in 1.0 L of water, to obtain a specific gravity of 1.18.

    b. To suspend cyst nematodes, dissolve 1,210 g of white sugar in 1.0 L of water, to obtain a specific gravity of 1.25.

8. Resuspend the nematodes in the sucrose solution by thoroughly mixing with a stirring rod or vibrator mixer.
9. Load the centrifuge as previously described. Centrifuge for 60 sec at 2,900 × *g* to extract vermiform nematodes and 1,150 × *g* to extract cyst nematodes. The nematodes will be at the top of the container.
10. Collect the nematodes by pouring the supernatant through a 300- to 625-mesh sieve (Figure 2.12F). Backwash the nematodes on the sieve into a 50-ml beaker or viewing dish.

CAUTION Do not permit nematodes to remain in the sucrose solution longer than 2 min. The osmotic pres-

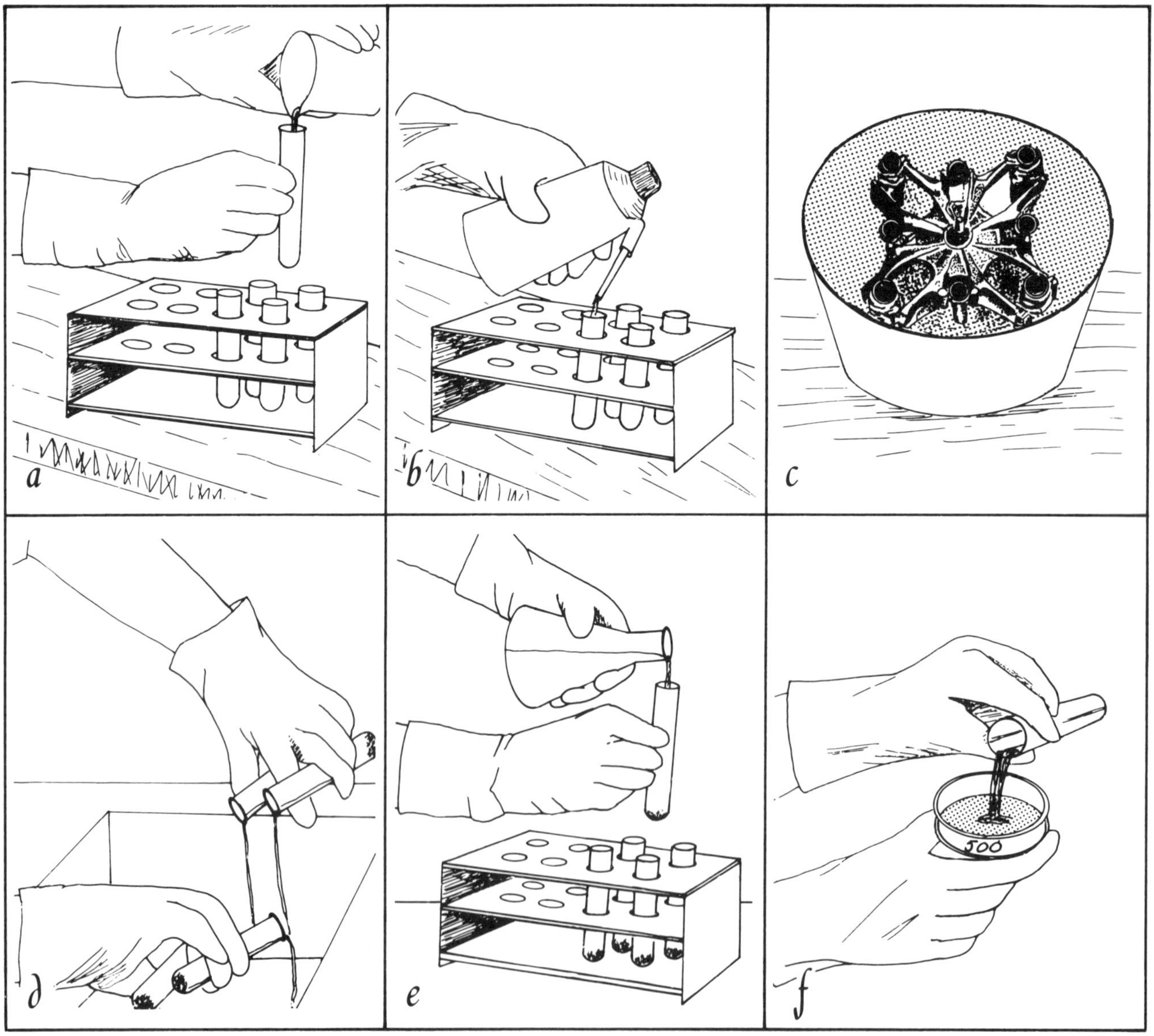


**Figure 2.12.** Extracting nematodes from soil samples by the centrifugal flotation method. See the text for a description of the procedure.

sure of the solution may distort or even kill them if they are left in it longer, making identification difficult.

#### MIST EXTRACTION

Mist extraction (Figure 2.13) is similar to the Baermann funnel method but has the advantage of not creating anaerobic conditions. Sap and toxic decomposition products from plant material, which inactivate nematodes, are washed away. This method is more efficient than the Baermann funnel method in collecting large numbers of nematodes and in recovering nematodes from a given quantity of soil and root samples, because toxic materials do not accumulate, oxygen remains at adequate levels, and nematodes are not lost during sieving of the samples. However, mist extraction requires more expensive equipment, often custom-built. A misting chamber is needed, but it is easy to construct one.

1. Set up Baermann funnels, one for each sample, as described above (Figure 2.9). Place the funnels in a rack, with a concave wire mesh in each funnel and double-thickness paper tissue over the wire mesh. Avoid tearing or puncturing the tissue. Place a large test tube (200 × 25 mm) under each funnel, with the bottom third of the funnel stem inserted in the test tube (Figure 2.13A). A spacer is required between the funnel and the test tube, to permit water in the test tube to overflow and not flood the funnel.
2. Remove pebbles, trash, etc., from the soil samples. Cut feeder roots into 0.5- to 2.5-cm lengths. Cut bulbs, corms, tubers, storage roots, foliage, stems, and woody tissues into cubes about 10 $cm^3$ or pieces 3 to 4 mm long.
3. Carefully place about 60 $cm^3$ (¼ cup) of moist soil or plant material, as prepared in step 2, uniformly over the tissue in each funnel. Label the funnels and test tubes with the sample number, time, and date. Place the labeled tubes in a test tube rack.
4. Place the loaded funnel rack in a mist chamber (Figure 2.13B), close the chamber door or curtain, turn on the misting system, and incubate for three to five days (or two days, for bud and foliar nematodes). Water under pressure is sprayed through small nozzle orifices, creating a fine mist that covers the material to be extracted. A solenoid in the line, connected to a time clock, provides

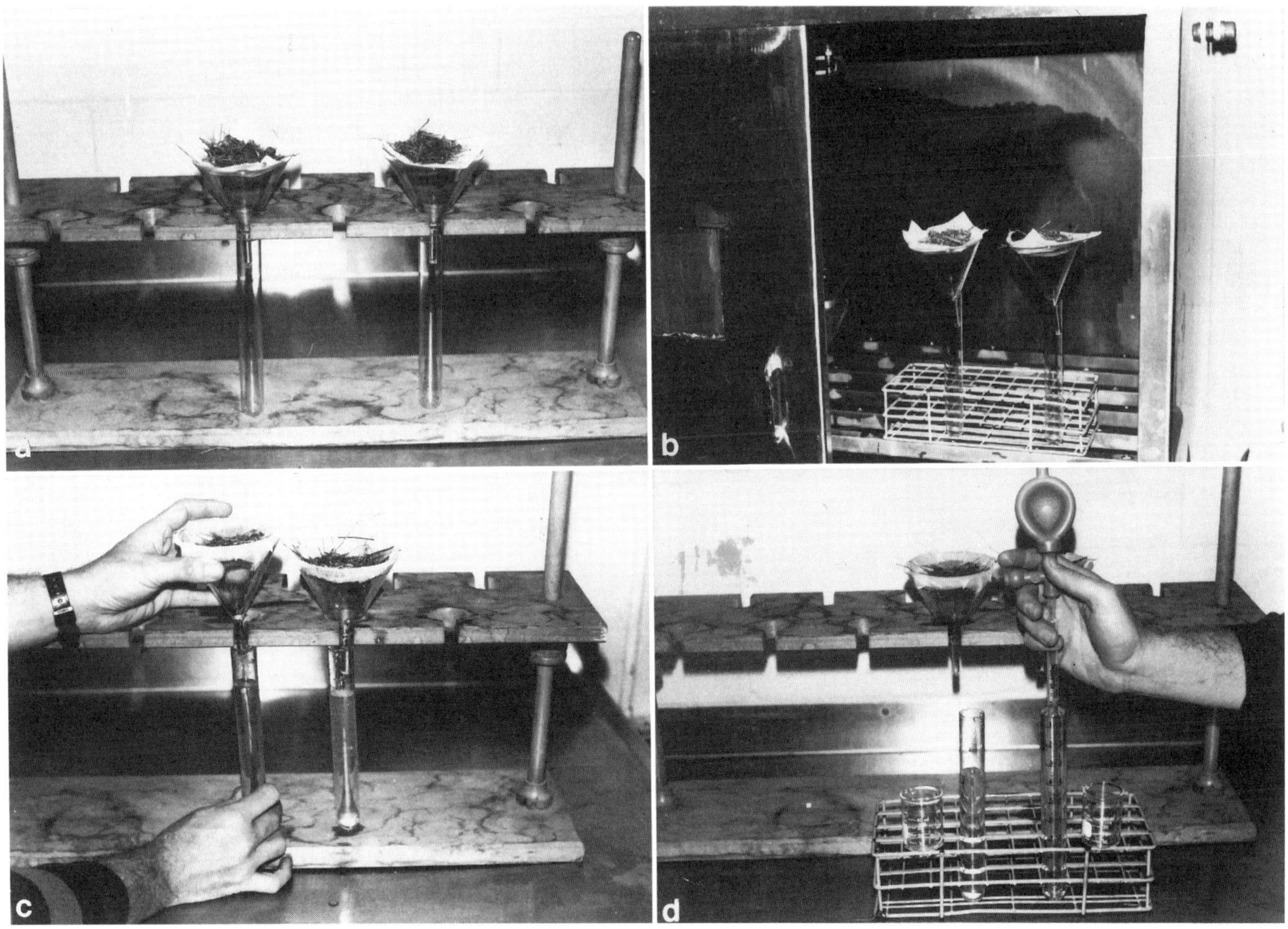


**Figure 2.13.** Extracting live vermiform nematodes from soil and plant tissue samples by the mist extraction method. See the text for a description of the procedure.

an intermittent misting cycle of 1.5 min on and 8.5 min off without dripping. Water percolating slowly through the funnel will gradually fill the test tube and overflow. The motion and density of nematodes (specific gravity greater than 1.0) cause them to settle and concentrate at the bottom of the test tube. A water pressure of about 2.8 kg/cm$^2$ (40 psi) is usually needed to supply a fine mist. Carefully regulate the supply of water to move the nematodes without washing them out of the funnel. Water at 27–29°C (80–85°F) is essential for good recovery.

5. After the extraction period, carefully remove the funnel stems from the test tubes (Figure 2.13C). With a large pipette, remove the nematodes in a few milliliters of water from the bottom of each tube (Figure 2.13D), and transfer them to a 50-ml beaker or viewing dish. Avoid stirring the water while removing the nematodes, or they will be resuspended.

   Use a clean pipette for each sample, or rinse the pipette inside and outside before reusing it, to avoid contamination of the next sample.

#### TRITURATION

Trituration (Figure 2.14) is excellent for extracting active nematodes from bulbs, cloves, corms, storage roots, crowns, leaves, and small plants (e.g., strawberry). However, the blending action may release toxic substances from the plant that can kill or immobilize nematodes. Some blended tissues also form a thick, sticky mass that is difficult to sieve and inhibits the movement of nematodes.

1. Place about 10 g of washed, diced plant material (in pieces 2 to 3 cm long) in the jar of an electric food blender (e.g., Waring, Sunbeam Domestic, Dormeyer, or M.S.E. Atomix), add 200 ml of water, and run the blender at medium speed for 20 to 50 sec to triturate the tissue without undue damage to the nematodes.

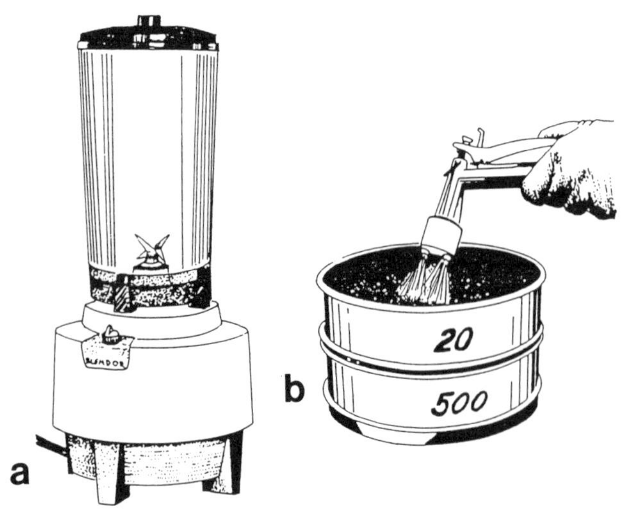


**Figure 2.14.** Extracting nematodes from plant parts by trituration.

2. Pour the slurried mixture through a 20-mesh sieve nested on top of a 500-mesh sieve. Thoroughly wash the slurry on the 20-mesh sieve with water from a fogger nozzle or hose to remove the nematodes (Figure 2.14B). Discard the slurry on the 20-mesh sieve, as the nematodes will be on the 500-mesh sieve.
3. Backwash the material on the 500-mesh sieve into a 50-ml beaker or viewing dish (Figure 2.11H).

#### OTHER METHODS OF EXTRACTION FROM PLANT TISSUES

Trituration and mist extraction are effective methods of recovering nematodes from plant tissues. Three other methods of extraction from plant tissue are described.

**Emergence in a viewing dish.** Cut infected leaf or stem tissue into 8-mm or smaller squares, and cut roots into pieces 8 mm long or less. Cover them with water in a shallow viewing dish. Endoparasitic migratory nematodes will wiggle out of infected tissue and can be viewed directly in the dish, often in less than an hour, or after soaking overnight. Species that can be extracted by this technique include foliar nematodes (*Aphelenchoides* spp.), the pinewood nematode (*Bursaphelenchus xylophilus*), burrowing nematodes (*Radopholus* spp.), and the coconut red ring nematode (*Rhadinaphelenchus cocophilus*).

**Soaking in water.** Migratory endoparasites (e.g., *Pratylenchus* and *Radopholus*), sedentary parasites in immature stages, and male sedentary parasites emerge from roots soaked in water over a period over several days.

1. Cut plant tissues such as leaves, stems, and roots (which should be free of debris and soil) into pieces 5 to 10 cm long. Fibrous roots should be no more than 3 mm in diameter. Roots of larger diameter and fleshy roots can be split longitudinally to help nematodes emerge.
2. Place moist plant tissues loosely in a container that can be tightly closed (e.g., a small bottle, screw-cap jar, petri plate, or reclosable plastic bag). Add enough water to cover the plant material. Seal and label the container.
3. Let the sealed container stand for three to four days at room temperature, about 20–25°C (68–77°F).
4. Remove nematodes and add fresh water daily, until no more nematodes emerge. Shake the container gently to wash the plant tissue, and decant the liquid over a coarse (48- or 60-mesh) sieve nested over a 325- or 500-mesh sieve. Discard roots or other plant material collected on the coarse sieve. Collect the nematodes on the fine sieve, holding it at an angle of 35 to 45° over a 50-ml beaker or collecting dish. Tilt the sieve and rinse it from the back with a small stream of water (10 to 20 ml) to wash the nematodes and debris down the side of the sieve and into the beaker or dish.

The addition of 1 to 3% hydrogen peroxide ($H_2O_2$) and antibiotics to the water improves extraction. Hydrogen peroxide raises the level of dissolved oxygen in the water, while antibiotics inhibit the growth of bacteria and fungi.

**Shaker agitation.** Migratory endoparasitic nematodes (e.g., *Pratylenchus*) also emerge from chopped roots in a flask or jar of water under continuous agitation in a shaker (motor-driven, wrist-action, or other type). Sufficient agitation is required for adequate aeration. The shaking may continue for several days. Extraction is improved if chopped roots are shaken in water with ethoxyethyl mercuric chloride (2 ppm) and dihydrostreptomycin sulfate (50 ppm).

#### WATER FLOTATION FOR RECOVERY OF CYST-FORMING NEMATODES

1. Place about 1,000 $cm^3$ of *dry* soil in a plastic bucket. Thoroughly stir the soil with a strong stream of water, or fill the bucket with water and stir mechanically or manually.
2. Pour all floating material through a 20-mesh screen nested over a 60-mesh sieve (an 80- or 100-mesh sieve may be needed to catch some species of small cyst nematodes, such as *Heterodera trifolii*). Discard the coarse material caught on the 20-mesh sieve.
3. With a gentle stream of water from a fogger nozzle, wash the material retained on the finer-mesh sieve, and backwash it into a 600-ml beaker. Let the suspension settle for 1 to 2 min. Label the beaker with the sample number.
4. Stir the suspension vigorously, and immediately pour it through the 60-mesh sieve and into the sink. Rotate the beaker while pouring to dislodge material from its sides. Leave heavy material (e.g., small pebbles) in the bottom of the beaker.
5. With a spatula, transfer the material on the sieve to a plastic scanning dish with partitions. Place no more than one-half teaspoonful of material in each section of the dish. Divide large amounts of material between separate scanning dishes.
6. With a wash bottle, rinse particles remaining on the spatula into the scanning dish. Then carefully add enough water to float the material to near the brim in each section.
7. Using a dissecting microscope, look for cysts in the floating material in each section of the scanning dish. Use a dissecting needle to spread clumps of material for a better look at cyst-like objects.

Another method of recovering cysts is the following:

1. Air-dry a soil sample, and then pass it through a coarse sieve (4-mm apertures).
2. Place 50 $cm^3$ of dried soil in a 1-L Erlenmeyer flask with 500 ml of water. Shake the flask vigorously, then fill it with water to just below its lip, and let it stand for 10 min. Cysts will float to the surface with other debris.
3. Collect cysts with a pipette or a fine-mesh sieve for further examination. Alternatively, pour the floating material onto filter paper in a funnel, let the water drain off, and then examine the filter paper under low magnification to detect cysts, which can usually be found along the high-water mark on the paper.

### Preparation and Mounting of Nematodes for Microscopic Observation

#### STORAGE OF NEMATODES IN SUSPENSION

Many nematodes remain active and in good condition for several days when stored in shallow, fresh tap water at about 5–10°C (40–50°F). Contaminating bacteria can be kept down by adding three drops of a 5% solution of streptomycin sulfate for each 5 ml of nematode suspension.

#### PREPARATION OF SPECIMENS FOR MICROSCOPIC OBSERVATION

The preparation of specimens (Figures 2.3, 2.4, and 2.6) is usually begun under a dissecting or stereoscopic microscope (with magnifications of 10× to about 100×), with the nematodes in distilled water in a small Syracuse dish or petri dish for viewing or counting (Figure 2.8). Details of morphology, such as stylet structure, head framework, esophageal lumen, excretory pore, and male gonads are best seen in living or freshly killed specimens. Microtools suitable for picking up individual nematodes are described below and shown in Figures 2.4 and 2.5. Individual nematodes are gently picked up (Figure 2.6) and transferred to a drop of water or anesthetic on a glass microscope slide for viewing morphological and anatomical features needed for identification.

Several anesthetics are available for immobilizing nematodes without killing them. A 0.5% aqueous solution of propylene phenoxital, made fresh every few days, is excellent. Dichlorodiethyl ether (two drops in 50 ml of water in a small stoppered bottle) also works, as does chlorobutanol (Chloretone, 2 g in 500 ml of water). Anaesthetized nematodes recover when placed in fresh water, or they can be killed and fixed for processing and mounting later.

The compound microscope should be of good to excellent quality (Figure 1.16), and the diagnostician must be skilled in its use. An oil immersion lens is valuable in seeing minute anatomical details. An excellent light source is needed for the microscope. Because nematodes are soft-bodied, special procedures are used to avoid distortions caused by pressure from the coverslip (Figure 2.7B and C). Mounts are useful for sending specimens to a nematologist for positive identification. Mounts may be temporary or permanent.

#### PREPARATION OF TEMPORARY MOUNTS

Temporary mounts in water are useful for a few hours or days if kept refrigerated or if the specimens are fixed in a 2% formaldehyde solution. Temporary mounts are commonly used in plant diagnostic laboratories.

1. Obtain the following equipment and supplies:

coarse fiberglass (glass wool), cut into pieces about 3 to 5 mm long and about the same diameter as the nematodes, or class IV glass microbeads, to be used as coverslip supports, stored in a small, covered viewing dish (e.g., a petri dish or Syracuse dish)

thin, sharp-pointed dissecting needles. Insect mounting pins (size 00, or the finest size) inserted in the eraser of a pencil are excellent. Needles or picks for picking up and transferring individual nematodes can also be made from a nylon toothbrush bristle or eyebrow hair cemented to the end of a glass or wooden rod; a dentist's pulp probe glued to the end of a needle holder; or a sharpened bamboo splinter, cut from a section of dry bamboo (2 to 3 cm in diameter and about 15–25 cm long) and shaped with a sharp knife to taper to a long, gradual point (Figure 2.5). The point is trimmed with a razor blade under a dissecting microscope, until at the tip consists of only two or three fibers of the outer bamboo cane.

micropipettes drawn in a gas flame from commercial, soft glass pipettes, controlled by mouth suction applied through rubber or soft plastic tubing

glass microscope slides (25 × 76 mm) and 18- to 19-mm, round, no. 0 or 1 glass coverslips. Cobb-type aluminum double-coverslip slides are preferable, if available. Their advantages are that they are more easily handled for storage and transit, they allow viewing of specimens from either side, they stack horizontally so that the glass does not touch other slides, and they commonly withstand being dropped on the floor. Glass slides with a thick card label at each end may also be stacked.

alcohol burner

small birthday candles

fine forceps

coverslip sealing material, kept in an airtight container with a fine brush attached to the stopper or cap. Silastic (Dow Corning), Permount (Fisher Scientific), and Silicone Seal (General Electric) in the United States and Glyceel (Hopkins & Williams) in the United Kingdom are excellent.

NOTE Silicone rubber should be diluted 30 to 40% with ligroin naphtha solvent (Caveness, 1969).

Clear fingernail polish applied with a fine brush is also satisfactory. The brush can be kept in the sealant or in a tube of solvent (e.g., *n*-butyl acetate). For a *permanent mount,* seal the edge of the coverslip with nail polish. When the nail polish is firm, ring the coverslip two or three times with a sealant (Glyceel, Permount, Silastic, or Silicone Seal).

A 50:50 mixture of paraffin (melting point about 60°C) and petroleum jelly also makes a satisfactory sealant but must be kept hot when it is being used. Paraffin wax is a satisfactory sealant and easily applied with the heated wick of a small birthday candle (Figure 2.7F).

2. Prepare several slides by applying a ring of sealing material, with a uniform depth of about 0.25 mm and breadth of 1 mm, to each slide with a soft, fine brush (Figure 2.7G). The diameter of the ring must be smaller than that of the coverslip. The sealant serves as a coverslip support to avoid pressing and distorting the specimen. Repeat the process after the first ring has dried, to give a long-lasting seal.
3. With an eyedropper place a drop of water, fixative (see p. 43), or mounting fluid on a clean glass microscope slide or a slide prepared with a dry ring of sealing material.
4. Transfer 10 to 15 extracted nematodes with a fine dissecting needle, toothbrush bristle, or bamboo pick to the drop of liquid on the slide as follows:
   a. Place a viewing dish containing the nematodes under a dissecting microscope. Focus on the nematodes at a magnification of 10 or 20×.
   b. Place the point of the needle, bristle, or pick under a nematode and gently raise it to the surface (Figures 2.4B and 2.6A). Adjust the microscope with the other hand to keep the nematode in focus as it is raised out of the water.
   c. Carefully lift the nematode out of the water and transfer it to the drop of liquid on the slide (Figure 2.6B).
   d. Repeat steps b and c until the desired number of nematodes (typically 10 to 15) has been transferred.
   e. If the nematodes are alive, gently kill (relax) them by passing the microscope slide back and forth several times over the flame of an alcohol burner (Figure 2.7A). *Do not overheat the slide,* as the water may boil or the slide may crack. Overheated nematodes are unsatisfactory for viewing.
5. Examine the nematodes under a magnification of 10 or 20× to make sure they are killed. Then gently move them toward the center of the drop without allowing them to touch one another.
6. With the tip of a moistened needle, place several pieces of fiberglass or a few glass microbeads evenly around the edges of the drop containing the nematodes (Figures 2.7B and C). The glass wool or beads should be about the same diameter as the nematodes or slightly thicker, so that the nematodes will not be flattened and distorted by the weight of the coverslip. These spacers are not needed if the slide was first ringed with sealing material (step 2).

   NOTE Pieces of fine nickel-chrome or stainless steel wire can also be used as spacers, particularly for thicker nematodes.
7. With a fine forceps, carefully pick up a clean coverslip and pass both sides over the flame to remove lint and moisture. Slowly and carefully lower the coverslip over the water drop, holding the coverslip at an angle to avoid

trapping air bubbles and displacing liquid containing nematodes (Figure 2.7D).

8. Slowly remove excess water with a piece of absorbent paper (Figure 2.7E), or adjust the size of the drop with a micropipette, while observing the specimen under the microscope to avoid drawing the nematodes from under the coverslip or removing too much liquid, which *must not protrude beyond the edges of the coverslip.* The drop should be smaller than the coverslip.
9. "Tack" the coverslip (place several drops of sealant around its edge) to prevent slippage during the sealing process. Then, using a fine brush, apply a 1- to 2-mm band of sealant around and slightly over the edge of the coverslip, to seal it to the slide. Wax from a birthday candle can also be used; melt the tip of the candle in a flame, without igniting the wick, and apply molten wax to the edge of the coverslip, using the wick as a brush (Figure 2.7F and G).
10. Allow the sealant to dry. Label the slide with a grease pencil or, preferably, a stick-on label. Include at least the date, the name of the host plant, and the locality where the specimen was collected (Figure 2.7H).

Well-sealed temporary mounts will keep in good condition for several hours or, if the nematodes are mounted in a fixative, for several weeks.

The importance of examining temporary mounts of living, anesthetized, or even freshly fixed nematodes cannot be overemphasized. Refractive structures, such as the stylet, head skeleton, esophagus lumen, excretory pores, and spicules, are much more distinct when fresh than when they have been fixed for some time or processed with glycerin.

#### PREPARATION OF PERMANENT MOUNTS

Permanent mounts of accurately identified nematodes are useful for reference in the identification of nematodes in plant diagnostic clinics. They are also used for nematodes that are to be shipped to a specialist. Permanent mounts will keep for years at room temperature. Specimens are first killed and then fixed to precipitate proteins, which preserve their lifelike appearance. Specimens may be kept indefinitely in fixatives, but they may undergo undesirable changes during prolonged storage. In fixed nematodes, much of the internal body contents, especially the gonads, may be difficult to see, because of the granular appearance of the intestine.

Specimens may be cleared by processing in hot (60 to 70°C) lactophenol, lactoglycerin, or glycerin (glycerol), which are also suitable mountants. Phenol is a dangerous poison, so use lactophenol with caution under an exhaust hood. Processing with lactoglycerin (equal volumes of lactic acid, glycerin, and distilled water, plus 0.05% acid fuchsin or 0.05% cotton blue [methyl blue] stain) is quicker than processing with glycerin. Well-sealed mounts in lactophenol or lactoglycerin should last for many years. Two methods are described.

**Glycerin (glycerol) method**

1. Kill the nematodes in distilled or deionized water by gentle heating (50 to 65°C). *Do not overheat the water.*
2. Transfer the nematodes to a viewing dish (e.g., a petri dish) containing a hot fixative (60 to 70°C), cover the dish, and let it stand for at least 12 h at room temperature. Nematodes can be stored indefinitely in fixative in vials. The vials should be labeled with the identity of the nematode or nematodes they contain (where known), the source of the specimen, the locality where it was collected, the fixative used, and the date of fixation. Five commonly used fixatives are described below.

Formalin

| | |
|---|---|
| Commercial formalin (37–40% formaldehyde) | 8 ml |
| Distilled water | 92 ml |

Formalin–acetic acid (FA)

| | |
|---|---|
| Commercial formalin (37–40% formaldehyde) | 10 ml |
| Glacial acetic acid* | 1 ml |
| Distilled water, to make up 100 ml | |

*Propionic acid may be substituted for acetic acid.

NOTE FA is best at boiling (100°C).

Place the nematodes in about 1 ml of water. Heat the fixative in a tube held in boiling water, and add about 5 ml of hot fixative to the nematodes.

Formalin-glycerin (FG)

| | |
|---|---|
| Commercial formalin (37–40% formaldehyde) | 10 ml |
| Glycerin | 6 ml |
| Distilled water | 84 ml |

Heat the fixative to boiling. Add a small volume of the boiling fixative to an approximately equal volume of nematodes in water. This causes the nematodes to straighten out for better viewing.

Formalin–acetic acid–alcohol (FAA)

| | |
|---|---|
| Commercial formalin (37–40% formaldehyde) | 2.4 ml |
| Glacial acetic acid | 1.6 ml |
| Ethyl alcohol, 95% | 60.0 ml |
| Distilled water | 80.0 ml |

Triethanolamine–formalin (TAF)

| | |
|---|---|
| Commercial formalin (37–40% formaldehyde) | 7 ml |
| Triethanolamine | 2 ml |
| Distilled water | 91 ml |

NOTE TAF is a commonly used fixative, as nematodes retain their lifelike appearance in it for several hours. However, it is *not a good long-term preservative,* as some degeneration of nematode cuticle can take place. For long-term preservation, transfer of nematodes to FA or an excess of FA plus 2% glycerin is recommended.

3. Remove most of the fixative with a clean, fine pipette or dropper. Take care that nematodes are not inadvertently drawn into the pipette.
4. Add 3 to 4 ml of a solution of glycerin, alcohol, and water, and cover the dish. This permanent mounting medium is prepared as follows:

   Glycerin, anhydrous ........ 2 ml
   Ethyl alcohol, 95% ........ 8 ml
   Distilled water ........ 90 ml

5. Let the sample stand at 50 to 60°C in an oven for a few days, or until the water and alcohol have all evaporated. This step is *critical.* If evaporation is too rapid, the nematodes shrink and become distorted. The rate of evaporation and the amount of time taken are controlled by the snugness of the lid covering the dish, the ambient temperature, and the amount of solution initially placed in the dish.

**Alternative method.** An alternative method of preparing permanent mounts takes about an hour.

1. Perform steps 4 and 5 of the glycerin method.
2. After the nematodes have been fixed, transfer them sequentially through a series of four hot solutions (50 to 55°C) with increasing amounts of glycerin and decreasing amounts of lactophenol.

| | Glycerin (ml) | Lactophenol (ml) | Formalin (ml) |
|---|---|---|---|
| Solution 1 | 25 | 75 | 0 |
| Solution 2 | 50 | 50 | 0 |
| Solution 3 | 72 | 25 | 3 |
| Solution 4 | 85 | 12.5 | 2.5 |

Allow the nematodes to remain in each solution for 10 min.

3. Place a drop of anhydrous glycerin on a clean microscope slide, transfer the nematodes to it, add a few fiberglass wool spacers to the edge of the drop, cover with a glass coverslip, seal the coverslip to the slide, and label the slide mount (for a description of these procedures, see Preparation of Temporary Mounts, pp. 41–43).

**Labeling and storage of permanent mounts.** It is very important to properly label and store slides for permanent collections. Always use waterproof ink, such as India ink. Permanently fix the label to the slide, giving the name of the genus and the species, where known; the growth stages and sex of the nematodes (♂, male; ♀, female; juv., juvenile); details of the habitat; the locality where the specimen was collected; the name of the collector or identifier; the method by which the nematodes were killed, fixed, and processed; and the mordant used. Slides with type specimens should be clearly marked as such. Store all slides horizontally in a cool, dry atmosphere and in fireproof cabinets whenever possible.

#### PRESERVING NEMATODES

Soil and plant samples and extracted nematodes are commonly stored for a few days to several months for later examination by a clinician or nematologist. Soil and plant samples can be kept several weeks if they remain cool and are protected from desiccation (see Care and Labeling of Samples, p. 24). Extracted nematodes in preservatives can be kept for years at room temperature. However, the killing and fixing must be carefully done to avoid distortion of the nematodes, which interferes with identification. Two methods are described below.

CAUTION All work with formaldehyde must be done under an exhaust hood.

**Hot formaldehyde method**

1. Prepare a 2% formaldehyde stock solution: pour 50 ml of commercial 38% formaldehyde into a 1,000-ml graduate cylinder, and add 950 ml of distilled water. The stock solution may be kept for several months but should be discarded if it becomes cloudy or a sediment develops. Keep the stock solution in a clean 4-L (1-gal) brown glass bottle with a screw cap, labeled "Formaldehyde 2% Solution," with the date of preparation. Store it in a dark place.
2. Heat 10–30 ml of the formaldehyde stock solution (10 ml for each sample) in a 250-ml Pyrex Erlenmeyer flask over a Bunsen burner or an electric hot plate, until the solution just begins to bubble.
3. Place 10 ml of a nematode suspension in a 50-ml beaker, and immediately add 10 ml of the hot formaldehyde stock solution. To process a series of samples at the same time, heat two flasks of formaldehyde solution. Use one flask for three consecutive samples and then alternate with the other flask, which has been kept hot.
4. Let the samples settle and cool in the beaker for about 10 min.
5. Carefully decant and discard about 10 ml of the supernatant into a sink. Be careful not to resuspend the nematodes by rough handling of the beaker.
6. Suspend the nematodes by gently swirling the beaker. Immediately pour the contents into a sturdy shipping vial. Cap it tightly, and tape the cap to the vial. Label the vial with a sample number and the date.
7. Repeat steps 3 to 6 until all samples are done.

**Hot water–formaldehyde method**

1. Heat water in a 250-ml Pyrex Erlenmeyer flask to about 100°C, until it just begins to boil.
2. Place a nematode suspension in a 50-ml beaker, and note the volume of water in the beaker.
3. Gently swirl the beaker to suspend the nematodes.
4. Slowly pour an equal volume of hot water from the Erlenmeyer flask into the beaker to kill the nematodes. To

process a series of samples at one time, reheat the water after preparing two or three samples, or heat two flasks of water and alternate the flasks for every two or three samples.

5. Let the samples settle and cool for about 30 min.
6. Gently decant and discard all but about 10 ml of the supernatant. Be careful not to resuspend the nematodes on the bottom by rough handling of the beaker.
7. Add 10 ml of formalin (37 to 40% formaldehyde) to the beaker containing the 10 ml of suspended nematodes in distilled water.
8. Proceed as outlined in steps 5 to 7 of the hot formaldehyde method.

#### MOUNTING ROOT-KNOT NEMATODES

Procedures for mounting root-knot nematodes (*Meloidogyne* species) to observe the perineal patterns (posterior cuticular markings around the vulva and anus) of females are described in the section on *Meloidogyne* in Chapter 3 and shown in Figure 3.35. The perineal patterns of six species of *Meloidogyne* are shown in Figure 3.34.

#### MOUNTING CYST NEMATODES

The vulval cones of cyst nematodes (*Globodera* and *Heterodera*), which include the vulva, fenestra, and associated internal structures, together with the general shape of the cysts, are used for identifying these nematodes.

Soak dry cysts in water overnight, or somewhat longer, before dissection. Place a moist cyst on a plastic slide on the stage of a dissecting microscope. Cut off the posterior end, so the fenestral area is in the center. With a very fine forceps and a dissecting needle, squeeze out the adhering body contents, including the eggs. Take care not to damage structures associated with the vulva. Thick-walled and heavily pigmented species can be bleached for a few minutes in hydrogen peroxide ($H_2O_2$) to more easily observe structures. Avoid overbleaching! Wash the cleared vulval cones in distilled water and then pass them successively through 70, 95, and 100% ethyl alcohol and then clove oil. Clear the cones in clove oil, and finally mount them in Canada balsam before transferring the specimens to a glass microscope slide. Support the coverslip with several pieces of fine glass rod or broken pieces of coverslip thick enough to prevent crushing the specimens. Vulval cones may also be mounted in Euparal, after passage through 70% ethyl alcohol and isobutanol, or they may be mounted directly in glycerin and sealed as described above.

## Mailing and Shipping Nematode Specimens

Plant diagnosticians sometimes send specimens to nematologists experienced with particular crops or nematodes to identify or confirm the identity of the specimens. Nematode samples can be mailed without permits from the U.S. Department of Agriculture or other agencies if the samples have been fixed or the specimens are on permanent slide mounts. Permits are required if live nematodes are shipped from areas under quarantine or shipped to other states or countries. Shipping permits are obtained from federal or state departments of agriculture. In the United States, application forms and federal permits may be obtained by writing to

Biological Assessment Support Staff
USDA, APHIS
Federal Building
6505 Belcrest Road
Hyattsville, MD 20782

#### NEMATODES MOUNTED ON GLASS SLIDES

Nematodes that have been killed and fixed and mounted on glass slides are easily packaged and shipped. Procedures for preparing slide mounts are described above (pp. 41–45). Package slides so that their coverslips are protected from abrasion and dust. A small slide box or tray, available from laboratory suppliers, is recommended. Specimens mounted on slides are preferable for shipping, since the package is small and lightweight, and specimens are not susceptible to deterioration due to aging or heat.

#### NEMATODES IN LEAKPROOF VIALS

Nematodes that have been killed and fixed can be stored in acidified glycerin in leakproof vials. Wrap each vial with absorbent paper toweling, and nest the vials snugly in a small box. Close the box, put it in a sealed plastic bag, and then package the bag in a sturdy box for shipment. This procedure should protect the samples from breakage in shipment. Also, if leakage occurs, the outer box should not get wet or release objectionable fumes of the preservative.

#### NEMATODES IN SOIL AND PLANT SAMPLES

Directions for packaging and shipping soil and plant samples are given above (pp. 24–25).

## Inoculating Plants to Demonstrate Pathogenicity

To establish that a nematode associated with diseased plants is the infectious agent causing the disease, four conditions must be fulfilled (Koch's postulates): the nematode must be consistently associated with the disease; a pure culture of the nematode must be obtained from an infected plant; healthy plants must be inoculated with the nematode and in time develop typical symptoms; and the nematode must be reisolated from the inoculated plants and found to be identical to the inoculum. Plant diagnostic clinics rarely perform inoculation trials to determine whether these conditions are met, because the process takes weeks or possibly months to complete. Inoculation tests to determine pathogenicity should be carried out by an experienced nematologist who clearly understands the biology of the nematode and the host plant and the pitfalls of the procedure. The general procedure is as follows:

1. Host plants are grown in a sterilized or pasteurized growing medium, usually a potting mix or a sandy soil, in clay or plastic pots (10 cm in diameter) on a porous greenhouse bench. Clay pots are preferred, because the environment in the pot is more conducive to nematode survival. The greenhouse temperature must be adjusted to suit the requirements of the plant and the nematode.

   CAUTION To prevent contaminating potted plants with nematodes and other pathogens from adjoining pots or water on the surface of the bench, place dividers between the pots, and elevate the pots above the bench surface with a saucer or inverted pot, or use a bench with a wire mesh top to prevent water from collecting on the bench surface.

2. Extract live nematodes from host plants or from the soil around them (see pp. 31–41). Take care to select plant tissue from which the nematodes are extracted. Remove individual nematodes from the mass extraction, and place them in water. Great care must be taken to avoid selecting species of nematodes other than the one under investigation.
3. Introduce the nematodes into the infection court:

   a. *Root-infecting nematodes* are often introduced by pouring water containing nematodes into small holes dug in the root zone of potted plants. The holes are then covered, and soil moisture is maintained within narrow limits. Drought or flood conditions may kill the nematodes.
   b. *Foliar nematodes* are introduced by pouring water containing nematodes over young tissues, such as apical buds or crowns. The inoculated plants are then placed in a mist chamber for several days, or at least kept under conditions that maintain a film of water on the plant shoot for several hours or more.

4. Initial inoculation often does not reproduce symptoms of the disease, because too few nematodes were introduced or because many nematodes in the inoculum were killed by handling. However, large numbers of nematodes are produced in the infected plants, and they can be used to adequately inoculate other plants and reproduce the disease.

Methods of inoculating plants for bioassay and to demonstrate pathogenicity of nematodes have been described by Barker et al. (1985a, 1986), Nickle (1991), and Zuckerman et al. (1990). A few methods are described here to identify nematodes to the level of genus, species, and race.

## Preparation of Inoculum

Several options are available for obtaining nematodes for inoculation studies or other purposes.

#### COLLECTION OF EGGS OF ROOT-KNOT NEMATODES AND RELATED NEMATODES

Eggs of root-knot nematodes (*Meloidogyne* spp.) and related nematodes (e.g., *Meloidodera*) can be collected from egg matrices by dissolution of the gelatinous matrix surrounding the eggs. Exposed egg sacs are abundant in heavily infected roots of favorable host plants.

1. Gently wash infected roots to remove adhering soil.
2. Vigorously shake pieces of roots for 4 min in a stoppered flask containing 5.25% liquid bleach (e.g., Clorox) diluted with four volumes of water, to make a 1% NaOCl solution.
3. Pour the liquid over nested sieves, and thoroughly wash the collected eggs with tap water to remove the bleach.
4. Suspend the eggs and adjust the concentration for inoculation.

This procedure can be combined with differential centrifugation in sucrose solutions (see pp. 37–39) to produce concentrated suspensions of eggs.

#### COLLECTION OF EGGS FROM CYSTS

Eggs from cysts (e.g., those of *Heterodera* and *Globodera* spp.) can be collected by trituration. The procedure is as follows (Acedo and Dropkin, 1982):

1. Collect cysts from soil on a 60-mesh sieve (250-μm apertures).
2. Wash the screenings onto a 100-mesh sieve (150-μm apertures).
3. Gently rub the cysts with a rubber stopper to release the eggs.
4. Wash the sieve over a bucket or large beaker with a gentle stream of water.
5. Concentrate the eggs by centrifugation at 800 × *g* for 5 min in 50-ml round-bottomed conical centrifuge tubes. A swinging horizontal-head rotor is best.
6. Carefully siphon off the supernatant, leaving about 2 ml of it above the pellet.
7. Stir the pellet, and layer the slurry over a sucrose gradient in 50-ml conical centrifuge tubes. The gradient consists of 10 ml of 50% sucrose overlaid with 10 ml of 40% sucrose and then 10 ml of 20% sucrose.
8. Centrifuge for 5 min as in step 5.
9. Clean embryonated eggs concentrated in a band in the 40% sucrose layer. This band may also contain hatched juveniles and some fungal spores.
10. Siphon off the band, and pass it through a 250-mesh nylon sieve (60-μm pores), which removes the juveniles and fungus spores. Catch the eggs on a 450-mesh sieve (31-μm pores).

#### ASEPTIC CULTIVATION OF NEMATODES

In aseptic culture, inoculum of plant-parasitic nematodes is reared on host plant callus, excised roots, or carrot slices in order to avoid interaction with microorganisms. Pure culture techniques must be used. Standard techniques of tissue culture are used to propagate host roots and callus. Aseptic culture is rarely used in diagnostic clinics.

## References


Acedo, J. R., and Dropkin, V. H. 1982. Technique for obtaining eggs and juveniles of *Heterodera glycines*. J. Nematol. 14:418–420.

Ayoub, S. M. 1980. Plant Nematology: An Agricultural Training Aid. Rev. ed. NemaAid Publications, Sacramento, Calif.

Barker, K. R., Carter, C. C., and Sasser, J. N., eds. 1985a. An Advanced Treatise on *Meloidogyne*. Vol. 2, Methodology. Department of Plant Pathology, North Carolina State University, and U.S. Agency for International Development, Raleigh, N.C.

Barker, K. R., Schmitt, D. P., and Noe, J. P. 1985b. Role of sampling for crop-loss assessment and nematode management. Agric. Ecosyst. Environ. 12:355–369.

Barker, K. R., Pederson, G. A., and Windham, G. L., eds. 1998. Plant and Nematode Interactions. American Society of Agronomy, Madison, Wisc.

Barker, K. R., Townshend, J. L., Bird, G. W., Thomason, I. J., and Dickson, D. W. 1986. Determining nematode population responses to control agents. Pages 283–296 in: Methods for Evaluating Pesticides for Control of Plant Pathogens. K. D. Hickey, ed. American Phytopathological Society, St. Paul, Minn.

Bridge, J., Page, S., and Jordan, S. 1982. An improved method for staining nematodes in roots. Rep. Rothamsted Exp. Stn. 1981, Part 1, p. 171.

Brown, R. H., and Kerry, B. R., eds. 1987. Principles and Practice of Nematode Control in Crops. Academic Press, San Diego.

Byrd, D. W., Jr., Kirkpatrick, T., and Barker, K. R. 1983. An improved technique for clearing and staining plant tissues for detection of nematodes. J. Nematol. 15:142–143.

Byrd, D. W., Jr., Nusbaum, C. J., and Barker, K. R. 1966. A rapid flotation-sieving technique for extracting nematodes from soil. Plant Dis. Rep. 50:954–957.

Caveness, F. 1969. Silicone rubber for ringing cover glasses. J. Nematol. 1:96.

Courtney, W. D., Polley, D., and Miller, V. L. 1955. TAF, an improved fixative in nematode technique. Plant Dis. Rep. 39:570–571.

Dropkin, V. H. 1989. Introduction to Plant Nematology. 2nd ed. John Wiley & Sons, New York.

Evans, D. A., Sharp, W. R., Amirato, P. V., and Yamada, Y. 1983. Handbook of Plant Cell Cultures. Macmillan, New York.

Evans, K., Trudgill, D. L., and Webster, J. M., eds. 1993. Plant Parasitic Nematodes in Temperate Agriculture. CAB International, Wallingford, Oxon., U.K.

Gerber, K., and Taylor, A. L. 1988. A simple technique for mounting whole root-knot nematode females. J. Nematol. 20:502–503.

Holbrook, C. C., Knauft, D. A., and Dickson, D. W. 1983. A technique for screening peanut for resistance to *Meloidogyne arenaria*. Plant Dis. 67:957–958.

Hooper, D. J. 1987. Observations on the curation of plant and soil nematodes. Nematologica 32:312–321.

Hussey, R. S., and Barker, K. R. 1973. A comparison of methods of collecting inocula of *Meloidogyne* spp., including a new technique. Plant Dis. Rep. 57:1025–1028.

Luc, M., Sikora, R. A., and Bridge, J., eds. 1990. Plant Parasitic Nematodes in Subtropical and Tropical Agriculture. CAB International, Wallingford, Oxon., U.K.

Mai, W. F., and Mullin, P. G. 1996. Plant-Parasitic Nematodes: A Pictorial Key to Genera. With photographs by H. H. Lyon and K. Loeffler. 5th ed. Comstock Publishing, Cornell University Press, Ithaca, N.Y.

Nickle, W. R., ed. 1991. Manual of Agricultural Nematology. Marcel Dekker, New York.

O'Bannon, J. H., Santo, G. S., and Nyczepir, A. P. 1982. Host range of the Columbia root-knot nematode. Plant Dis. 66:1045–1048.

Premachandran, D., Von Mende, N., Hussey, R. S., and McClure, M. A. 1988. A method for staining nematodes, secretions and structure. J. Nematol. 20:70–78.

Sasser, J. N. 1989. Plant-Parasitic Nematodes: The Farmer's Hidden Enemy. Department of Plant Pathology and Consortium of International Crop Protection, North Carolina State University, Raleigh.

Sasser, J. N., and Jenkins, W. R., eds. 1960. Nematology, Fundamentals and Recent Advances with Emphasis on Plant Parasitic and Soil Forms. University of North Carolina Press, Chapel Hill.

Shurtleff, M. C., and Averre, C. W., III. 1997. The Plant Disease Clinic and Field Diagnosis of Abiotic Diseases. American Phytopathological Society, St. Paul, Minn.

Southey, J. F., ed. 1986. Laboratory Methods for Work with Plant and Soil Nematodes. 6th ed. Ministry of Agriculture, Fisheries and Food, Tech. Ref. Book 402. Her Majesty's Stationery Office, London.

Taylor, A. L. 1971. Introduction to Research on Plant Nematology: An FAO Guide to the Study and Control of Plant-Parasitic Nematodes. Rev. ed. Food and Agriculture Organization of the United Nations, Rome.

Thorne, G. 1961. Principles of Nematology. McGraw-Hill, New York.

Zuckerman, B. M., Mai, W. F., and Krusberg, L. R., eds. 1990. Plant Nematology Laboratory Manual. Rev. ed. University of Massachusetts Agricultural Experiment Station, Amherst.

CHAPTER 3

# Classification and Descriptions of Plant-Parasitic Nematodes

Differences in morphology, anatomy, and fine structures as seen through a compound microscope have traditionally been used to identify genera and species of plant-parasitic nematodes. Plant diagnosticians sometimes diagnose diseases caused by nematodes on the basis of host symptoms and nematode habitat and later confirm their conclusions by observing nematodes microscopically.

Technological advances, such as scanning electron microscopy (SEM), and molecular techniques, such as serological methods (enzyme-linked immunosorbent assay, or ELISA, and techniques using monoclonal antibodies), polymerase chain reaction (PCR), restriction fragment length polymorphism (RFLP), DNA probes, and protein electrophoresis, have introduced the possibility of more accurate identification of nematodes at the species and subspecies level, beyond what is possible with light microscopy. New biochemical techniques are constantly being studied for better identifying nematode species.

More than 200 genera of known or suspected plant-parasitic nematodes have been described in the literature since 1975. Only the most important genera are discussed in detail in this volume. Certain recently described or rare genera found in association with plant roots have been widely accepted by plant nematologists, but the validity and taxonomic placement of others remain subjects of discussion.

## Abbreviated Taxonomic Classification of Important Plant-Parasitic Nematodes

Nematode taxonomy is in a state of flux, and several systems are in use for the classification of nematodes. None of these systems is well adapted for routine diagnoses, because many of the taxonomically distinctive structures of nematodes are minute, obscure, or difficult to see.

Taxonomic keys to mature females are needed. Positive identification of a nematode genus depends on observation of adult females; in most genera, males and juveniles can only confirm an identification based on characteristics of the females. The use of taxonomic keys, however, is beyond the scope of this manual. The reader is directed to the references listed at the end of the chapter for appropriate texts, especially those by Mai and Mullin (1996) and Barker et al. (1998).

An abbreviated taxonomic key to plant-parasitic nematodes is presented in this section (see also Tables 3.1–3-6) in order that the diagnostician can appreciate nematode relationships and possibly characterize nematodes not described in this manual. This key is largely based on the summary of Mai and Mullin (1996).

Nematodes belong to the kingdom Animalia, in the phylum Nematoda (or Nemata), which is often divided into two classes, the Adenophorea and the Secernentea. Plant-parasitic nematodes are members of three orders: the Dorylaimida, in the class Adenophorea, and the Tylenchida and Aphelenchida, in the class Secernentea (Table 3.1). Most of the economically important plant-parasitic nematodes are species in the order Tylenchida.

In the system of nomenclature commonly used for nematodes, the names of the higher categories are formed with the following standard endings:

| | |
|---|---|
| phylum | -a |
| class | -ea |
| subclass | -ia |
| order | -ida |
| suborder | -ina |
| superfamily | -oidea |
| family | -idae |
| subfamily | -inae |

In the following sections, genera marked with an asterisk are described and illustrated in the text.

### Class Adenophorea (Aphasmidea)

Nematodes in the class Adenophorea have a generally smooth cuticle. The amphid aperture is usually postlabial and well developed. There is no median bulb in the esophagus. The one-cell excretory system lacks a lateral canal. Phasmids and deirids are absent. In most species, males have two testes and ventral, preanal supplementary papillae and do not have a bursa (*Paratrichodorus* spp. are an exception in having a bursa). Other important characteristics are the relatively small number of eggs produced; the odontostyle, with an extension, in plant-parasitic species; setae on the lips; and the presence of hypodermal glands.

### Order Dorylaimida

Most individuals in the order Dorylaimida are free-living nematodes inhabiting soil and fresh water; others are nematode predators and plant parasites. Some dorylaims have an odontostyle with a dorsal aperture and a slanted tip, and without rounded basal knobs; others have a mural tooth in the stoma. The amphids are never spiral. Cuticle annulations are usually not visible with a compound microscope. The dorsal esophageal gland duct opening is posterior to the nerve ring. The esophagus consists of two parts: a slender anterior portion, sometimes with small muscular swellings, and an expanded posterior portion, which may be reduced to a simple, valveless bulb. Setae and a caudal spinneret are not present, and the excretory pore is usually absent or rudimentary.

**Table 3.1.** Plant-parasitic nematodes according to the classification used in this book[a]

- Class Adenophorea (Aphasmidea)
  - Order Dorylaimida
    - Family Longidoridae
      - Genera *Longidorus,* Paralongidorus, Xiphinema**
    - Family Trichodoridae
      - Genera *Paratrichodorus,* Trichodorus**
- Class Secernentea (Phasmidea)
  - Order Tylenchida
    - Family Tylenchidae
      - Genera *Aglenchus, Anguina,* Basiria, Cephalenchus, Clavilenchus, Cynipanguina, Ditylenchus,* Lelenchus, Malenchus, Miculenchus, Neopsilenchus, Ottolenchus, Paranguina, Pleurotylenchus, Psilenchus, Subanguina, Tylenchus*
    - Family Tylenchorhynchidae
      - Genera *Macrotrophurus,* Paratrophurus, Trophurus, Tylenchorhynchus**
    - Family Dolichodoridae
      - Genera *Brachydorus, Dolichodorus**
    - Family Belonolaimidae
      - Genera *Belonolaimus,* Carphodorus, Histotylenchus, Merlinius, Morulaimus, Telotylenchoides, Telotylenchus, Trichotylenchus*
    - Family Pratylenchidae
      - Genera *Apratylenchoides, Hirschmanniella,* Hoplotylus, Nacobbus,* Pratylenchoides, Pratylenchus,* Radopholus,* Zygotylenchus*
    - Family Hoplolaimidae
      - Genera *Acontylus, Antarctylus, Aorolaimus, Aphasmatylenchus, Helicotylenchus,* Hoplolaimus,* Peltamigratus, Rotylenchoides, Rotylenchulus,* Rotylenchus,* Scutellonema**
    - Family Heteroderidae
      - Genera *Cactodera, Cryphodera, Globodera,* Heterodera,* Meloidodera,* Punctodera, Sarisodera, Sherodera, Zelandodera*
    - Family Meloidogynidae
      - Genera *Meloidoderella, Meloidogyne,* Meloinema, Nacobbodera*
    - Family Criconematidae
      - Genera *Bakernema, Blandicephalanema, Caloosia, Criconema, Criconemella (Mesocriconema),* Criconemoides,* Crossonema, Discocriconemella, Hemicriconemoides, Hemicycliophora,* Lobocriconema, Macroposthonia, Nothocriconema, Nothocriconemoides, Ogma, Xenocriconemella*
    - Family Paratylenchidae
      - Genera *Cacopaurus, Gracilacus, Paratylenchoides, Paratylenchus**
    - Family Tylenchulidae
      - Genera *Sphaeronema,* Trophonema, Trophotylenchulus, Tylenchulus**
  - Order Aphelenchida
    - Family Aphelenchidae
      - Genus *Aphelenchus**
    - Family Aphelenchoididae
      - Genera *Aphelenchoides,* Bursaphelenchus,* Rhadinaphelenchus,* Tylaphelenchus*

[a] Genera marked with an asterisk are described and illustrated in the text.

#### Family Longidoridae

Nematodes in the family Longidoridae are long (usually over 2 mm) and slender and have a very long stylet (more than 3 body widths at the stylet base) without basal flanges. The guiding ring is prominent, and its location is important in subfamily and genus identification. The esophagus is bottle-shaped, with a long, narrow, muscular anterior portion and a short, larger-diameter, glandular posterior portion, which contains the salivary glands. The vulva is usually median and slitlike. The genital tracts are paired, opposed, reflexed, and short. The tail is bluntly rounded to roundly conoid. The male tail has massive, banana-shaped, ventrally curved (arcuate) spicules with lateral accessory pieces and prominent ventromedian copulatory papillae (supplements). There is no gubernaculum or bursa. The cuticle is smooth, without lateral fields. There is no excretory pore. Juvenile longidorids are easily recognized by their functional odontostyle with a replacement just posterior to it.

**Genera.** *Longidorus,* Paralongidorus, Xiphinema.**

#### Family Trichodoridae

Nematodes in the family Trichodoridae are short (0.5–1.4 mm) and cigar-shaped and have a thick cuticle and a bluntly rounded head. The simple stylet lacks basal knobs and is attenuated, short (about 2 body widths at the stylet base), conspicuously curved ventrally, and grooved rather than hollow. The esophagus is narrow and cylindrical in the anterior region and expands into a posterior bulb, which sometimes overlaps the intestine. An excretory pore is present in the esophageal region, which in males contains one to four ventral papillae. The ovaries are paired, opposed, and reflexed. The vulva is usually median. The anus is close to the tail terminus. Males have a single testis, directed anteriorly; elongate and mostly straight spicules; and a gubernaculum. The tail is short (less than 1 anal body width), blunt, and round, with a bursa (in *Paratrichodorus*) or without one (in *Trichodorus*). The esophageal glands form a basal bulb. The body cuticle is thin and loose, lacks ornamentation, and often wrinkles when the nematode moves.

**Genera.** *Paratrichodorus,* Trichodorus.**

## Class Secernentea (Phasmidea)

Nematodes in the class Secernentea have a many-celled, tubular excretory system with one or more lateral canals. Phasmids and deirids are usually present but may be difficult to see. The amphid apertures are usually small and porelike or are small slits located dorsolaterally on the lateral lips or the anterior extremity. There are no setae on the lips. The cuticle is annulated. The spear aperture is ventral oblique. The stylet (a stomatostylet) is usually straight, with three basal knobs. The esophageal tube is attached to the stylet. The median bulb has a valve. There are no hypodermal glands. Other key characteristics are a single testis in males of most species; the large number of eggs produced; and papillae on the lips.

### Order Tylenchida

The order Tylenchida contains forms that feed on fungi, higher plants, and insects. They have a moderately to strongly sclerotized, narrow, pointed stylet, usually with three conspicuous, rounded basal knobs, without a sclerotized valvular apparatus, to which muscles are attached. The cuticle has distinct annulations, which usually are interrupted by incisures on the lateral fields. The esophagus is divided into three parts: the anterior procorpus; an enlarged metacorpus (median bulb) with crescentic plates (or valves) and a narrow isthmus encircled by the nerve ring; and a posterior glandular bulb or set of lobes. The dorsal esophageal gland orifice is usually near the stylet base, in the procorpus. The excretory pore is conspicuous and most often located near the latitude of the nerve ring. Females have one or two ovaries. Sexual dimorphism is common.

#### Family Tylenchidae

Nematodes in the family Tylenchidae are small to moderate-sized and are mostly vermiform. The lip region is distinct and marked by fine transverse striations. Phasmids or phasmid-like structures are located in a submedian position, near the vulva in females and near the lateral field but outside and dorsal to it. Both males and females are active and elongate. The lip region is rarely offset from the body contour. The stylet is small; basal knobs are present in some genera. The cuticular annulations are usually indistinct. The esophageal glands distinctly abut the intestine; if there is a short overlap with the intestine, then head sclerotization is indistinct. The vulva is posterior (between the middle of the body and the anus); the female genital tract is usually single and anterior. The length of the tail is more than 2 anal body widths. The female tail is pointed. Males have caudal alae, which do not reach the tail tip.

**Genera.** *Aglenchus, Anguina,* Basiria, Cephalenchus, Clavilenchus, Cynipanguina, Ditylenchus,* Lelenchus, Malenchus, Miculenchus, Neopsilenchus, Ottolenchus, Paranguina, Pleurotylenchus, Psilenchus, Subanguina, Tylenchus.*

#### Family Tylenchorhynchidae

Males and females in the family Tylenchorhynchidae are active, vermiform nematodes about 0.8 to 1.5 mm long. The stylet is well developed and has basal knobs. The esopha-

**Table 3.2.** Plant-parasitic nematodes found in all stages of development in soil samples, including ectoparasites and immature endoparasites

| Length (mm) | Remarks | Genus |
|---|---|---|
| 0.2–0.5 | Short to very short body<br>Stylet medium-sized in females, smaller in males<br>One ovary | *Paratylenchus* |
| 0.2–1.0 | Short, thick body with wide, large annules<br>Medium-sized stylet<br>One ovary<br>Males rarely present | *Criconemella* and related genera |
| 0.3–1.0 | Short, strong stylet<br>One ovary<br>Present in all stages in the cortex of roots with lesions and rot | *Pratylenchus* |
| 0.4–0.9 | Short stylet<br>Two ovaries<br>Present in all stages in the cortex of roots with lesions and rot<br>Vulva near midsection | *Radopholus* |
| 0.4–1.8 | Curved stylet in two parts, without knobs<br>Esophagus lacking a median bulb<br>Two ovaries | *Paratrichodorus* |
| 0.4–1.2 | Body curved or spiral<br>Medium-sized to heavy stylet<br>Two ovaries<br>Vulva posterior to median | *Helicotylenchus* |
| 0.5–1.7 | Short to medium-sized stylet<br>Two ovaries<br>Esophagus with posterior, non-overlapping basal gland | *Tylenchorhynchus* |
| 0.6–2.0 | Medium-sized, heavy stylet with prominent, cup-shaped knobs<br>Two ovaries<br>Rounded tail enveloped by bursa<br>Vulva in midsection | *Rotylenchus* |
| 0.4–2.0 | Females with cuticular sheath and medium-sized, slender stylet<br>One ovary<br>Vulva posterior | *Hemicycliophora* |
| 0.9–2.0 | Lip region distinctly set off<br>Heavy stylet with tulip-shaped knobs<br>Two ovaries<br>Rounded tail | *Hoplolaimus* |
| 0.9–4.2 | Found in rice paddies and very wet soils<br>Short stylet<br>Two ovaries, equally developed | *Hirschmanniella* |
| 1.3–5.5 | Very long stylet with basal flanges<br>Body lacking annulations<br>Esophagus lacking a median bulb<br>Vulva usually near midsection | *Xiphinema* |
| 2.0–3.0 | Very long, slender stylet<br>Two ovaries<br>Long, rounded tail in females | *Belonolaimus* |
| 1.5–3.0 | Long, slender body with coarsely annulated cuticle<br>Very long stylet<br>Lips set off<br>Rounded, pointed tail | *Dolichodorus* |
| 4.4–11.0 | Very long, narrow body<br>Guiding ring near lips<br>Long stylet in two parts, without flanges and knobs<br>Esophagus lacking a median bulb<br>Vulva near midsection | *Longidorus* |

**Table 3.3.** Plant-parasitic nematodes found in soil and roots

| Length (mm) | Stages found in soil | Remarks | Genus |
|---|---|---|---|
| 0.2–0.5 | Immature females, males, juveniles, eggs | Often found in soil under citrus<br>Mature females partially embedded in citrus roots<br>Body covered with egg mass<br>Vulva at posterior end | *Tylenchulus* |
| 0.3–0.9 | Immature females, males, juveniles, eggs | Distance from stylet knobs to dorsal gland orifice is 0.6–1.9 times stylet length<br>Mature females typically reniform (kidney-shaped), with neck embedded in roots, sometimes lying partially within roots<br>Body covered with egg mass | *Rotylenchulus* |
| 0.3–0.5 | Juveniles, eggs | Associated with galls on roots<br>Mature females globose, entirely embedded in roots<br>Two ovaries | *Meloidogyne* |
| 0.3–0.6 | Juveniles, cysts, eggs | Stylets of juveniles 20–30 μm long<br>Mature females lemon-shaped to globular, yellow or white, with neck embedded in roots, sometimes lying partially within roots, forming brown cysts in the soil | *Heterodera* |
| 0.4–0.6 | Juveniles, cysts, eggs | Stylets of juveniles 20–30 μm long<br>Mature females spherical, usually white, with neck embedded in roots, forming yellow or white cysts, which turn golden to brown when mature | *Globodera* |

geal glands are in a distinct basal bulb. Females have two ovaries; the vulva is in the middle of the body. The tail length is 2 to 6 anal body widths. The female tail is rounded or acute. Phasmids are distinct on the midtail of females and extend into the bursal wings of males. The bursa is prominent, enveloping the tail.

**Genera.** *Macrotrophurus,* Paratrophurus, Trophurus, Tylenchorhynchus.**

#### Family Dolichodoridae

Nematodes in the family Dolichodoridae are medium-sized to large (1 to 3 mm long). The lip region is heavily sclerotized. The cuticle is distinctly annulated. The vulva is median, and the genital tracts are paired and opposed. The female tail is round to elongate-conoid. Males have a trilobed bursa. Distinct phasmids are present in the middle of the lateral fields, mostly on the tail. The esophageal glands are enclosed in a terminal bulb.

**Genera.** *Brachydorus, Dolichodorus.**

#### Family Belonolaimidae

Males and females in the family Belonolaimidae are elongate and have well-marked annules. The lip region is offset with a moderate cephalic framework. The stylet is long and slender and has basal knobs. The cuticular annulation is distinct. The esophageal glands are in a lobe overlapping the intestine. The vulva is median. The genital tracts are paired, opposed, and outstretched. The tail is cylindrical to elongate-conical, with a length of much more than 2 anal body widths, and is enveloped by a long bursa. Distinct phasmids are present near the middle of the tail.

**Genera.** *Belonolaimus,* Carphodorus, Histotylenchus, Merlinius, Morulaimus, Telotylenchoides, Telotylenchus, Trichotylenchus.*

#### Family Pratylenchidae

Males and females in the family Pratylenchidae are active, elongate nematodes (except mature females of *Nacobbus*). The males resemble the females. The lip region is low, broad, and rounded to flat. The cephalic region is distinctly sclerotized. The stylet is stout and has large basal knobs. These nematodes have three esophageal glands, in a lobe overlapping the intestine. Females have one or two ovaries. The tail length is usually more than 2 anal body widths. The phasmids are postanal and located near the center of the tail. The bursa is subterminal to terminal.

**Genera.** *Apratylenchoides, Hirschmanniella,* Hoplotylus, Nacobbus,* Pratylenchoides, Pratylenchus,* Radopholus,* Zygotylenchus.*

#### Family Hoplolaimidae

Males and females in the family Hoplolaimidae are vermiform nematodes with a high, convex-conoid or broadly rounded lip region and distinct cephalic sclerotization. The cuticular annules are prominent. The stylet is stout (usually over 20 μm long) and has prominent and distinctive basal knobs. The esophageal gland lobes overlap the intestine. Females have one or two ovaries. The phasmids are usually distinct and variable in position. The female tail is short, usually with a length of less than 2 anal body widths. A distinct bursa overlaps the male tail.

**Table 3.4.** Plant-parasitic nematodes found in roots[a]

| Length (mm) | Stages found in roots | Female shape | Remarks | Genus |
|---|---|---|---|---|
| 0.3–1.0 | All | Slender | Associated with root lesions and rot<br>One ovary<br>Vulva posterior | *Pratylenchus* |
| 0.4–0.9 | All | Slender | Associated with root lesions and rot<br>Two ovaries, outstretched | *Radopholus* |
| 0.4–1.2 | All | Slender | Curved or spiral body<br>Two ovaries<br>Occasionally associated with root lesions and rot | *Helicotylenchus* |
| 0.2–0.5 | All | Enlarged, with pointed tail | Females with head and neck embedded in citrus roots<br>Body covered with egg mass | *Tylenchulus* |
| 0.3–0.9 | All | Kidney-shaped, with pointed tail | Found on roots of many plants<br>Females with neck embedded in roots<br>Body covered with egg mass | *Rotylenchulus* |
| 0.4–1.3 | Females, juveniles | Spherical, short-necked | Females and juvenile stages fully embedded in root galls<br>Found on many plants | *Meloidogyne* |
| 0.3–1.0 | Females, juveniles | Lemon-shaped to globular | Females white or yellow and short-necked, with neck embedded in roots<br>Short neck<br>Juveniles fully embedded | *Heterodera* |
| 0.3–0.9 | Females, juveniles | Spherical | Same as *Heterodera*<br>No vulvular cone | *Globodera* |
| 0.6–1.6 | Females | Enlarged | Females completely embedded in roots | *Nacobbus* |
| 0.9–4.2 | All | Slender | Found in roots of rice and other plants in very wet soils<br>Short stylet<br>Two ovaries, equally developed | *Hirschmanniella* |

[a] Also found in soil, in some stages of development.

**Genera.** *Acontylus, Antarctylus, Aorolaimus, Aphasmatylenchus, Helicotylenchus,* Hoplolaimus,* Peltamigratus, Rotylenchoides, Rotylenchulus,* Rotylenchus,* Scutellonema.**

#### Family Heteroderidae

Mature females in the family Heteroderidae are swollen and pear-shaped, globular, or lemon-shaped. They are sessile in or on roots. Some form cysts. The cephalic framework of females is weak and not sclerotized; it is more developed in males. The stylet is more than 17 μm long. The vulva is terminal or behind the middle of the body; the genital tracts are paired. The anus is terminal. Males are elongate, vermiform, and active. They have a well-developed stylet and cephalic framework, a single testis, and a well-developed, ventrally curved spicule with a small gubernaculum; they have no bursa.

**Genera.** *Cactodera, Cryphodera, Globodera,* Heterodera,* Meloidodera,* Punctodera, Sarisodera, Sherodera, Zelandodera.*

#### Family Meloidogynidae

Mature females in the family Meloidogynidae are swollen to subspherical. They do not form cysts. The cuticle is somewhat annulated, especially in the posterior region. The tail is absent or very short. The anus and vulva are terminal to subterminal. The excretory pore is opposite or anterior to the median esophageal bulb. Males have a well-developed stylet and cephalic framework. The amphid aperture is slitlike and is normally covered by enlarged lateral lips.

**Genera.** *Meloidoderella, Meloidogyne,* Meloinema, Nacobbodera.*

#### Family Criconematidae

The classification of the family Criconematidae is somewhat confused. There is strong sexual dimorphism. Both sexes are usually small (under 1 mm long) and vermiform, but females may be quite robust. The cuticular annulations may be coarse and strong. In some species the annules have overlapping scales, fringes, or plates. The shape of the lip region is variable, sometimes with four extra submedial lobes. The cephalic framework is strong; the stylet is large and robust (50–100 μm long) and has well-developed basal knobs. The large, oval median bulb, with a well-developed prominent valve, is broadly fused with the procorpus into a single division of the esophagus. The crescentic plates are elongate. The isthmus is short and narrow. The esophageal glands are in a small basal bulb. The vulva is well posterior (V = 80–90), and the genital tract is single and anterior. Lateral fields are not present; sometimes irregular lines form by anastomosing annules. The female tail is generally short and blunt to conical. Males have a degenerate esophagus, a weak stylet or no stylet, and a tail with or without a bursa. Juveniles, although smaller, are somewhat similar to the females.

**Genera.** *Bakernema, Blandicephalanema, Caloosia, Criconema, Criconemella (Mesocriconema),* Criconemoides,* Crossonema, Discocriconemella, Hemicriconemoides, Hemicycliophora,* Lobocriconema, Macroposthonia, Nothocriconema, Nothocriconemoides, Ogma, Xenocriconemella.*

#### Family Paratylenchidae

Nematodes in the family Paratylenchidae are generally small and vermiform, but females of some forms may become obese. The labial sclerotization is generally weak (except in *Paratylenchoides*). The body annules are rounded and fine, not retrorse, ornamented, or overlapping. The female stylet is well developed and has rounded basal knobs. The metacorpus (median bulb) is poorly offset from the short procorpus but has a prominent valve. The vulva is posterior, and the genital tract is single and anterior. In males, the stylet is reduced or absent, the esophagus is degenerate, and the bursa is very small or absent.

**Table 3.5.** Plant-parasitic nematodes found in or on aboveground parts of plants[a]

| Length (mm) | Remarks | Genus |
|---|---|---|
| 0.8–5.0 | Found in galls on leaves, stems, and seed, especially of cereals and grasses<br>Short stylet<br>Median bulb with distinct valve<br>One ovary, anterior<br>Tail tapering to a point | *Anguina* |
| 0.6–1.5 | Endoparasitic or ectoparasitic on bulbs, stems, leaves, and seed of many plants<br>Short stylet<br>Median bulb with weak valve<br>Vulva in posterior quarter<br>One ovary, anterior<br>Bursa and spicules present<br>Elongate-conoid tail | *Ditylenchus* |
| 0.4–1.2 | Ectoparasitic or endoparasitic on buds, leaves, and seed of many plants<br>Short stylet<br>Large median bulb with conspicuous valve<br>Conoid tail<br>Spicules shaped like rose thorns<br>Gubernaculum and bursa absent | *Aphelenchoides* |
| 0.4–1.2 | Endoparasitic and migratory on woody tissues in trunk and branches of pines and palms<br>Associated with slight necrosis<br>Female tail tapered, with rounded tip<br>Vulva with wide, overlapping flap | *Bursaphelenchus* |

[a] Adults and juveniles, especially those of *Anguina, Aphelenchoides,* and *Ditylenchus,* may be found in soil at the surface when infected plant parts fall to the ground and host tissues begin to decompose.

**Genera.** *Cacopaurus, Gracilacus, Paratylenchoides, Paratylenchus.**

#### Family Tylenchulidae

Both sexes in the family Tylenchulidae are usually small. Mature females are stout to very swollen (saccate or subspherical) and partially buried in roots, with their swollen posterior portion protruding from the surface. The cuticular annulations are fine. The stylet is short and well developed in females but reduced or absent in males. The esophageal isthmus (between the median bulb and the basal glands) is

**Table 3.6.** Stylet-bearing, non-plant-parasitic nematodes found in soil, the rhizosphere, and necrotic plant tissues[a]

| Length (mm) | Remarks | Family or genus |
|---|---|---|
| <0.85 | Common in decaying plant tissues and the rhizosphere<br>Rounded tail<br>Short stylet, without basal knobs<br>Prominent median bulb with valve<br>Prodelphic ovary<br>V = 70–80<br>Similar to *Aphelenchoides* | *Aphelenchus* |
| <0.90 | Family of numerous genera and species found in soil around plants and in lesions on underground plant parts<br>Short, well-developed stomatostylet (<15 μm long) with basal knobs<br>Esophagus lacking a bulb and valve<br>Prodelphic, outstretched ovary<br>V > 60 | Neotylenchidae |
| <1.5 | Frequently found in cultivated soils<br>Long, slender body (a = 33–38)<br>Esophagus with unusually long procorpus, rounded median bulb with valve, and long, narrow isthmus<br>Slender stylet, sometimes with knobs<br>Very long, pointed tail (c = 7.6), usually ending with a clavate swelling<br>Two outstretched ovaries<br>V = $^{32}47^{32}$<br>Similar to *Tylenchus* (one ovary) and often recovered in the same sample | *Psilenchus* |
| 1.2 | Abundant around plant roots<br>Odontostylet with a strong dorsal stiffening piece and three flanges, similar to that of *Xiphinema*<br>Esophagus a long narrow tube with an ovoid terminal bulb<br>Single ovary, posterior and reflexed<br>V = $^{7}30^{30}$ | *Tylencholaimus* |
| 0.3–1.5 | Frequently the most numerous nematode in the rhizosphere<br>Well-developed, slender stylet with strong basal knobs<br>Rounded median bulb with prominent refractive valve<br>Very long, pointed (filiform) tail<br>Prodelphic (anterior), outstretched ovary with rudimentary, small postuterine branch<br>V = 55–65<br>Some juveniles similar to those of *Meloidogyne* spp. | *Tylenchus* (this group is now divided into many genera) |

[a] Also found in soil, in all stages of development.

long, and the median bulb, with a prominent valve, is distinctly set off from the procorpus. The basal bulb is weakly defined. The vulva is posterior, and the genital tract is single and anterior. Males are slender and have a degenerate esophagus, and there is no bursa. All members of this family are obligate, sedentary semiendoparasites of roots.

**Genera.** *Sphaeronema,* Trophonema, Trophotylenchulus, Tylenchulus.**

### Order Aphelenchida

The order Aphelenchida includes forms that feed on fungi and higher plants (generally on aboveground parts), predators, and obligate parasites of insects. It is sometimes classified as the suborder Aphelenchina of the order Tylenchida.

Members of this group are generally small (1 mm long or less) and vermiform. The lip region is often set off by a constriction. The stomatostylet is weakly sclerotized and has small or no knobs. The esophagus consists of three parts: a narrow procorpus, a broad metacorpus, and a gland lobe. Both dorsal and subventral glands open into the lumen in the metacorpus (median bulb). The lumen of the procorpus is a straight tube without bends, posterior to the stylet. The dorsal esophageal gland orifice is located in the metacorpus and empties into a large, muscular median valve just anterior to the valve or subventrally posterior to it. The esophageal glands overlap the anterior part of intestine. Females have one ovary.

#### Family Aphelenchidae

Nematodes in the family Aphelenchidae have a low lip region, which is rounded to flattened anteriorly. The stylet lacks basal thickenings. The esophageal glands overlap the intestine dorsally and laterally, often in a long lobe. The lateral field has six to 14 incisures in the midbody region. The tail is short and usually bluntly rounded. The vulva is posterior; the genital tract is single and anterior. Males have slender, ventrally curved spicules, a gubernaculum, and a prominent bursa, which envelops the tail and is supported by ribs.

**Genus.** *Aphelenchus.**

#### Family Aphelenchoididae

Both sexes in the family Aphelenchoididae are usually vermiform. The stylet does not have conspicuous knobs. The lateral field has fewer than six incisures. The dorsal esophageal gland opens into a (usually large) median bulb anterior to the crescentic plates. There is no isthmus in the esophagus. The esophageal glands are in a long lobe overlapping the intestine, usually dorsally. The vulva is posterior; the genital tract is single and anterior. Males have rosethorn or irregularly shaped spicules; the gubernaculum is indistinct or absent.

**Genera.** *Aphelenchoides,* Bursaphelenchus,* Rhadinaphelenchus,* Tylaphelenchus.*

## Descriptions of Plant-Parasitic Nematodes

### *Anguina:* Seed and Leaf Gall Nematodes

#### Class Secernentea, order Tylenchida, family Tylenchidae

##### HOSTS

The principal hosts of *Anguina* species are many members of the family Poaceae (Gramineae) worldwide, along with some plants in other families, including alkaligrass (*Puccinellia distans*), American dunegrass (*Leymus mollis,* syn. *Elymus mollis*), *Apera, Arctagrostis,* barley, bentgrasses, bermudagrass, bluegrasses, bluejoint (*Calamagrostis canadensis*), bromegrasses, *Dactylis,* dropseed (*Sporobolus* spp.), fescues, hairgrass (*Deschampsia*), Harding grass (*Phalaris stenoptera*), June grass (*Koeleria*), weeping lovegrass (*Eragrostis curvula*), nigerwood (*Guizotia abyssinica*), oat, orchids, Oregon sunflower (balsamroot, *Balsamorhiza sagittata*), *Phalaris, Plantago aristata,* primroses, redtop, rye, ryegrasses, *Stipa occidentalis,* sunflower, timothy, *Trisetum,* vetch (*Vicia* sp.), wheat, wheatgrass (*Agropyron*), wild sugarcane (*Saccharum spontaneum*), *Wyethia,* yarrow (*Achillea millefolium*), and others.

##### PARASITISM AND HABITAT

Males and females of *Anguina* species are predominantly migratory endoparasites, inducing galls in leaves, stems, and seeds (Figure 1.3A–C and Plate 1A). The galls contain adults, large numbers of eggs, and developing juveniles. Juveniles are ectoparasites, invading seedlings, entering young floral tissue, and feeding on developing ovules to form galls within which nematodes mature to become adults. Juveniles can remain dormant in dry galls for at least 30 years. They usually emerge in the spring when the galls imbibe water. They survive poorly in soil, possibly up to one year. Development proceeds through four stages. The first molt may occur in the egg. In some species only one generation develops within a gall during a growing season, with a single female laying up to several hundred eggs, which hatch as second-stage juveniles. Other species may produce more than one generation in a season.

Juveniles in the infective stage (the second or third stage, depending on the species) invade rapidly growing tissues of seedlings and are carried up as the plant grows. Leaves of infected wheat are small and twisted. A single flower gall may contain up to 40 adults of each sex and 30,000 or more eggs and juveniles. Infected flowers develop into small, dark galls, which become mixed with seed during harvest or fall to the ground. Some galls on leaves, stems, and inflorescences are characteristic in size, shape, color, and location. Infected plants may be stunted or dwarfed, with crinkled, twisted, and distorted leaves and curled shoots in

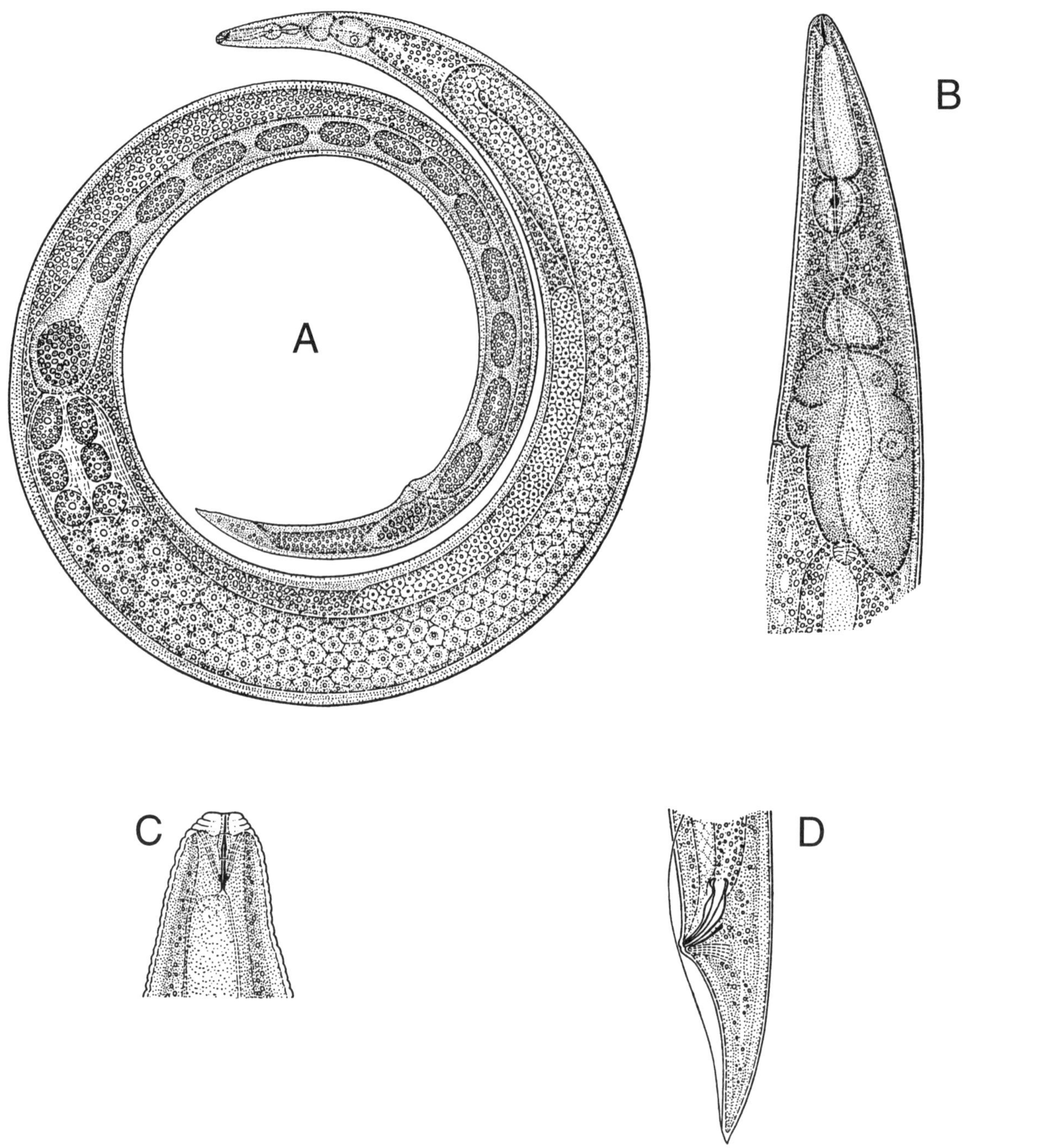


**Figure 3.1.** *Anguina tritici.* **A,** Adult female (120×). **B,** Adult female head (680×). **C,** Head (1,020×). **D,** Male tail (340×). (Reprinted from Thorne, 1949, with the permission of *Comparative Parasitology,* formerly *Journal of the Helminthological Society of Washington*)

the early stages (Figure 1.3A); later they have reduced and irregular heads with infected kernels (cockles), which are dark, stunted, distorted, and filled with hundreds or thousands of juvenile nematodes (Figure 1.3B and Plate 1A). Severely infected plants may die without heading.

### BODY

Species of *Anguina* vary considerably in body morphology (Figure 3.1). These nematodes are large (1.5–5.0 mm long), robust, and stout (a = 20–30). The ratio of overall length to greatest width is 20 or less in some species; others are more slender. Males are smaller (1.0–2.5 mm) and much more slender than females and have a swollen midsection. Females are obese, tapered toward both ends, and usually strongly curved (coiled), whereas males are only slightly curved. On heat relaxation females die coiled ventrally in a spiral, and males are ventrally or dorsally arcuate. The head is of the same contour as the body (not offset). In all stages, the lips are low, flattened, and distinctly set off, and the labial framework is weak. The cuticular annulations are fine in both sexes and difficult to see on adults except on the neck and lip region. The lateral fields may be plain bands or may bear four or more minute incisures, which are usually not visible on mature adults. The amphid apertures at the tips of the lateral lips are visible only from a face view. Deirids and six minute papillae surround the oral opening. Four other papillae are located on the outer margins of the submedian lips.

#### STYLET

The stylet is short (8–10 μm) and weak, with small, well-developed, rounded knobs.

#### ESOPHAGUS

The median bulb has a distinct valve. Three large esophageal glands overlap the intestine in a basal gland, in most but not all species. The glandular tissues may become greatly swollen and irregular in form. The isthmus is crossed by the nerve ring.

#### REPRODUCTIVE STRUCTURES

Females have a single ovary, anteriorly reflexed two or more times, with a small, rudimentary postuterine branch; numerous oocytes and spermatocytes, in multiple rows; and a short postvulval sac. Spermatheca may be present or absent. The ovary ends in a cap cell.

Males have a single, well-developed testis, with one or more anterior flexures and a cap cell. The spicules are robust, wide, paired, arcuate, and not fused. The gubernaculum is small, troughlike, and slightly curved.

#### VULVA

The vulva is near the posterior end of the body (V = 90–95) and has prominent lips.

#### TAIL

The tail constitutes 3 to 5% of the total body length. It is conoid and tapered to an obtuse or rounded tip in *A. tritici* and an acute or finely mucronate tip in *A. agrostis;* it is finely rounded or minutely lobed in *A. graminis.* The bursa arises just anterior to the spicules and extends nearly to the terminus or in some cases envelops it.

#### GENERA RESEMBLING *ANGUINA*

*Aphelenchoides*
*Cynipanguina*
*Ditylenchus*
*Subanguina.* Adults of this genus occur only in plant galls on roots of cereals and grasses. They resemble *Anguina* but are not quite as swollen.

#### COMMENTS

There are some 20 host-specific species of *Anguina,* several of which include more than one race.

When placed in water and crushed, galls caused by *Anguina* species release large numbers of eggs, moving juveniles, and adults, which are visible to the naked eye as a cottony mass.

*A. tritici* lays eggs in seed galls, and all of its juvenile stages and adults are produced in galls. Each gall contains 10,000 to 30,000 eggs. Second-stage juveniles can survive in galls for up to 30 years. This species is a common pest of wheat in many regions where galls are not removed from seeds by modern seed-cleaning methods.

*A. agrostis* causes seed galls on bentgrasses and other grasses in many temperate regions.

Species of *Anguina* are also associated with toxin-producing bacteria on certain grasses eaten by livestock. In India, *Anguina* species interact with *Clavibacter toxicus* (*Corynebacterium rathayi*), which causes yellow ear rot of wheat. In wheat infected with the nematode but not the bacterium, the typical symptom is ear cockle, in which the kernels are replaced by dark galls and the ear is small and malformed. Wheat ears infected with both the nematode and the bacterium develop a disease called yellow ear rot, or *tundu,* in which they exude a yellow slime. The bacterium by itself causes no symptoms.

A similar association occurs in Australia and South Africa, where certain species of *Anguina,* especially *A. funesta,* serve as vectors transmitting *C. toxicus* to seeds of annual ryegrass (*Lolium rigidum*). The bacterium is often infected with a bacteriophage, which induces it to produce glycolipid compounds (corynetoxins). These compounds are extremely toxic, and short exposures often cause death in sheep, cattle, horses, pigs, and other animals that feed on infected seed heads or grain. The bacterium apparently develops only in nematode galls.

## *Aphelenchoides:* Foliar or Leaf Nematodes and Bud Nematodes

### Class Secernentea, order Aphelenchida, family Aphelenchoididae

#### HOSTS

*Aphelenchoides* species have very extensive host ranges, including aconite (Turk's-cap, *Aconitum napellus*), African violet, alfalfa, almond, alumroot (*Heuchera*), amaryllis, *Anchusa,* anemone, *Annona cherimola* (cherimalla, cherimoya), *Anthurium,* apple, arrowhead (*Sagittaria*), aspen, asters, *Astilbe simplicifolia,* azaleas, bamboo, *Barleria cristata,* barley, beans (*Phaseolus* spp.), beet (sugar beet), begonia, *Bergenia cordifolia,* Oriental bittersweet (*Celastrus orbiculatus*), *Bletilla hyacinthina,* boysenberry, *Buddleja,* buttercup, cabbage, *Calceolaria, Caltha,* camellia, carnation, carrot, celery, cherries, chickpea (*Cicer arietinum*), China aster, China-grass (ramie, *Boehmeria nivea*), *Chionodoxa,* chrysanthemum, *Cimicifuga simplex,* cineraria, clover (*Trifolium*), coconut palm (*Cocos nucifera*), *Colchicum,* coleus, columbine, corn (maize), *Cornus canadensis,* cot-

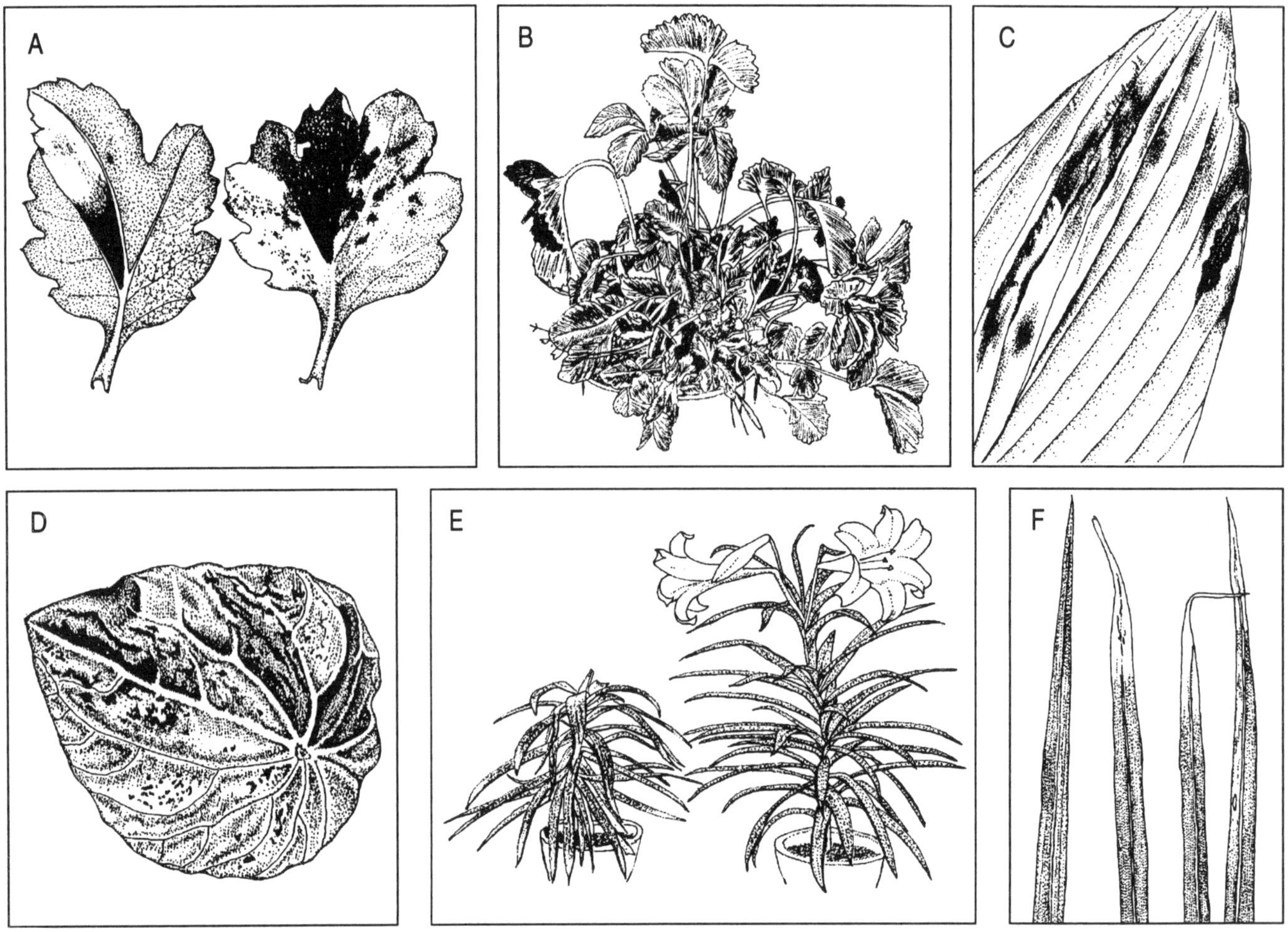


**Figure 3.2.** Damage caused by species of *Aphelenchoides.* **A,** Chrysanthemum leaves infected with *A. ritzemabosi.* **B,** Strawberry plants with distortion caused by *A. fragariae* or *A. besseyi.* **C,** Leaf blotches in a *Hosta* leaf invaded by *A. ritzemabosi.* **D,** Lower surface of a begonia leaf infected with *A. ritzemabosi.* **E,** Lily plant severely stunted by *A. ritzemabosi* (left) and a healthy plant. **F,** Healthy rice leaf (left) and leaves with white tip, caused by *A. besseyi.*

ton, crocus, *Crossandra,* currants, cyclamen, dahlia, date palm (*Phoenix dactylifera*), delphinium, dogwood, Douglas fir, elder (*Sambucus*), *Eremurus stenophyllus,* everlasting flower (*Helichrysum*), ferns, *Ficus elastica,* firs, flax, *Fritillaria imperialis, Galanthus,* gardenia, garlic, geranium (*Pelargonium*), *Gerbera, Geum,* ginger (*Zingiber*), ginseng (*Panax*), gloxinia, gooseberry, grapefruit, grape-hyacinth, grasses (pasture grasses, turfgrasses, and wild species), *Helenium,* hemlock, *Heuchera,* hosta, hyacinth bean (*Lablab purpureus*), *Hyacinthoides,* hydrangea, Indian sorrel (*Hibiscus sabdariffa*), iris, junipers, *Lamium, Lantana,* eastern larch (tamarack), lavender, lettuce, *Ligularia,* lilies, lily-of-the-valley, lupines, magnolia, mesquite (*Prosopis*), millets, *Mimulus,* mints, mushrooms, narcissus, *Nerine,* oaks, oat, onions, orchids, mountain papaya (*Carica candamarcencis*), passionfruit, peach, peanut, peonies, peperomia, phlox, pines, poppy, potato, primroses, *Ranunculus,* rhododendrons, rhubarb, rice, rubber plant (*Ficus elastica*), sage, *Saxifraga, Scabiosa, Scilla,* sego-lily (*Calochortus nuttallii*), shadscale (*Atriplex confertifolia*), silver-dollar plant (*Lunaria annua*), sorghums, soybean, Colorado blue spruce (*Picea pungens*), strawberry, strawflower, *Streptocarpus,* sugar beet, sugarcane, sunflower, sweet potato, taro (*Colocasia esculenta*), teasel (*Dipsacus*), *Thalassia testudinum,* tobacco, tomato, *Trollius,* tuberose (*Polianthes tuberosa*), tulip, verbena, hairy vetch (*Vicia villosa*), viburnums, *Viola,* violets, wallflower, wheat, yams (*Dioscorea* spp.), zinnia, and others.

Weed hosts include dandelion, ground ivy (*Glechoma hederacea*), milkweed (*Asclepias*), and red goosefoot (*Chenopodium rubrum*). *Aphelenchoides* species also feed on lichens and many fungi.

#### PARASITISM AND HABITAT

*Aphelenchoides* species are common and widespread migratory ecto- and endoparasites of leaves, buds, stems, and occasionally corms. Necrotic tissue is often extensive. Males and females of some species live in soil; other species do not. The feeding habits of different species vary greatly. Some feed and also reproduce on developing leaves, flower buds, and fungi.

These nematodes invade plants primarily through leaf stomates. They spread from the base of the plant upward, generally infecting young leaves more severely than older leaves. Feeding destroys cells in buds and growing points, causing malformation of leaves, stunting of plants, and reduced flowering and fruit production.

Infection of rice causes white tip disease (Figure 3.2F and Plate 1C), in which leaves turn white at the tip and later turn brown or black. Affected plants may be stunted, and their grain is poorly filled or unfilled and chaffy.

Chrysanthemum, begonia, and peperomia leaves develop necrotic dried areas, which are often wedge-shaped and usually delimited by veins (Figure 1.3D and E, Figure 3.2A and D, and Plate 1B).

Fern and hosta leaves develop elongated brown or black blotches between the leaf veins (Figure 3.2C).

Easter lilies develop bunchy top and die back; their leaves turn yellow to bronze and soon droop, then turn blackish brown, and later die; stems are stunted, and infected buds fail to open (Figure 3.2E).

*A. besseyi* is an important ectoparasite of strawberry, in which it causes summer crimp, or summer dwarf disease, and in rice, in which it causes white tip. It survives between crops in a quiescent state under the hulls of rice kernels. It can remain in this state for at least several years but survives only four months on grain left in the field.

*A. fragariae* and *A. ritzemabosi* cause spring crimp, or spring dwarf disease, in strawberry (Figure 3.2B) and many ornamentals. Leaves of affected plants are twisted and crinkled or puckered with rough surfaces or may be crimped, cupped, and abnormally yellow. Flowers are abnormal and may have greenish petals. Fruit yield is reduced, fruit buds are "blind," and the plants may be killed. The nematodes migrate to growing points of leaves and stems when films of moisture are present.

Plants invaded by *Aphelenchoides* species may be more seriously damaged by other pathogens associated with it, including the bacteria *Xanthomonas phaseoli* pv. *begoniae* and *Rhodococcus fascians* (syn. *Corynebacterium fascians*) and the fungus *Sclerotium oryzae*. *A. fragariae* and *A. ritzemabosi* may also interact with *R. fascians* to produce cauliflower disease of strawberry, in which the crown of the plant is reduced to a stunted, fleshy rosette.

*Aphelenchoides* species are resistant to desiccation and commonly survive winter in buds, growing points at the bases of plants, or dried leaves. When the temperature rises and shoots are wet, they ascend in films of water to enter growing leaves, buds, and floral parts, causing distortion.

#### BODY

Females are elongate (0.5–1.2 mm long, and usually less than 1 mm) and very slender (a = 32–65) (Figure 3.3). The lip region is not striated and may be wider than the neck or continuous with the body contour. The head is flattened to rounded anteriorly, with a weak cephalic framework, and is slightly offset. The amphid apertures at the apex of the lateral lips are minute. The lateral field has up to four incisures. The cuticle is marked by fine transverse annulations. Males are very similar to females in general size and form, except for secondary reproductive characteristics. When relaxed with heat, females lie straight or bowed ventrally; male tails curl sharply ventrally. The sexes are separate in many species, but males are rare in some species. *A. besseyi* may reproduce parthenogenetically.

#### STYLET

The stylet is slender and very short (9 to 17 μm), with a needle-like anterior portion and small to minute but distinct basal knobs.

#### ESOPHAGUS

The procorpus is long, slender, and cylindrical. The median bulb is large (about three-fourths the body width) and has a conspicuous sclerotized, spherical to rounded-rectangular valve. The esophageal gland overlaps the intestine dorsally in a long lobe. The dorsal esophageal gland orifice is in the median bulb, just anterior to the valve plates. The excretory pore opens at the level of the nerve ring or close behind it. The nerve ring encircles the anterior end of the esophageal glands and the intestine. The hemizonid is posterior to the excretory pore. The male esophageal region is similar to that of the female.

#### REPRODUCTIVE STRUCTURES

Females have a single ovary, which is outstretched anteriorly, may be reflexed, and has one or more rows of oocytes. A long postuterine branch usually extends behind the vulva.

Males have a single, outstretched testis. The spicules are paired, well developed, and curved ventrally, resembling rose thorns. There is no gubernaculum or bursa.

#### VULVA

The vulva is a transverse slit located in the posterior third of the body (V = 60–75). The vagina usually slopes forward from the vulva.

#### TAIL

The tails of both sexes are conoid (never filiform), long (up to 5 anal body widths), and uniformly tapered to a blunt or acute tip. One to four mucros (minute spikes) may be present at the tip, or mucros may be absent. The male tail is frequently curved ventrally, usually with three pairs of ventro-submedian papillae, posterior to the anus. The phasmids are usually subterminal and are minute and difficult to observe.

#### GENERA RESEMBLING *APHELENCHOIDES*

*Aphelenchus.* Nematodes in this genus, like members of *Aphelenchoides,* occur in soil and plant parts and feed on fungi. In *Aphelenchus,* the body is smaller (0.54 mm long)

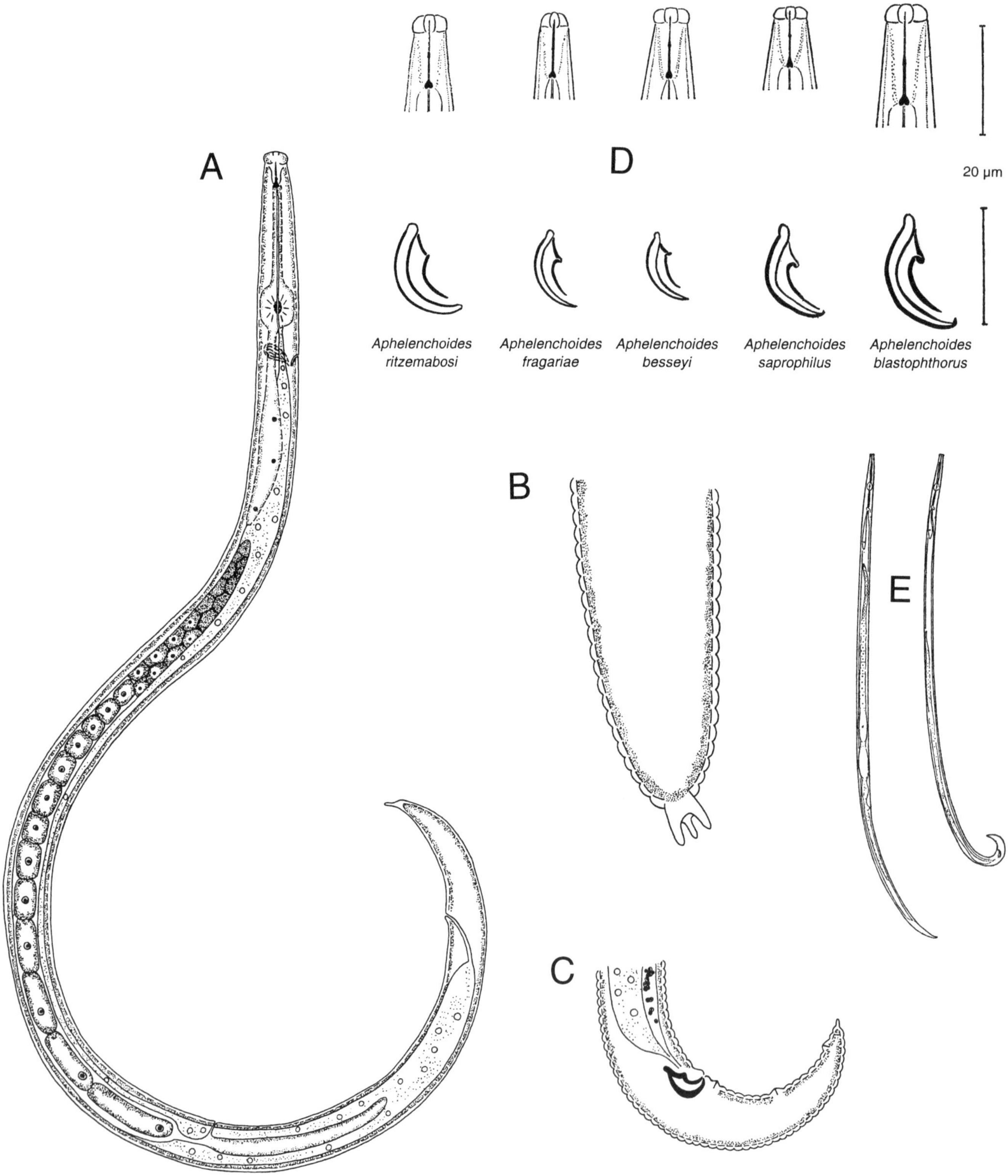


**Figure 3.3.** *Aphelenchoides.* **A,** Adult female. **B,** Female tail with mucros (terminal spikes). **C,** Male tail. **D,** Heads and spicules of five species. **E,** *A. ritzemabosi* adult female and male. (**A–C** reprinted from Jenkins and Taylor, 1967; **D** and **E,** British Crown copyright material reproduced with the permission of the Controller of Her Majesty's Stationery Office, from Franklin, 1978)

than in *Aphelenchoides,* the female tail is rounded and lacks mucros, and males (which are infrequent) have a gubernaculum and bursa.

*Rhadinaphelenchus.* This genus is similar in most respects to *Aphelenchoides,* with females having a very long postvulval sac; a very long, slightly tapering tail with a rounded tip; and a vulval flap. Males have a small, circular bursa at the tip of the tail.

#### COMMENTS

There are 180 to more than 225 species of *Aphelenchoides,* many of which are marine and soil inhabitants that feed readily on various fungi, algae, lichens, and mosses. Major plant-parasitic species include *A. arachidis, A. besseyi, A. blastophthorus, A. composticola, A. fragariae* (which has more than 250 hosts), *A. parietinus,* and *A. ritzemabosi* (which has 190 hosts). *A. composticola* is an economic pest of mushrooms (*Agaricus bisporus*) throughout the world, and some 16 other species of *Aphelenchoides* are of potential economic importance.

The life cycle of these nematodes can be completed in a week to 14 days. The optimum temperature for development is 21–25°C (70–77°F), with the life cycle taking 10 days at 21°C and eight days at 23°C. Females may lay 25 to 30 or more eggs at a time. The eggs hatch in three to four days, and the nematodes mature in six to 12 days. Ten generations may be produced during a single season. These nematodes can be vigorous "swimmers," traveling by rapid undulating movements. They are often observed at all levels in a counting dish.

Some species (e.g., *A. fragariae*) that feed on fungi are also parasitic in higher plants. *A. ritzemabosi* may be found in association with infectious bacteria (e.g., *Pseudomonas, Rhodococcus,* and *Xanthomonas*) and fungi (e.g., *Phytophthora*). Several species of *Aphelenchoides* feed on mycorrhizal fungi and inhibit the growth of these beneficial organisms.

To observe *Aphelenchoides* species, tease apart leaf tissue in water with needles and observe nematodes swarming out. Avoid badly decayed tissue. As many as 15,000 nematodes have been reported from a single leaf.

*Aphelenchoides* species are often associated with *Ditylenchus dipsaci* in alfalfa, currants, and gooseberries.

## *Aphelenchus*

### Class Secernentea, order Tylenchida, family Aphelenchidae

#### HOSTS

Nematodes in the genus *Aphelenchus* feed on various fungi, including species of *Armillaria, Fusarium,* mushrooms, mycorrhizal and nematode-trapping fungi, *Neurospora, Periconia, Pyrenochaeta, Pythium, Rhizoctonia, Sclerotium,* and *Thielaviopsis.*

These nematodes are cosmopolitan in the soil rhizosphere, feeding on fungi in decaying roots, bulbs, tubers, rhizomes, leaf sheaths, plant crowns, and the cortex of some roots, where they appear to feed on the cell contents of living tissues. Reports of parasitism on vascular plants are generally discounted.

#### PARASITISM AND HABITAT

*Aphelenchus* species are cosmopolitan, migratory, ectoparasites on fungi and possibly other microorganisms in the soil. These mycophagous nematodes are easily reared on a number of fungi growing in petri plates. They may occur in large numbers in moist, decaying plant tissues both above- and belowground.

#### BODY

Females are almost cylindrical, slender (a = 27–33), and elongate (0.5–1.2 mm long), tapering only slightly anteriorly and posteriorly, sometimes abruptly behind the vulva (Figure 3.4). Males are somewhat smaller but similar to females, except for sexual dimorphism. Second- to fourth-stage juveniles are similar to mature females in their esophageal region, tail, and overall shape. The lip region is bluntly rounded to flattened and may be slightly offset from the body, with a weak cephalic framework. Cuticular annulations, 1 to 2 μm wide, cover the body except the head. The lateral fields are widest in the midbody region, where they have eight to 14 minute incisures. Anteriorly, they are narrower and have fewer incisures; posteriorly, the lateral field often widens just in front of the vulva and narrows just posterior to it. Deirids are present at about the level of the excretory pore. These nematodes lie almost straight when killed by heat.

#### STYLET

The stylet is very short (less than 17 μm), with slight thickenings at the base, but without distinct basal knobs (Figure 1.15).

#### ESOPHAGUS

The procorpus is long, cylindrical, and slightly constricted anteriorly. The metacorpus (median bulb) is rectangular to oval, large (three-quarters of the body width), and well developed, and it contains a prominently sclerotized valve. The esophageal isthmus is short (1.5 body widths), about half the bulb length behind the median bulb, and is surrounded by the nerve ring. The dorsal lobe of the esophageal glands is usually well developed and usually overlaps the intestine. The excretory pore is opposite the nerve ring. The hemizonid is well developed, posterior to the excretory pore, separated from it by three or four annules. The esophageal region of the male is similar to that of the female.

#### REPRODUCTIVE STRUCTURES

Females have a single anterior, outstretched (prodelphic) ovary with a rather obscure postvulval sac, which reaches about halfway from the vulva to the anus. The oocytes are in a partly single, partly double row. The eggs are roundly rectangular and 60–88 × 20–30 μm.

Males have a single, outstretched testis. The spicules are paired, slender, and ventrally slightly arcuate, 20–30 μm long, resting on a simple, almost straight gubernaculum about one-third as long as the spicules.

#### VULVA

The vulva is a prominent transverse slit with slightly protruding lips, in the posterior third of the body ($V = 70^{70}–81^{91}$). The vagina has thickened walls, sloping forward from the vulva.

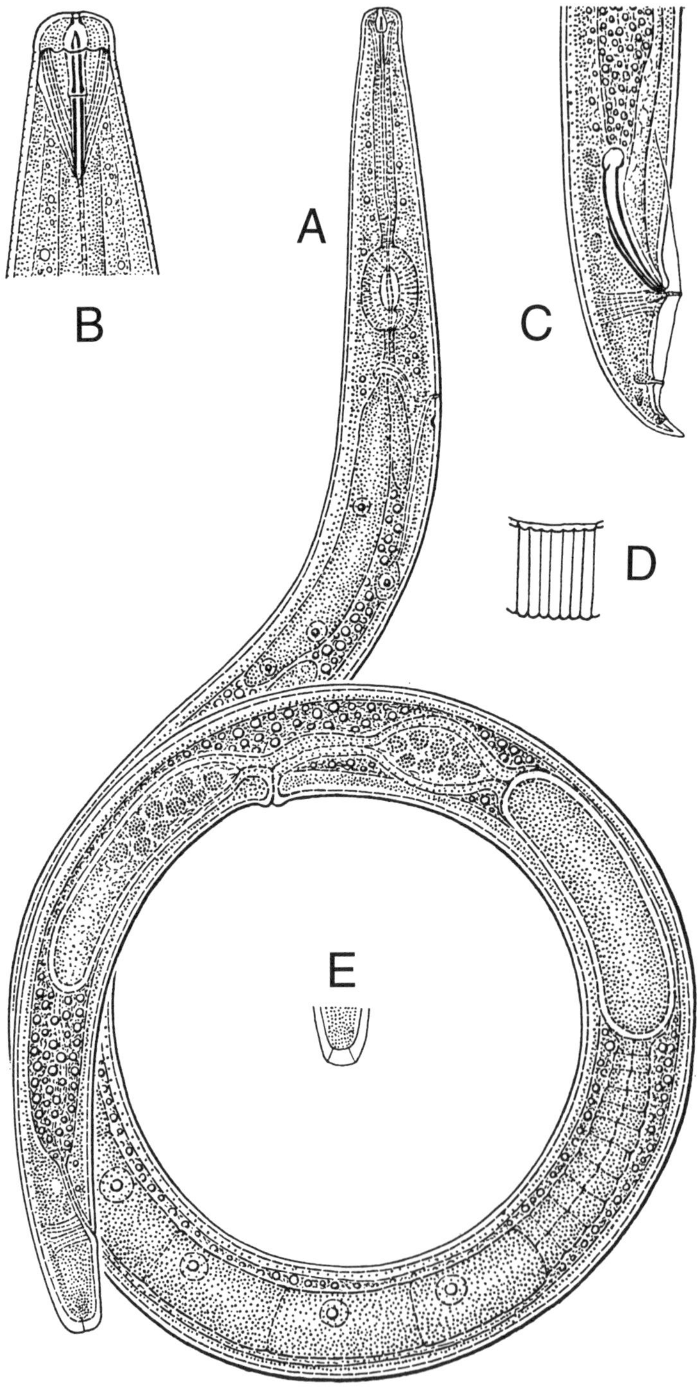


**Figure 3.4.** *Aphelenchus.* **A,** Adult female. **B,** Head. **C,** Adult male, posterior part. **D,** Lateral field. **E,** Phasmids, dorsal view. (Reprinted, by permission, from Thorne, 1961)

#### TAIL

The female tail is cylindrical and bluntly rounded, with a length of 1 to 4 anal body widths, and lacks mucros. The phasmids are subterminal. The male tail is short and conoid, with a bursa supported by three or four pairs of ribs, arising opposite the head of the spicules and extending to the terminus. The gubernaculum is about half as long as the spicules. The rectum has a length of about 1 or 2 body widths. The male phasmids are located in the lateral field near the terminus.

#### GENUS RESEMBLING *APHELENCHUS*

*Aphelenchoides.* Nematodes in this genus have two to four incisures in the lateral field; males lack a bursa and gubernaculum; and females usually have a conoid tail, often with one or more mucros (sharp points) at the tip.

#### COMMENTS

*Aphelenchus* species are frequently found in plant and soil samples. Males are seldom seen, being absent or extremely rare in many populations and cultures. Reproduction of *A. avenae,* the major species, is generally considered to be parthenogenetic; it is easily cultured on a nutrient agar medium, in which it feeds on the hyphae of various fungi. There are some 12 other species.

## *Belonolaimus:* Sting Nematodes

### Class Secernentea, order Tylenchida, family Belonolaimidae

#### HOSTS

*Belonolaimus* species (sting nematodes) have very extensive host ranges, including alfalfa, *Ardisia,* barley, beans (*Phaseolus* spp.), beet (garden beet, sugar beet), bush and tree fruits, cabbage, cantaloupe, carrot, celery, chrysanthemum, citrus, clovers (*Trifolium*), conifers (e.g., pines), corn (maize), cotton, cowpea (*Vigna unguiculata*), cucumber, eggplant, elms, garlic, *Gerbera,* gherkin, grape, grasses (pasture grasses, turfgrasses [Plate 1D], and wild species), holly, lettuce, lima bean, lupines, maples, millets, morning-glory, mung bean (*Vigna radiata*), muskmelon, oat, onions, okra, pea, peach, peanut, pepper (garden or bell pepper and ornamentals, *Capsicum*), pines, planetree, potato, pumpkin, rye, soybean, squashes, strawberry, Sudangrass, sugarcane, sunflower, sweet potato, sweetgum (*Liquidambar*), sycamore, tomato, turnip, wheat, and many others.

Weed hosts include ball nightshade (*Solanum carolinense*), mild water-pepper (*Polygonum hydropiperoides*), nutgrass (*Cyperus rotundus*), and sea-oats (*Uniola paniculata*).

**Figure 3.5.** Corn seedlings severely damaged by *Belonolaimus longicaudatus.*

#### PARASITISM AND HABITAT

*Belonolaimus* species are elongate, polyphagous, migratory ectoparasites. They feed on root tips and along the sides of succulent young roots of various plants, mostly in sandy soils in the southeastern and some western states in the United States, where they sometimes completely destroy a crop. Poor stands with erratic, uneven growth, severe stunting, and chlorosis with a tendency to wilt under dry conditions are common (Plate 1D). Shoots show symptoms of nutrient and water deficiencies. Affected plants often have little root growth below the top 3 in. of soil but excessive root development near the soil surface (Figure 3.5). Affected roots appear to have been clipped, with the ends slightly swollen. Root systems are small, coarse, and devitalized, with large, dark, sunken, lesions. Surface lesions and cavities in roots do not extend beyond the inner cortex, some spreading lengthwise or laterally, girdling the root. Feeder roots are short and stubby, with necrotic, knobby tips (Plate 1E). Taproots are often malformed. During feeding, these nematodes destroy host cells in the outer portion of the root, producing dark, necrotic lesions. Root growth is severely restricted when nematode populations are high, and shoot growth and yield are consequently reduced.

*B. longicaudatus* is a destructive pest, reproducing at soil temperatures up to 34°C (93°F) but hindered at higher temperatures. It is the common species in the southeastern United States, where it is restricted to soils containing more than 85% sand.

#### BODY

The sexes are separate, and the males and females are similar in appearance and size (Figure 3.6). Females are long (2.0–3.0 mm) and slender ($a \approx 52$ in *B. gracilis;* $a =$

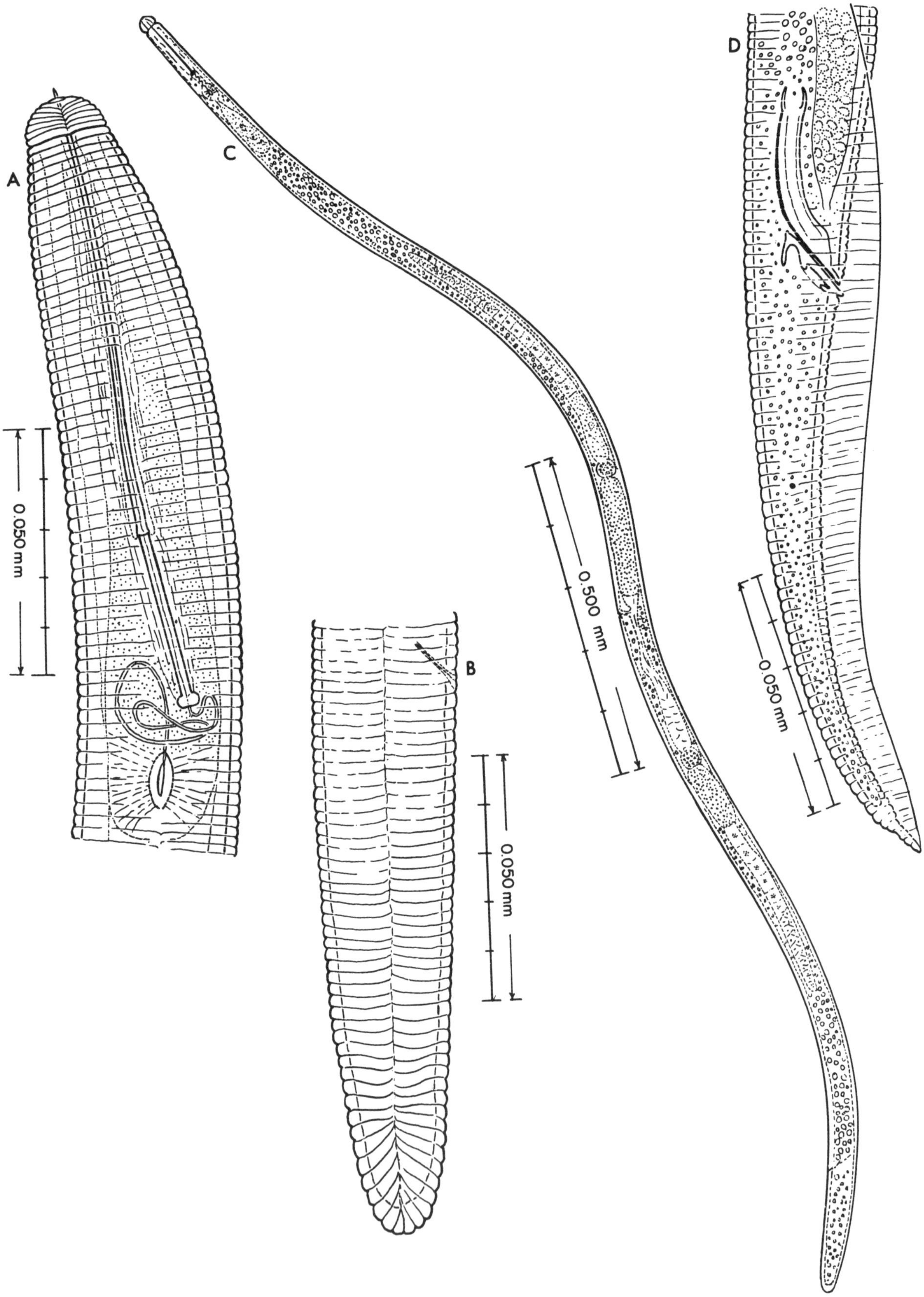


**Figure 3.6.** *Belonolaimus longicaudatus.* **A,** Female, anterior part. **B,** Female, posterior part. **C,** Female, full length. **D,** Male, posterior part. (Reprinted, by permission, from Taylor, 1971)

56–75 in *B. longicaudatus*). Males are slightly smaller (1.5–2.5 mm long). The lip region is hemispherical and distinctly offset; it is divided into four distinct lobes and two smaller ones and has six to 10 or more annules. Viewed *en face,* six distinct lips can be seen, with amphid openings near the edges of the lateral lips. There is a terminal head cap. The cephalic framework is moderately sclerotized. The cuticle is coarsely striated. A single lateral line or groove is evident on each side in both sexes.

#### STYLET

The stylet is typically very long (100–160 µm), slender, and flexible, with small, rounded basal knobs, in both males and females. The cone is much longer than the shaft.

#### ESOPHAGUS

The procorpus is long and swollen. It is attached to the base of the stylet knobs and lies coiled between the base of the knobs and the valve of the median bulb when the stylet is retracted, straightening when the stylet is protruded. The median bulb is ovate. The isthmus is short, expanding to a long, spatulate glandular lobe that overlaps the intestine lateroventrally. There is an obscure muscular valve at the junction of the esophagus and the intestine. The esophageal glands overlap the anterior end of the intestine latero-ventrally. The intestine extends almost to the terminus. The dorsal esophageal gland opens close to the spear base. The excretory pore is opposite the middle of the glandular lobe.

#### REPRODUCTIVE STRUCTURES

Females have two opposed, outstretched ovaries, joining at the midbody, and distinct spermathecae.

Males have a single, outstretched testis and well-developed, curved (arcuate) spicules with ventral flanges. The gubernaculum is well developed; it varies with the species but in most species has a posterior projection.

#### VULVA

The vulva is a transverse slit with lips not protruding, located in the midsection (V = 48–52 in *B. gracilis;* V = 46–54 in *B. longicaudatus*).

#### TAIL

The tail length is 3 to 6 anal body widths. The female tail is cylindrical, with a rounded tip. The male tail is long and pointed and enveloped by a long, narrow bursa.

#### GENUS RESEMBLING *BELONOLAIMUS*

*Dolichodorus.* In this genus, females have sharply pointed tails, and males have short tails.

#### COMMENTS

There are nine recognized species of *Belonolaimus,* of which the most important are *B. gracilis* and *B. longicaudatus.* These elongate, voracious nematodes are among the most damaging ectoparasites known. Both nematodes act together with pathogenic fungi (e.g., *Fusarium, Pythium,* and *Verticillium*) to cause severe damage to roots. Geographic variants occur in *B. longicaudatus* and may result in a splitting of the species; there are several biotypes or races. This species reproduces only in soils with a high sand content, and most individuals are found in the upper 30 cm of the soil. A generation takes 28 days at an optimal soil temperature of 28–30°C (82–86°F). *B. gracilis* has a shorter tail and a longer stylet and is wider than *B. longicaudatus.*

Other species of sting nematodes are *B. euthychilis, B. hastulatus, B. lineatus, B. maritimus,* and *B. nortoni.*

*B. hastulatus* and *B. lineatus* have incisures in the lateral fields, distinguishing them from the other species.

## *Bursaphelenchus:* Pinewood or Pine Wilt Nematode and Related Species

### Class Secernentea, order Aphelenchida, family Aphelenchoididae

#### HOSTS

The pinewood or pine wilt nematode, *Bursaphelenchus xylophilus* (syn. *B. lignicolus*), infects the wood of most species of pines and, rarely, a few other related conifers, including balsam fir (*Abies balsamea*), Atlas cedar (*Cedrus atlantica*), deodar (*C. deodara*), Alaska yellow-cedar (Nootka cypress, *Chamaecyparis nootkatensis*), eastern larch (tamarack, *Larix laricina*), European larch (*L. decidua*), western larch (*L. occidentalis*), black spruce (*Picea mariana*), Colorado blue spruce (*P. pungens*), white spruce (*P. glauca*), Douglas fir (*Pseudotsuga menziesii*), American arborvitae or northern white-cedar (*Thuja occidentalis*), western hemlock (*Tsuga heterophylla*), and western red cedar (*Thuja plicata*).

The most susceptible pines include Austrian (*Pinus nigra*), jack (*P. banksiana*), Japanese black (*P. thunbergiana*), Japanese red (*P. densiflora*), loblolly (*P. taeda*), maritime (cluster, *P. pinaster*), red (*P. resinosa*), sand (*P. clausa*), Scotch (scots, *P. sylvestris*), and southern red (*P. massoniana*).

Moderately susceptible pines include lodgepole (*P. contorta*), ponderosa (*P. ponderosa*), and western white (*P. monticola*).

Pines that appear resistant include eastern white (*P. strobus*), Jeffrey (*P. jeffreyi*), pitch (*P. rigida*), and Virginia (*P. virginiana*).

*B. xylophilus* is a serious pest in Japan and parts of China, Taiwan, and South Korea and on exotic pine species in the United States.

#### PARASITISM AND HABITAT

*Bursaphelenchus* species are migratory endoparasites in woody tissue (petioles, branches, trunks, and roots). Males are abundant. Most species have a very close relationship

**Figure 3.7.** **A** and **B,** Scotch pines infected with the pinewood nematode, *Bursaphelenchus xylophilus.* Diseased trees generally turn completely from healthy dark green to pale yellowish green and finally reddish brown. **C,** Female (left) and male (right) long-horned Carolina pine sawyer beetles (*Monochamus carolinensis*), vector of the pinewood nematode. **D,** Male *M. carolinensis* feeding on a Scotch pine twig, where nematode transmission takes place. **E,** Felled pine with wedge-shaped blue stains in the wood. Blue stain fungi enter the tree following invasion by nematodes. **F,** Disks cut from a pine branch and placed in water in a petri dish overnight to extract nematodes from infected tissue. **G,** Tail of male *B. xylophilus.* (**C** and **D** courtesy J. E. Appleby)

with insects, especially bark beetles and wood-boring insects, though some are known to feed on fungi. Juveniles (fourth-stage dauer larvae) are disseminated by bark beetles, long-horned beetles, wood borers, and weevils, mostly in the families Cerambycidae, Curculionidae, and Scolytidae. Dauer larvae are well adapted to survive desiccation. Most species feed on fungi (e.g., *Ceratocystis* spp.) growing in insect tunnels and are carried externally on the insects.

*B. xylophilus* alternates between pines and long-horned wood-boring beetles, mostly *Monochamus alternatus, M. carolinensis* (Figure 3.7C and D), *M. marmorator, M. nitens, M. obtusus, M. saltuarius, M. scutellatus,* and *M. titillator.* From spring into summer, adult beetles generally emerge from dead, infected pines (Figure 3.7A and B and Plate 1F), carrying large numbers of nematodes (an average of 15,000 per beetle) in their tracheae. The beetles fly to healthy pines, where they strip the outer bark from tender shoots (Figure 3.7D). Nematodes leave the beetle tracheae and enter resin canals and cambial tissues of the tree, where they feed, mature, and reproduce. Tremendous populations build up throughout the tree when host tissues are moist. Reproduction is rapid; a complete generation takes four to five days at 25°C (77°F), and the nematodes migrate throughout the tree. Populations as high as 1,000 nematodes per dry gram of wood have been recorded.

Infected trees die, often during hot and dry summer weather. Dying pines are invaded by blue stain fungi (Figure 3.7E and Plate 1F), on which the nematodes feed. During summer, female beetles lay eggs only on stressed (weakened) or dead pines, in which nematode populations are soaring. Emerging insect larvae tunnel into the wood and emerge as adults the following spring and summer. Resistant, fourth-stage juveniles (dauer larvae) aggregate around developing pupae and enter the mesothoracic spiracles of newly molted adults. The nematodes continue their life cycle when the beetles feed on a healthy tree.

Nematode populations from different geographic regions may have different host ranges. The incidence of the pinewood nematode in North America is limited mostly to species of pines planted in warm to hot areas outside their normal or expected range and to areas where pines are stressed by sustained high temperatures combined with water stress.

#### BODY

Both sexes are relatively long (in *B. xylophilus,* the average length is 0.81 mm) and slender (a = 37–40) (Figure 3.8). Males are somewhat shorter than females but otherwise resemble them, except for secondary reproductive characteristics. The lips are high and angular to rounded. The adult head is offset with massively sclerotized labial arches. The cuticle is finely annulated with four lateral incisures. The hemizonid is about 1 body width posterior to the excretory pore.

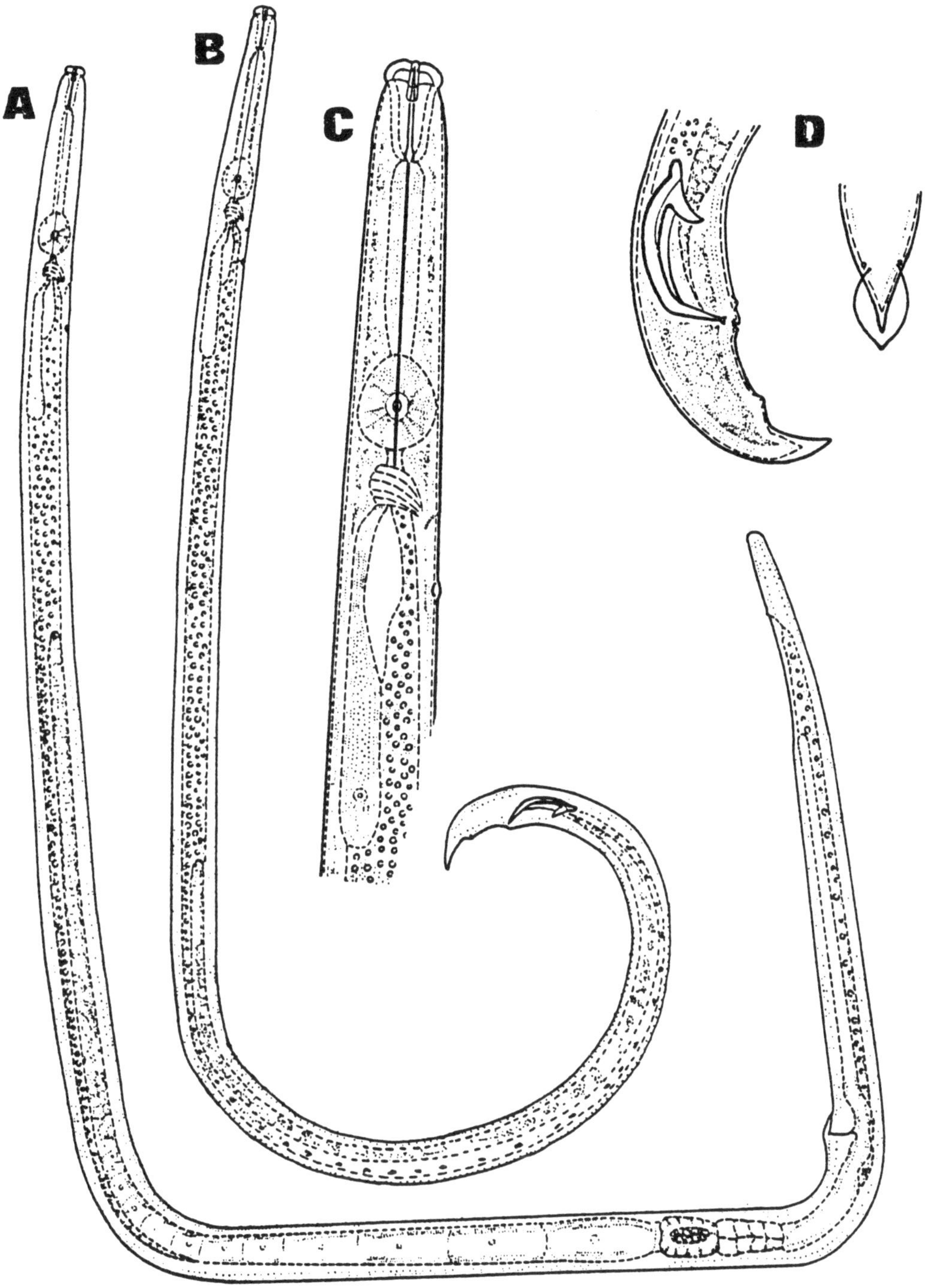


**Figure 3.8.** *Bursaphelenchus xylophilus.* **A,** Adult female. **B,** Adult male. **C,** Anterior region. **D,** Male tail, lateral and dorsal views. (Reprinted, by permission, from Mamiya and Kiyohara, 1972)

#### STYLET

The stylet is short (10.7–16 μm) and slender, with small, weak, rounded basal knobs, especially in immature specimens.

#### ESOPHAGUS

The procorpus is elongate, cylindrical, and slender. The median bulb is ovate, two-thirds the body width, and has a conspicuous oval valve. The dorsal esophageal gland orifice is in the median bulb. The esophageal glands are long and slender and overlap the intestine dorsally for 5 to 6 body widths. The excretory pore is at the level of the esophago-intestinal junction or just behind the nerve ring (a distance of two-thirds of the body width behind the median bulb) and anterior to the conspicuous hemizonid, which is about the width of the body and three to five annules long.

#### REPRODUCTIVE STRUCTURES

Females have a single, anterior ovary, which is outstretched and not reflexed. The uterus is thick-walled and compact. The oocytes are large, several times longer than wide, and in a single file, with a long postuterine sac about three-quarters of the distance to the anus.

Males have a single, anteriorly outstretched testis. The spicules are paired, generally large, thorn-shaped (free or partially fused), and arcuate, usually with a prominent, sharply pointed rostrum with disklike expansions and an apex that tends to continue the line of the transverse bar rather than that of the dorsal limb (Figure 3.8D). They usually lack a gubernaculum.

#### VULVA

The vulva is slitlike, sclerotized, and curved, with a wide overlapping flap, posterior to the midsection (V = 64–75). The vagina slopes anteriorly.

#### TAIL

The female tail is elongate-subcylindrical, narrowing to a usually broadly rounded, unstriated terminus. The male tail is pointed, acute, curved ventrally, and sclerotized; it bears two to four pairs of ventro-submedian papillae. A circular to oval bursa reaches the tail. The terminus appears talon-like in lateral view. The juvenile tail is sharply pointed.

#### GENERA RESEMBLING *BURSAPHELENCHUS*

*Aphelenchoides*

*Rhadinaphelenchus.* The coconut red ring nematode (*R. cocophilus*) is placed in *Bursaphelenchus* by some authorities.

#### COMMENTS

There are about 50 species of *Bursaphelenchus,* of which *B. xylophilus* is the major species. Other plant-pathogenic species in this genus are *B. fraudulentus,* which infects deciduous trees in Germany, and *B. kolymensis,* which infects larch in far eastern Russia.

Specimens of *B. xylophilus* can be obtained from small pieces of bark-free wood (chips, slivers, increment borer plugs, etc., about 5–15 mm long) from an infected trunk and branches, placed in shallow water in a petri dish, and left to stand for several hours or overnight (Figure 3.7F). The nematode can be readily cultured on media containing various fungi, including *Alternaria, Botrytis, Ceratocystis, Fusarium, Gliocladium, Graphium,* and *Pestalotia.*

*B. xylophilus* reproduces in the resin canals of pine trees. The first visible pathological response to infection (invasion by dauer larvae, transmitted by insects), several weeks after nematode invasion, is the death of parenchyma cells, followed by reduction and cessation of resin flow (oleoresin exudation). Needles turn from bright green to dull green or yellow and finally become reddish brown. Susceptible pines appear to die suddenly.

## *Criconemella, Criconemoides,* and *Mesocriconema:* Ring Nematodes

### Class Secernentea, order Tylenchida, family Criconematidae

#### HOSTS

*Criconemella, Criconemoides,* and *Mesocriconema* species (ring nematodes) are widespread and associated with many plants, including *Aeonium haworthii,* African violet, alfalfa, almond, *Aloe,* amaryllis, apple, apricot, *Aralia, Artemisia,* ashes, aster, avocado, banana, barley, American basswood (*Tilia americana*), beans (*Phaseolus* spp.), beech (*Fagus*), beet, begonia, black pepper (*Piper nigrum*), blackberry, blueberry, boxwood, cabbage, cacao (cocoa, *Theobroma cacao*), camellia, carnation, *Carpenteria* sp., cashew (*Anacardium occidentale*), cassava (*Manihot esculenta*), *Chamaedorea,* cherries, chrysanthemum, cinnamon (*Cinnamomum verum*), citrus, clove (*Syzygium aromaticum*), clover (*Trifolium*), coffee, coleus, conifers (e.g., pines), corn (maize), cotton, eastern cottonwood (*Populus deltoides*), cowpea (*Vigna unguiculata*), cranberry, dahlia, date palm (*Phoenix dactylifera*), daylily (*Hemerocallis*), deodar (*Cedrus deodara*), dogwood, *Echeveria, Echinopsis,* eggplant, elms, ferns, fig, firs, fuchsia, grape, grasses (pasture grasses, turfgrasses, and wild species), hemlock, hickories, holly, huckleberry (*Gaylussacia*), junipers, kenaf (*Hibiscus cannabinus*), lawnleaf (*Dichondra carolinensis*), *Lobelia,* loquat (*Eriobotrya japonica*), lupine, magnolia, mango, maples, marigold, *Mesembryanthemum,* mesquite, mints (*Mentha*), mulberry, mung bean (*Vigna radiata*), nectarine, oat, oaks, oleander (*Nerium*), *Osmanthus,* palmetto (*Sabal palmetto*), papaya (*Carica papaya*), pea, peach, peanut, pear, pecan, pines, *Pittosporum undulatum,* planetree, plums, potato, raspberry, rhododendrons, rice, roses, rye, sedges (*Carex* spp.), sorghums, soybean, spruces, strawberry, sugarcane, swamp taro (*Cyrtosperma chamissonis*), sweet potato, sweetgum (*Liquidambar*), tobacco, tuberose (*Polianthes*), turmeric (*Curcuma domestica*), hairy vetch (*Vicia villosa*), walnut, wheat, wineberry (*Aristotelia*), wisteria, and yam (*Dioscorea*).

Weed hosts include ball nightshade (*Solanum carolinense*), nutgrass (*Cyperus rotundus*), plantain (*Plantago*), *Polygonum,* ragweed (*Ambrosia*), and *Rumex.*

#### PARASITISM AND HABITAT

Ring nematodes are slow-moving, migratory ectoparasites that become sedentary on the roots of many hosts, especially woody plants. Females and juveniles are found in soil; males are rare or never found. High populations can severely damage roots, especially in sandy soils where moisture is retained. New growth of roots is destroyed, and moisture stress is increased as a result. The nematodes feed on the epidermal cells of small feeder roots, mostly near the root cap. Small, necrotic lesions are formed, which may en-

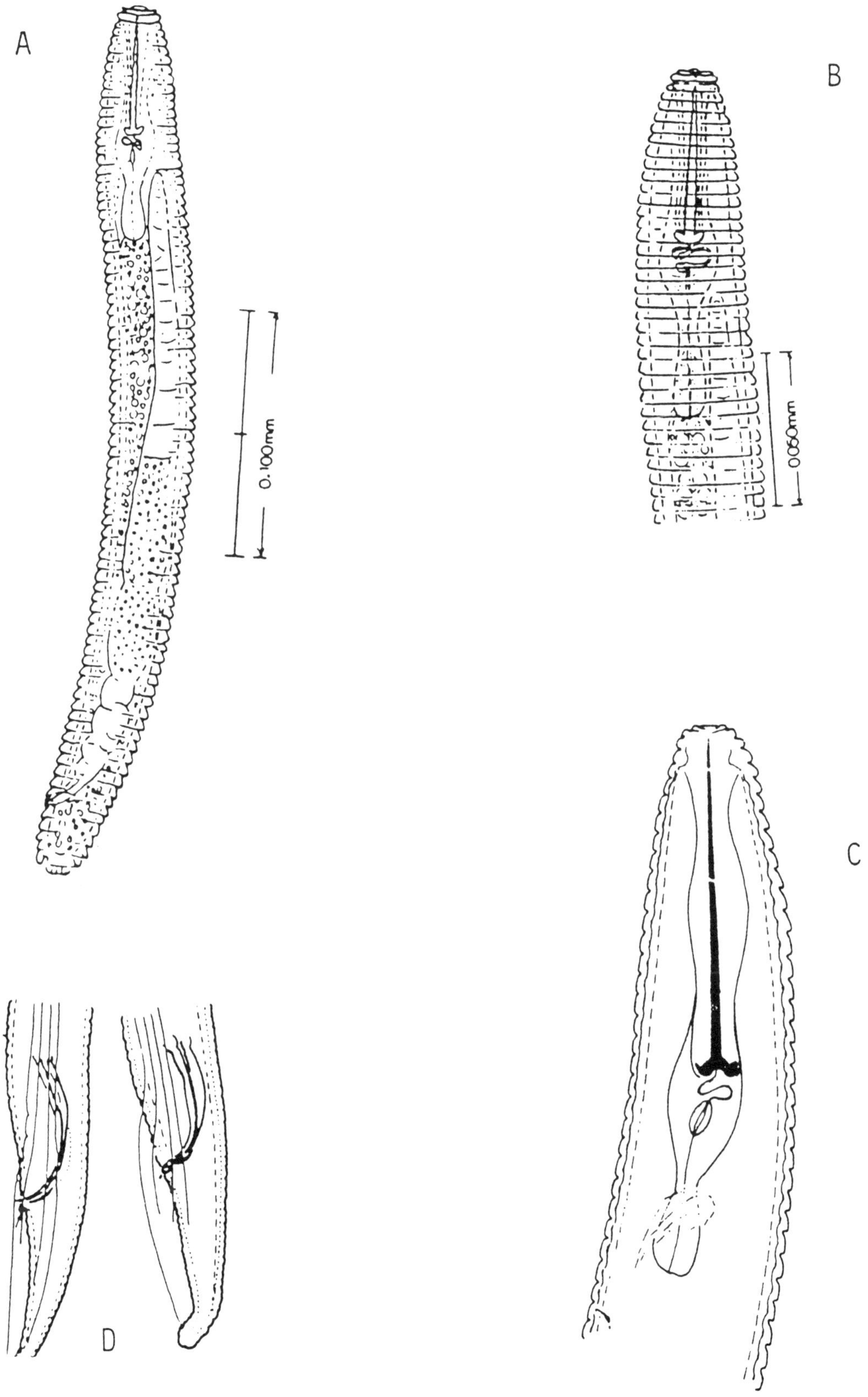


**Figure 3.9.** *Criconemoides.* **A,** Adult female. **B,** Female, anterior part. **C,** Neck region of *C. morgensis.* **D,** Male tail region of *C. morgensis,* lateral view. (**A** and **B** reprinted, by permission, from Taylor, 1971; **C** and **D** reprinted, by permission, from Loof, 1974)

large, merge, girdle, and kill the tips of roots. The lesions are invaded by other microorganisms (e.g., *Fusarium* and *Pythium* spp. and bacteria), which cause stunting, discoloration, and low vigor in the root system with few feeder roots. Aboveground symptoms include stunting and yellowing (chlorosis) of plants, early defoliation, decline, and reduction of flower and fruit yield. Ring nematode populations decline rapidly in the presence of poor hosts.

#### BODY

Members of *Criconemella* and related genera are called ring nematodes because of the large transverse annulations that appear like rings around the body (Figure 3.9). These nematodes are typically short (0.4–1.0 mm), fusiform, stout ($a \approx 10$), and cylindrical over most of the length of the body. The annules, numbering from 33 to 200, are very broad, prominent, ringlike, and retrorse in females, with smooth or finely serrate posterior margins; they are much smaller in males. Annules on juveniles nearly always have an angular posterior edge. The head area in females is rounded, not well separated from the rest of the body, and marked by one or two thinner annules. Males are slender and short, with a rounded anterior end; they are very rare. These nematodes lie straight or slightly curved when relaxed by heat.

#### STYLET

The stylet is long (about 30–90 µm) and stout, with anchor-shaped knobs directed forward, typically of medium size. In most species, males do not feed and do not have a stylet.

#### ESOPHAGUS

The procorpus and metacorpus are fused. The esophageal tube is typically in coils above the median bulb valve when the stylet is not extended. The median bulb is large and broad, with a distinct triradiate lining. The posterior bulb is short and narrow. The esophagus and intestine are degenerate in males.

#### REPRODUCTIVE STRUCTURES

Females have a single, anterior ovary, which is outstretched and may reach beyond the esophageal base. Most species are parthenogenic.

Males have a single, outstretched testis; short, slender, slightly curved spicules; and a linear gubernaculum. The bursa is generally weakly developed, extending almost to the tail tip (terminus), and rarely absent.

#### VULVA

The vulva is a transverse slit, difficult to see, with the anterior lip not overhanging, in the posterior part of body ($V = 80–95$). The annules are modified at the vulva.

#### TAIL

The female tail is short and conical to bluntly rounded. The male tail tapers to a rounded or pointed tip.

#### GENUS RESEMBLING *CRICONEMELLA*

*Criconema.* Nematodes in this genus are distinguished by having annules ornamented with spines, scales, plates, or stalked appendages on the posterior margins.

#### COMMENTS

*Criconemella, Criconemoides,* and *Mesocriconema* are the only genera of ring nematodes now generally recognized. These nematodes were previously classified in eight genera, including *Discocriconemella, Lobocriconema, Macroposthonia, Nothocriconema, Nothocriconemoides,* and *Xenocriconemella,* in addition to *Criconemella* and *Criconemoides.*

There are over 90 species of *Criconemella* (*Mesocriconema*). Pathogenicity has been demonstrated for only a few of them. The major species are *C. curvata, C. ornata, C. rustica, C. sphaerocephala,* and *C. xenoplax.* There are about 280 other species in the Criconematidae.

These nematodes move in a sluggish, earthworm-like fashion, not with the serpentine motion characteristic of most genera. Only a small percentage move through tissue paper in a Baermann funnel (Figure 2.9) or tray apparatus, and extraction by this method is therefore not recommended. The centrifugal flotation technique (Figure 2.12) is much more effective for extraction of these nematodes.

Most species are parthenogenic. Only a few species have been proven to cause serious damage. *Criconemella xenoplax* limits the number of feeder roots on peach and has been implicated in peach tree short life. Root infection of almond and plum trees by *C. xenoplax* may result in more extensive cankers by *Pseudomonas syringae* pv. *syringae.* Pecan roots parasitized by *Criconemella rusium* are more likely to be severely diseased in the presence of *Pythium irregulare* and *Fusarium solani.*

## *Ditylenchus:* Stem and Bulb Nematodes (Potato Rot Nematode and Related Species)

### Class Secernentea, order Tylenchida, family Tylenchidae

#### HOSTS

Plants damaged by *Ditylenchus* species include alfalfa, *Allium,* amaryllis, American potato bean (*Apios americana*), *Amsinckia,* anemone, apple, apricot, *Astrantia,* azaleas, baby's-breath, barley, beans (*Phaseolus* spp.), beet (garden beet, sugar beet), begonias, *Bergenia,* blueberry, broad bean (faba bean, *Vicia faba*), broccoli, cabbage, *Calceolaria, Campanula,* carnation, *Carpinus laxiflora,* carrot, cauliflower, celery, celeriac, *Cheiranthus,* chicory, China aster (*Callistephus chinensis*), *Chionodoxa,* chives, *Cimicifuga, Clerodendrum,* clovers (*Trifolium*), cluster bean (*Cyamopsis tetragonoloba*), *Colchicum, Collomia,* corn (maize), cotton, cowpea (*Vigna unguiculata*), cranberry, crocus,

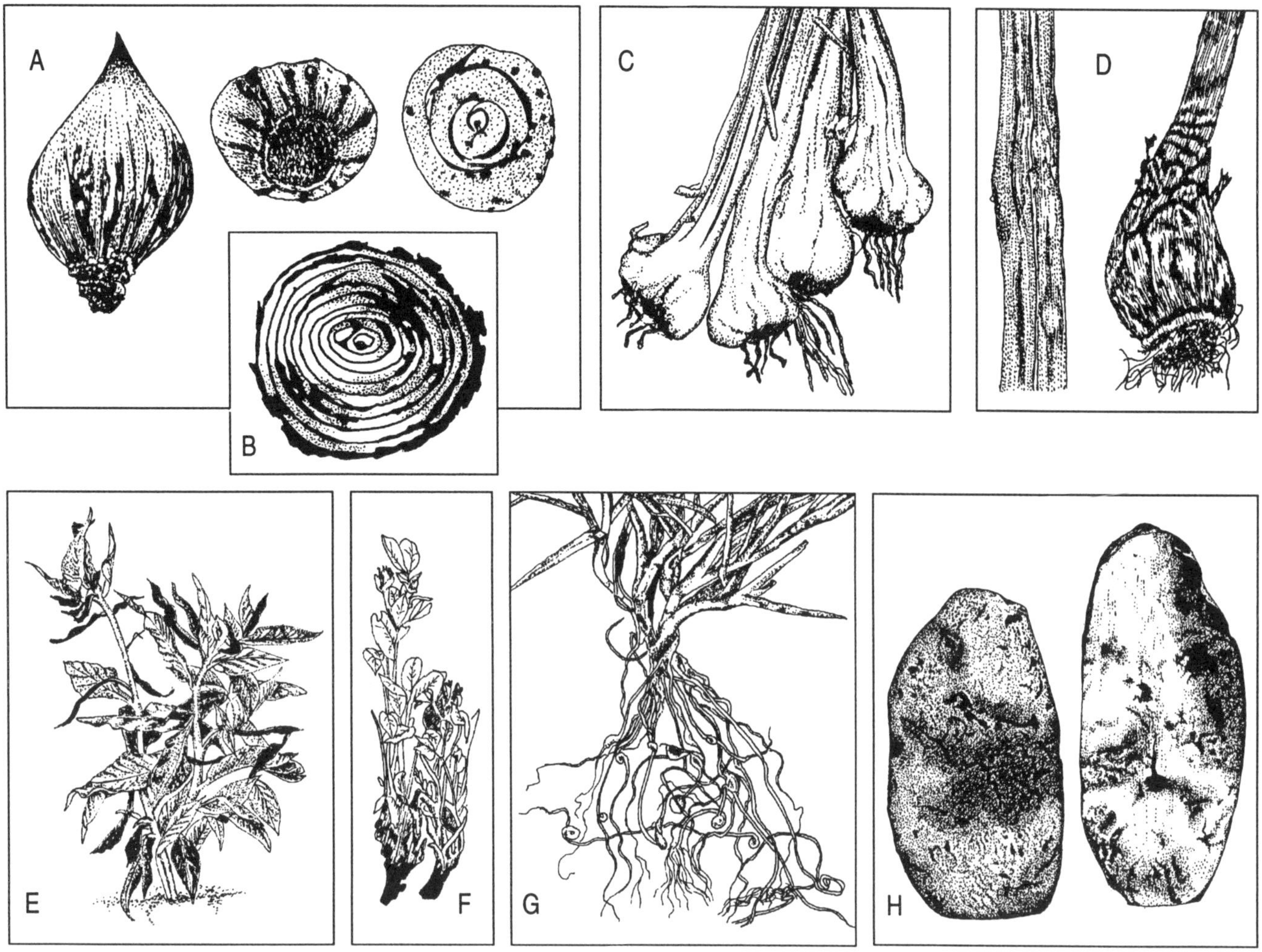


**Figure 3.10.** Damage to various plants caused by species of *Ditylenchus.* **A,** Wedgewood iris infected with *D. dipsaci.* **B,** Cross section of a hyacinth bulb infected with *D. dipsaci* and subsequently by secondary fungi and bacteria. **C,** Garlic bulbs with basal plate damage caused by *D. dipsaci.* **D,** Narcissus leaf section with characteristic "spikkles" (left) and narcissus bulb with masses of nematode wool on the basal plate (right), caused by *D. dipsaci.* **E,** Phlox infected with *D. dipsaci.* **F,** Alfalfa plants with crowns infected with *D. dipsaci.* **G,** Wheat roots with galls caused by *D. radicicolus.* **H,** Russet potato tuber infected with *D. destructor,* the potato rot nematode or tuber nematode.

cucumber, cyclamen, dahlia, dandelion, dogwood, dryad (*Dryas integrifolia*), *Erysimum,* fall daffodil (*Sternbergia lutea*), *Galanthus, Galtonia,* garlic, gherkin, *Gilia,* ginger, gladiolus, grape-hyacinth, grasses (forage grasses, turfgrasses, and wild species), *Helenium, Heuchera,* hyacinth, *Hyacinthoides,* hydrangea, iris, kale, kohlrabi, *Lamium,* leek, *Leersia hexandra,* lentil (*Lens culinaris*), lettuce, lilac, lilies, lily-of-the-valley, lupines, *Lychnis, Lyonia, Lysimachia,* mangel (mangold, *Beta vulgaris* subsp. *vulgaris*), mints, *Monarda,* mushrooms, narcissus, *Nerine,* oat, *Oenothera,* onions, *Ornithogalum,* parsley, parsnip, pea, peach, peanut, pear, *Penstemon,* phlox, pigeon pea (red gram, *Cajanus cajan*), pineapple (*Ananas comosus*), pinks, plantain (*Musa* × *paradisiaca*), potato, primroses, *Puschkinia,* radish, *Ranunculus,* rape (colza, *Brassica napus* var. *napus*), raspberry, rhododendrons, rhubarb, rice, rutabaga, rye, sage (*Salvia*), salsify, sanfoin (*Onobrychis viciifolia*), *Schizanthus, Scilla,* shallot, *Solidago,* soybean, strawberry, sunflower, sweet potato, teasel (*Dipsacus*), thyme (*Thymus vulgaris*), *Tigridia,* tobacco, tomato, tulips, turnip, vetches (*Vicia angustifolia;* hairy vetch, *V. villosa*), wheat, witloof (*Cichorium intybus*), yucca, and others.

These nematodes also feed on a wide range of fungi and many weeds, including *Bassia muricata* and plantain (*Plantago major*).

#### PARASITISM AND HABITAT

*Ditylenchus* species are migratory ecto- and endoparasites in parenchymatous plant tissues—stems, bulbs, tubers, leaves, flowers, and (rarely) roots.

*D. dipsaci* is a serious pest of more than 500 plants worldwide. It invades all aboveground parts of its hosts and is present in seed at harvest. It causes stunting of plants and swelling and often twisting of stems and other aerial parts. Symptoms include leaf deformation (Figure 3.10E and F) and swellings or blister-like areas on the leaf surface (Figure

3.10D and Plate 1G). Leaves grow in a disorderly fashion and often hang as if wilted (Figure 3.10E and F), and they may later become chlorotic. Severely infected seedlings and young plants may die.

Onions and hyacinth infected with *D. dipsaci* become swollen (bloat) and may rot during storage. The inner scales are more severely attacked than the outer scales (Figure 3.10A). Later in the season the bulb becomes soft, and the inner scales are brown in concentric circles when viewed in cross section (Figure 3.10B and Plate 1G).

In garlic, *D. dipsaci* causes leaf yellowing and death from damage to the basal plate (Figure 3.10C) but not deformation or swelling.

Masses of dauer juveniles of *D. dipsaci* form at the base of infected narcissus bulbs. The outermost nematodes die and form a crust around the remainder of the juveniles, thus slowing the drying of those in the interior of the mass. This aggregation is called nematode wool (Figure 3.10D).

In strawberry, *D. dipsaci* produces symptoms very similar to those caused by foliar nematodes (*Aphelenchoides* spp.), so nematodes must be extracted and identified from infected plant tissue to confirm a diagnosis.

In some alfalfa varieties, *D. dipsaci* may increase the severity of bacterial wilt, caused by *Clavibacter michiganensis* subsp. *insidiosus*. Fusarium and Verticillium wilts are also increased in the presence of the nematode.

*D. dipsaci* is very tolerant of desiccation. It can survive for 20 years in a dormant state in soil or in soil particles on unclean seed. The fourth-stage juvenile can persist in clay soils for many years without a host plant. The life cycle of this species is completed in an average of 19 to 25 days at 15°C (59°F).

*D. destructor,* the potato rot or tuber nematode, is a serious pest of potato worldwide and more than 70 other hosts. It enters lenticels and eyes on potato tubers, forming small, off-white, mealy spots just below the surface. These spots later turn dark and are woolly in texture. Infected areas merge into lesions of dry, granular tissue as nematode numbers increase. Secondary organisms (bacteria, fungi, and mites) enter the wounds, causing a generalized dark brown decay (Figure 3.10H). *D. destructor* can be serious in potatoes in storage, as the infection spreads from tuber to tuber. It also feeds on roots of other plants and at least 70 species of fungi, whereas *D. dipsaci* is an obligate parasite of higher plants. *D. destructor* overwinters in soil as adults or juveniles and possibly as eggs. Unlike *D. dipsaci,* it cannot stand excessive desiccation and thus is usually important only in cool, moist soils. *D. destructor* does not form nematode wool. It probably survives in field soil by colonizing weed hosts.

*D. radicicolus* produces galls on roots of wheat (Figure 3.10G), barley, rye, and numerous grasses.

*D. angustus,* the rice stem nematode, occurs in rice-growing areas of Bangladesh, Burma, India, Thailand, Vietnam, and other parts of Asia. It is an ectoparasite and grows on tender tissues, where it can destroy developing rice leaves and flowers. It can survive dry conditions over a year in a dormant, coiled position, resuming activity when fields are flooded. *D. angustus* spreads from field to field by means of irrigation water and rain.

*D. myceliophagus* is an important pest of mushrooms worldwide. It feeds on and destroys mushroom mycelia, causing yield reductions in commercial mushrooms. Large populations may aggregate to form clumps. The individuals in such clumps are very resistant to drying. Insects carry *D. myceliophagus* to healthy mushroom beds, where it causes new infections.

#### BODY

*Ditylenchus* species are moderate in length (0.6–1.5 mm) and slender (a = 30–62) (Figure 3.11). The sexes are separate and are generally similar. The lips are low, flattened anteriorly, without obvious annules, and slightly offset or not offset from the body, with weak cephalic sclerotization. The cuticle is finely annulated. The lateral fields are marked by four or six incisures. The body lies straight or slightly curved when relaxed by heat.

#### STYLET

The stylet is small (usually 10–14 µm long) and slender, with distinct, well-developed basal knobs.

#### ESOPHAGUS

The median bulb is ovoid to fusiform, with a weak valve located anterior to the midpoint of the esophagus. The isthmus gradually expands to form a distinct bulb, which may extend back as a short, club- or spoon-shaped lobe over the anterior end of the intestine. The deirids are small but usually visible. The phasmids are extremely minute, if visible at all.

#### REPRODUCTIVE STRUCTURES

Females have a single, anterior, outstretched ovary with a rudimentary postvulval uterine branch extending about halfway or more from the vulva to the anus. The gonad cells are in one or two lines, not arranged around a rachis.

Males have a single, outstretched testis; rather weak, ventrally curved, anteriorly expanded spicules; and a short, simple gubernaculum.

#### VULVA

The vulva is a distinct transverse slit located in the posterior quarter of the body (V = 75–80).

#### TAIL

The tail is elongate-conoid in both sexes; its length is usually 4 to 7 anal body widths. It often has a sharply pointed or narrowly rounded terminus. In males, the bursa extends from the spicules to envelop one-fourth to three-quarters of the tail.

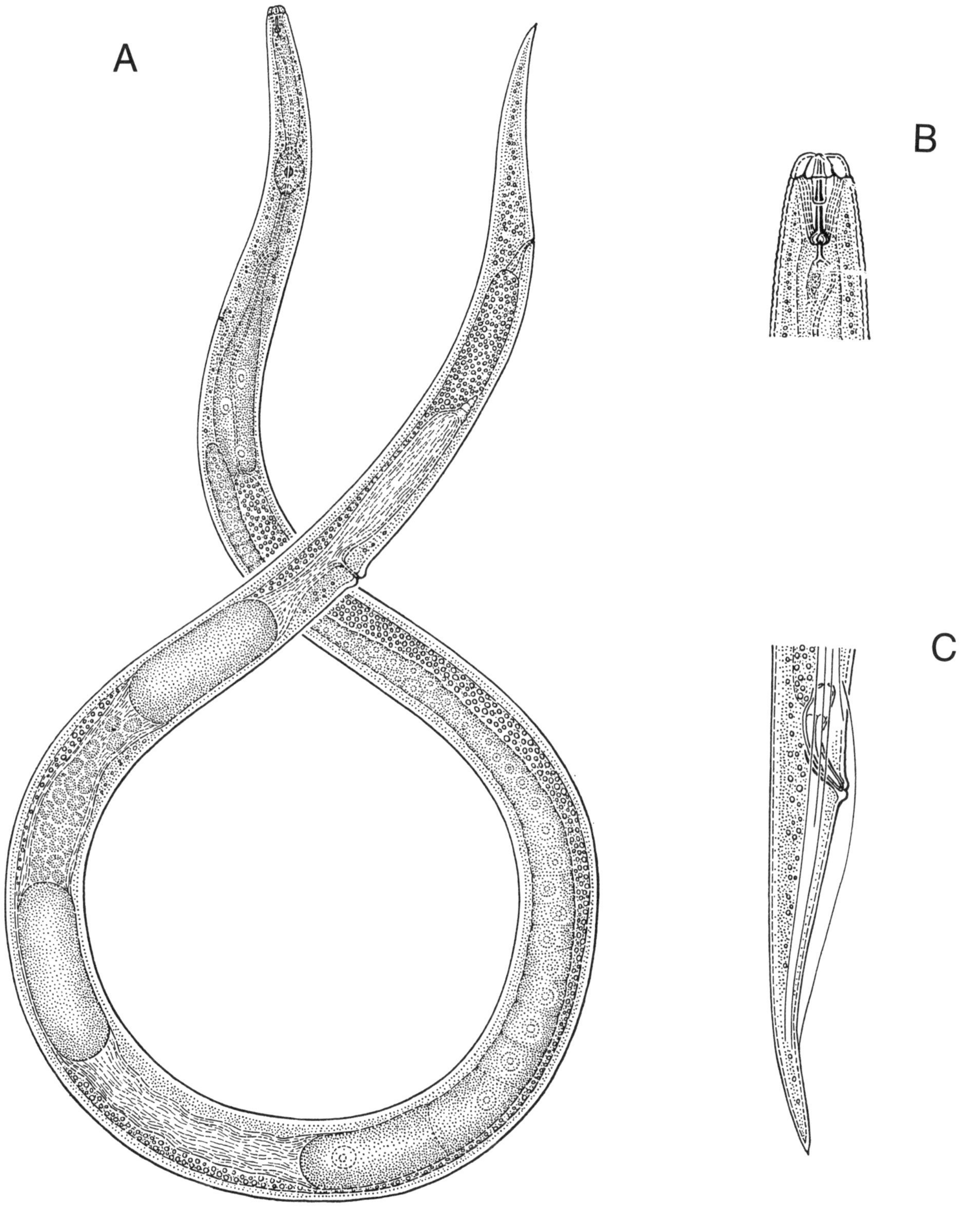


**Figure 3.11.** *Ditylenchus dipsaci.* **A,** Adult female. **B,** Head. **C,** Adult male tail. (Reprinted, by permission, from Thorne, 1961)

#### GENERA RESEMBLING *DITYLENCHUS*

*Anguina.* This genus is closely related to *Ditylenchus* but causes galls on aboveground parts of various plants.

*Aphelenchoides*

*Nothotylenchus*

*Pseudhalenchus*

*Subanguina.* This genus was established in 1967, and the former *D. radicicola* was transferred to it, as *S. radicicola.* This nematode causes root galls in members of the family Poaceae, such as annual bluegrass (*Poa annua*) and quackgrass (*Elytrigia repens,* syn. *Agropyron repens*).

*Tylenchus*

#### COMMENTS

There are approximately 85 species of *Ditylenchus*. They are distributed worldwide but are poorly adapted to subtropical and tropical areas. The major species are *D. dipsaci*, a serious pest of more than 500 plants worldwide; *D. angustus*, the rice stem nematode; *D. destructor*, the potato rot or tuber nematode; *D. myceliophagus;* and *D. radicicolus*. A number of species comprise distinct biological races (several to 20 or more), which differ in host preference.

These nematodes will swarm out of infected leaf or bulb tissue teased apart in water and thus become available for observation. Certain species of *Ditylenchus* can be grown in media containing various fungi.

## *Dolichodorus:* Awl Nematodes

### Class Secernentea, order Tylenchida, family Dolichodoridae

#### HOSTS

Hosts of *Dolichodorus* species (awl nematodes) include beans (*Phaseolus* spp.), black pepper (*Piper nigrum*), blueberry, *Bromus*, cabbage, cacao (cocoa, *Theobroma cacao*), celery, chestnut (*Castanea*), Chinese water chestnut (*Eleocharis dulcis*), clove (*Syzygium aromaticum*), coconut palm (*Cocos nucifera*), corn (maize), cowpea (*Vigna unguiculata*), garden balsam (*Impatiens balsamina*), grasses (forage grasses, turfgrasses, and wild species), *Juncus roemerianus*, Parry manzanita (*Arctostaphylos manzanita*), red maple (*Acer rubrum*), pea, pepper (garden or bell pepper and ornamentals, *Capsicum annuum*), pines, *Sparganium greenei*, tomato, water chestnut (*Pachira aquatica*), and others.

#### PARASITISM AND HABITAT

*Dolichodorus* species are widespread ectoparasites in the eastern United States, especially in Florida. They feed principally at the tips of roots, with some feeding on the sides of succulent roots and at the base of the hypocotyl. The number of fine feeder roots is severely reduced. Injury symptoms are stubby root or coarse root, or both, often with considerable discoloration and necrosis of tissues. Large numbers of nematodes can completely destroy the roots. Terminal galls are occasionally formed. Severe stunting accompanies a depleted root system. Planted seeds may fail to germinate if they are attacked.

These nematodes feed in all stages and are found in all stages in wet soils. They resemble sting nematodes (*Belonolaimus* species) in feeding habits as well as size, shape, and general appearance (see the section on *Belonolaimus*).

#### BODY

Awl nematodes are long (1.5–3.0 mm) and slender (a = 35–60) and have a coarsely annulated cuticle (Figure 3.12). Males are similar to females, except for sexual dimorphism. The lips have eight to 10 minute annules and are set off from the body by a deep constriction. The head is distinctly set off from the remainder of the body. In lateral view, the head is rounded, with a massive sclerotized framework in the lip region. The amphids open at the lateral margins of the lip cap. The lateral fields are marked with three or four deep incisures at the middle of the body. The body lies almost straight when relaxed by gentle heat.

#### STYLET

The stylet is very long (50–160 μm), slender, and flexible in both sexes, with conspicuous, backward-sloping basal knobs. The cone is much longer than the shaft.

#### ESOPHAGUS

The esophageal lumen is folded and swollen in the procorpus when the stylet is retracted. The median bulb is ovate and conspicuous, with a large valve and a distinct triradiate lining. The isthmus is short and slender, widening to an elongate, pyriform terminal bulb, which contains the esophageal gland nucleus. The terminal bulb does not overlap the intestine. A postanal intestinal sac is present. The hemizonid is posterior to the excretory pore.

#### REPRODUCTIVE STRUCTURES

Females have paired, opposed, outstretched ovaries accompanied by spermathecae.

Males have large, slightly curved spicules. The gubernaculum is linear and has a reflexed part, the tip protruding from the body.

#### VULVA

The vulva is a transverse slit at or near the midsection (V = 50–60).

#### TAIL

The female tail is bluntly rounded to pointed, ending in a conoid spike. The male tail is short (about 1 anal body width) and pointed at the tip, which is surrounded by a well-developed, deeply cleft bursa projecting beyond the terminus.

#### GENUS RESEMBLING *DOLICHODORUS*

*Belonolaimus.* There is only one incisure in the lateral field in this genus, whereas in *Dolichodorus* there are three or four incisures. Distinguishing features of *Dolichodorus* females are the abruptly acute tail and the basal esophageal gland, which does not overlap the anterior part of the intestine.

#### COMMENTS

There are some 16 valid species of *Dolichodorus*. The type species is *D. heterocephalus*. Biotypes of this species have been identified. The host range is probably large, especially in wet, sandy soils.

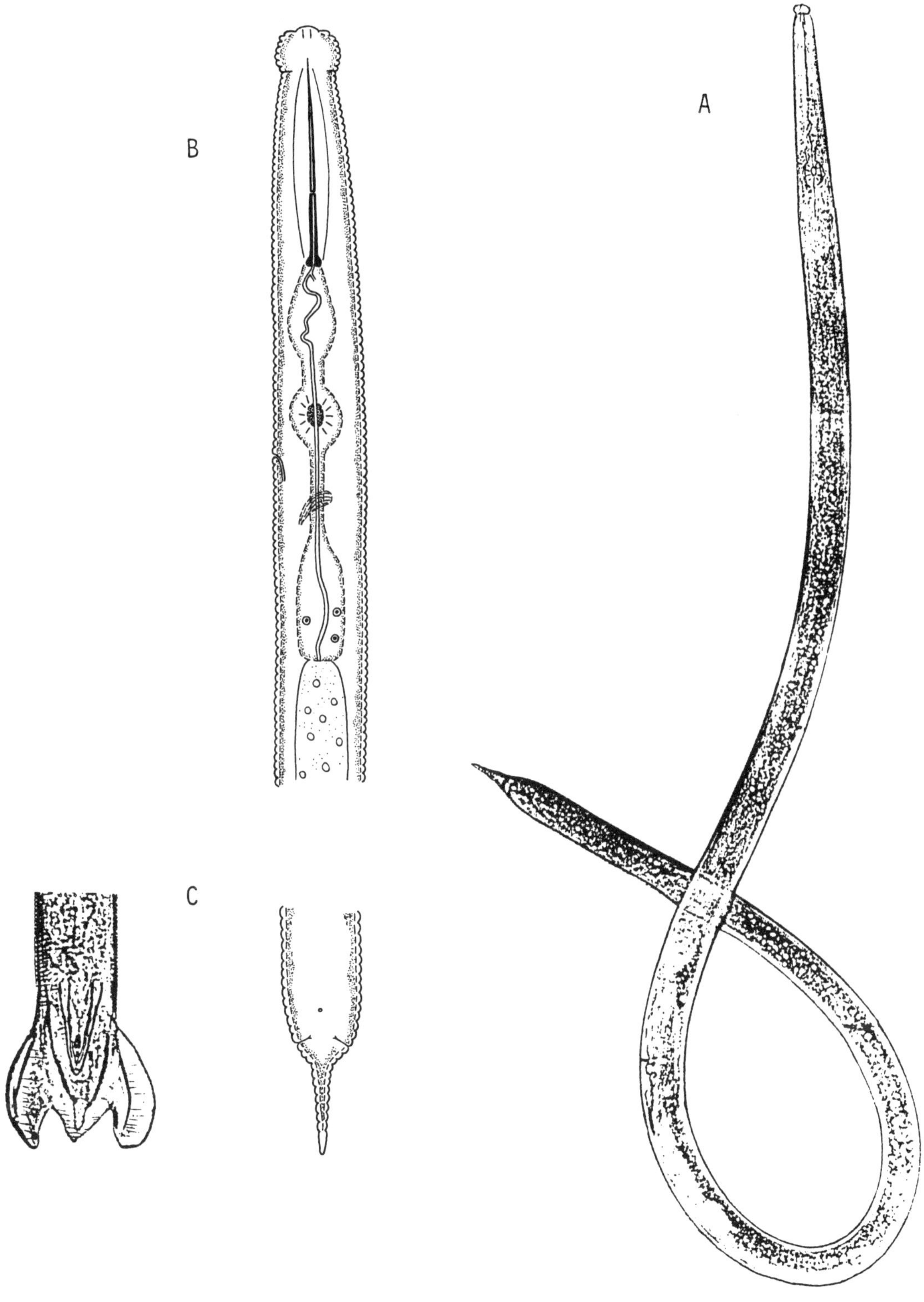


**Figure 3.12.** *Dolichodorus.* **A,** Adult female. **B,** Female, anterior part. **C,** Tails of adult male (left) and female (right). (**A** and **C** [left] reprinted from Mai and Mullin, 1996, © 1960 by W. F. Mai and H. H. Lyon, and used by permission of the publisher, Cornell University Press; **B** and **C** [right] reprinted from Jenkins and Taylor, 1967)

## *Globodera:* Round Cyst Nematodes

### Class Secernentea, order Tylenchida, family Heteroderidae

#### HOSTS

Different species of *Globodera* feed on different hosts, including broad bean (faba bean, *Vicia faba*), *Cestrum,* eggplant, fruit trees, fuchsia, gardenia, potato, *Sansevieria,* most *Solanum* species, snapdragon, *Swietenia,* tobacco, tomato, and yarrow (*Achillea millefolium*).

These nematodes also feed on many weeds, including bird rape (*Brassica rapa*), bittersweet nightshade (*Solanum dulcamara*), curly dock (*Rumex crispus*), and European nettle (*Urtica dioica*).

#### PARASITISM AND HABITAT

Females are sedentary endoparasites in roots. Males and preparasitic second-stage juveniles are migratory and inhabit soil and root tissue. Males are common. All species of *Globodera* and *Heterodera* induce syncytia, consisting of partially fused cells without nuclear division (see the section on *Heterodera*). In severe infections, plant growth is suppressed, and plants become yellow; wilting occurs during the heat of the day, and plants may die. Roots are short, brown, and excessively branched. The minute, pearly white bodies of enlarged females, which later turn dark brown and become cysts, can be seen attached to the surface of roots (Figure 3.13 and Plate 1H) and in soil. Symptoms of disease usually appear seven or eight years after the nematode has been introduced into a field. Annual diagnostic surveys are important.

#### BODY

Females are virtually spherical, with a projecting neck containing the esophagus and part of the associated glands (Figure 3.14). There is no protruding vulval cone. The cuticle surface is lacelike. The anus is subterminal. About three to four weeks after entering a root, the female enlarges to a dark, thick-skinned, resistant, spherical or subspherical cyst, 0.5–0.8 mm in diameter, containing several hundred eggs with infective juveniles in a quiescent state. The female, which lacks the terminal cone of *Heterodera,* ruptures the root cortex and protrudes from the root, exposing the vulval region. The head and neck remain inserted in the root. The enlarged female is white when it first emerges from the root cortex. Cysts of *G. rostochiensis* later turn golden yellow and finally dark brown when they die. Cysts of *G. pallida* are white, turning cream-colored in some populations, and finally glossy brown when they die. The average distance from anus to vulva is 60 μm in cysts of *G. rostochiensis* and 44 μm in *G. pallida. G. rostochiensis* has more cuticular ridges than *G. pallida.*

Adult males are 0.7–1.7 mm long and vermiform (a ≈ 30–50) and have regular cuticular annulations. The lateral

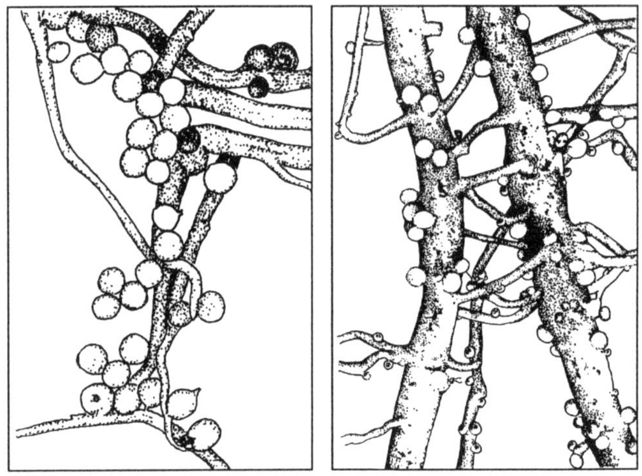

**Figure 3.13.** *Globodera pallida,* the white or pale potato cyst nematode (left), and *G. rostochiensis,* the round or golden cyst nematode (right), on potato roots. The pearly white cysts later turn dark brown.

field has four incisures, which terminate on the tail. The male head is rounded and offset and has a large oral disk surrounded by six small lips.

Juveniles are 0.4–0.6 mm long and vermiform (a = 22), with a tail tapering to a finely rounded tip. The head region is rounded.

#### STYLET

The stylet is slightly less than 30 μm long, stout, and heavily sclerotized, with well-developed, rounded basal knobs that slope backward (except in second-stage juveniles of *G. pallida,* in which the knobs project forward). In juveniles of *G. rostochiensis,* the stylet is typically shorter (about 21 μm) and the basal knobs are more rounded than in *G. pallida* (in which the stylet is about 23 μm long).

#### ESOPHAGUS

The median bulb is very large and almost circular and has a prominent valve and crescentic valve plates. The esophageal glands are in a broad lobe, often displaced forward in females by the large, paired ovaries. In males and second-stage juveniles, the esophageal gland lobe overlaps the intestine ventrally.

#### REPRODUCTIVE STRUCTURES

Females have two highly convoluted ovaries, often occupying most of the body cavity.

Males have a single, anteriorly directed testis. The spicules are stout and arcuate and terminate distally in single pointed tips very near the tail terminus.

#### VULVA

The vulva is near the terminus and anus, not on a terminal vulvular cone. It lies between finely papillated crescentic areas occupying most of the vulval slit and surrounded

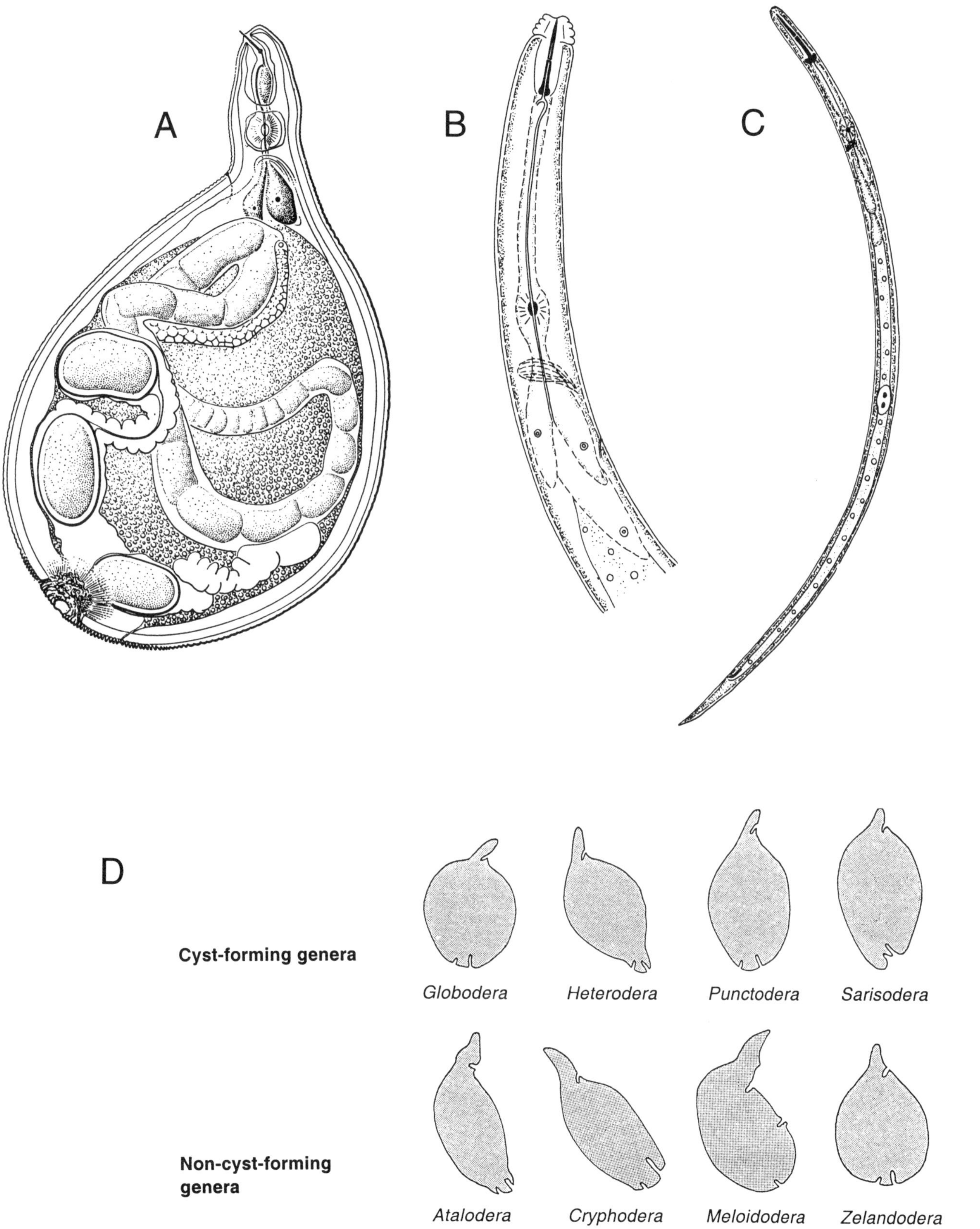


**Figure 3.14.** *Globodera.* **A,** Adult female. **B,** Adult male, anterior part. **C,** Second-stage juvenile. **D,** Body shapes of cyst-forming genera. (**A** reprinted, by permission, from Thorne, 1961; **B** and **C** reprinted from Jenkins and Taylor, 1967; **D** adapted from Hesling, 1978)

by a single, circular fenestra. In most species, all the eggs are retained within the mature cyst, but a few species also lay eggs in an external gelatinous matrix.

#### TAIL

The adult female tail is much reduced. The male tail is short and blunt and lacks a bursa. The juvenile tail tapers to a pointed tip with a clear terminus.

#### GENERA RESEMBLING *GLOBODERA*

**Cyst-Forming Genera**

*Cactodera.* Nematodes in this genus form spheroid to lemon-shaped cysts, with a small posterior cone; the vulva is terminal, becoming a circumfenestrate opening in the cyst.

*Heterodera* and *Punctodera.* In 1976 the genus *Heterodera* was split into *Heterodera, Globodera,* and *Punctodera.* Cyst nematodes have three basic shapes (Figures 1.3F and 3.14D):

spherical, with an anterior necklike projection, as in *Globodera*
pyriform (elongate-ovoid), the length of the body being about 1.5 times its width, with a protruding neck and no vulval cone, as in *Punctodera*
lemon-shaped, with both anterior and posterior projections, as in *Heterodera*

*Sarisodera.* This genus differs from *Heterodera* in having hypertrophied vulval lips, and it lacks structures associated with the vulva in *Heterodera* and *Punctodera.* Males of *Sarisodera* have a pointed tail with straight spicules at the very tip. *Sarisodera* species induce a single huge syncytium with a nucleus that is much larger than normal.

**Non-Cyst-Forming Genera**

*Atalodera*
*Cryphodera*
*Meloidodera*
*Thecavermiculatus.* Females in this genus are pearl white and round to oval; they have a prominent neck and do not form cysts. The vulva and anus are close together and flush with the body contour.
*Zelandodera*

#### COMMENTS

There are about 14 species of *Globodera,* each having a narrow host range. The most important species are the golden potato cyst nematode, *G. rostochiensis;* the white or pale potato cyst nematode, *G. pallida;* and the tobacco cyst nematodes, *G. tabacum* and *G. solanacearum.* Subspecies and pathotypes (races) have been distinguished in some species. Pathotypes of the potato cyst nematodes are recognized by reproduction on a series of *Solanum* clones.

The life history of *Globodera* is similar to that of *Heterodera,* but *Globodera* species retain all their eggs in the cyst, where they remain viable for several years, until induced to hatch in soil or water. Some cysts have been known to remain infective in cool soils for up to 28 years. Root exudates from host plants stimulate the hatching of juveniles, but other factors may be involved. Cysts are readily distributed by water, by tilling the soil, and with seed tubers, etc. Usually there is one generation per crop per season.

Potato cyst nematode infection enhances attack by the wilt pathogens *Verticillium dahliae* and *Pseudomonas solanacearum.*

*G. tabacum* and *G. solanacearum* increase the severity of black shank of tobacco, caused by *Phytophthora nicotianae* var. *parasitica.*

## *Helicotylenchus:* True Spiral Nematodes

### Class Secernentea, order Tylenchida, family Hoplolaimidae

#### HOSTS

*Helicotylenchus* species feed on numerous plants worldwide, including abacá (*Musa textilis*), *Aechmea,* African violet, alfalfa, *Amaranthus,* amaryllis, apple, apricot, arborvitae, *Ardisia,* ashes, avocado, azaleas, bamboo (*Bambusa*), banana, barberry, beans (*Phaseolus* spp.), beet (garden beet, sugar beet), begonia, black pepper (*Piper nigrum*), blackberry, blueberry, boxwood, breadfruit (*Artocarpus altilis*), cacao (cocoa, *Theobroma cacao*), camellia, cardamom (*Elettaria cardamomum*), carrot, cassava (*Manihot esculenta*), cauliflower, cedar (*Cedrus*), ceiba (silk-cotton-tree, *Ceiba pentandra*), celeriac, celery, *Centrosema, Cephalotaxus,* cherries, chickpea (*Cicer arietinum*), citrus, clove (*Syzygium aromaticum*), clovers (*Trifolium*), cockscomb (*Celosia argentea* var. *cristata*), coconut palm (*Cocos nucifera*), coffee, corn (maize), cotton, cowpea (*Vigna unguiculata*), cranberry, cucumber, custard-apple (*Annona cherimola* and *A. squamosa*), date palm (*Phoenix dactylifera*), dieffenbachia, dogwood, eggplant, elms, fenugreek (*Trigonella foenum-graecum*), fig, forest trees, garlic, bedding geranium (*Pelargonium* × *hortorum*), gherkin, ginger (*Zingiber officinale*), gladiolus, gourd (*Lagenaria*), grape, grasses (forage grasses, turfgrasses, and wild species), guava (*Psidium*), hairy indigo (*Indigofera hirsuta*), holly, European hop (*Humulus lupulus*), jack bean (*Canavalia ensiformis*), jackfruit (*Artocarpus heterophyllus*), jasmine, night jessamine (*Cestrum nocturnum*), junipers, jute (*Corchorus capsularis* and *C. olitorius*), kenaf (*Hibiscus cannabinus*), lantana, leek, lentil (*Lens culinaris*), lespedezas, lilies, *Lippia,* lupines, *Mammillaria,* mango, maples, marguerite, millets, mints, *Mucuna,* mulberry, mung bean (*Vigna radiata*), muskmelon,

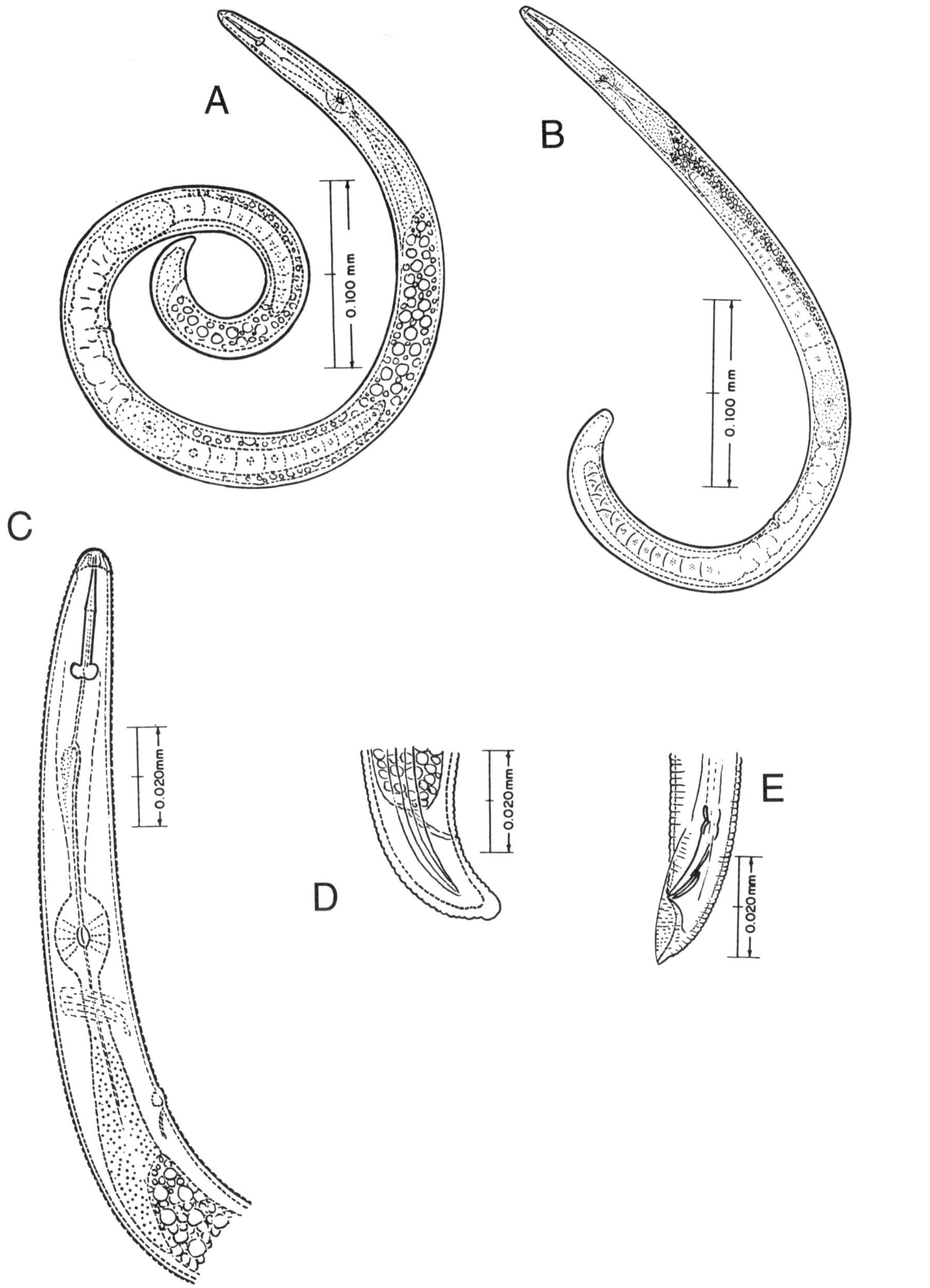


**Figure 3.15.** *Helicotylenchus.* **A** and **B,** Adult females. **C,** Female, anterior part. **D,** Female, posterior part. **E,** Male, posterior part. (Reprinted, by permission, from Taylor, 1971)

oaks, oat, okra, olive, onions, palms, papaya, parsley, passionfruit, pawpaw, pea, peach, peanut, pear, pecan, peony, pepper (garden or bell pepper and ornamentals, *Capsicum annuum*), pigeon pea (red gram, *Cajanus cajan*), pineapple (*Ananas comosus*), pines, pistachio-nut (*Pistacia vera*), planetree, plantain (*Musa* × *paradisiaca*), plums, pomegranate (*Punica granatum*), poplars, potato, California privet (*Ligustrum ovalifolium*), pumpkin, radish, raspberry, rhododendrons, rice, rose, rubber-tree (*Hevea brasiliensis*), rutabaga, rye, *Sansevieria,* sapodilla (*Manilkara zapota*), sedge (*Carex*), *Sedum,* sorghums, soybean, spinach, spruces, squashes, strawberry, sugarcane, sunflower, swamp taro (*Cyrtosperma chamissonis*), sweet potato, sycamore, *Talinum,* tea, *Tillandsia,* tobacco, tomato, turmeric (*Curcuma domestica*), turnip, vanilla, hairy vetch (*Vicia villosa*), walnut, watermelon, wheat, willows, *Xanthosoma,* yams (*Dioscorea* spp.), yew (*Taxus*), Japanese zelkova (*Zelkova serrata*), and many others.

Weed hosts include ball nightshade (*Solanum carolinense*), *Crotalaria, Oxalis, Polygonum,* and ragweed (*Ambrosia*).

#### PARASITISM AND HABITAT

*Helicotylenchus* species are cosmopolitan, polyphagous, migratory endo-, semiendo-, and ectoparasites. They are found in all stages in soil, roots, and other underground parts of plants. Like other spiral nematodes, they penetrate partly into roots; sometimes an entire nematode enters a root to feed. These nematodes cause small, localized, light to dark brown or reddish brown necrotic lesions in the cortex of roots and in other underground parts, as a result of the death of cells on which they feed. Favored points of entry are behind the root tips and at the junction of the lateral and main roots. When damage is severe, primary roots are shriveled, short, and discolored and have few lateral roots, and the cortical tissue sloughs off. Infected plants show mild to moderate stunting, chlorosis, reduced vigor, and decline, but the disease is rarely severe.

#### BODY

Females are 0.5–1.2 mm long (a = 21–34), typically arcuate, and somewhat spiral in form or C-shaped when dead or relaxed by gentle heat (Figure 3.15). The males resemble the females, except for secondary reproductive characteristics. The lip region is hemispherical and more or less continuous with the body contour, with weak to moderate sclerotization. The lip annulations are not subdivided by longitudinal striations. The head region is conoid-rounded, rarely truncate, with a strongly sclerotized labial framework. The cuticular annulations are well marked and extend almost to the tail tip (terminus). The lateral field has four prominent incisures, usually areolated only anteriorly. Phasmids, located in the anal region, are small and dotlike. The sexes are separate in some species; males are unknown in others.

#### STYLET

The stylet is well developed and moderately long (20–35 μm), with strongly developed, rounded or slightly cup-shaped basal knobs.

#### ESOPHAGUS

The median bulb is round to oval. The esophageal glands are often distinct, overlapping the anterior end of the intestine dorsally, laterally, and ventrally for a distance of 1 or 2 body widths. The dorsal esophageal gland opening is located 6 to 16 μm (one-quarter or more of the stylet length) posterior to the stylet knobs. There is an obscure muscular valve at the junction of the esophagus and the intestine.

#### REPRODUCTIVE STRUCTURES

Females have paired ovaries, which are opposed (anterior and posterior, occasionally more posterior) and outstretched.

Males have a single, anteriorly outstretched testis. The spicules are well developed and arcuate, resting on a moderately developed and slightly curved gubernaculum.

#### VULVA

The vulva is prominent, a depressed transverse slit posterior to the middle of the body (V = 55–70). It is occasionally more posterior (V = 75–95) when the posterior gonad is degenerate.

#### TAIL

The female tail is usually more curved dorsally; its length is 1 to 2.5 anal body widths. The tip is usually bluntly conical but is occasionally rounded. Some species have a short terminal projection on the ventral side. The male tail is short, conical, and enveloped by a bursa.

#### GENERA RESEMBLING *HELICOTYLENCHUS*

*Aorolaimus*
*Hoplolaimus*
*Peltamigratus*
*Rotylenchus.* Nematodes in this genus have an overlapping dorsal gland lobe, and the dorsal gland duct opening is more anterior than in *Helicotylenchus.*

#### COMMENTS

There are some 180 species of *Helicotylenchus.* The major species are *H. cavenessi, H. crenacauda, H. digonicus, H. dihystera, H. erythrinae, H. indicus, H. microcephalus, H. microlobus, H. multicinctus, H. nannus,* and *H. pseudorobustus.*

These nematodes are often recovered from soil samples near plant roots, but they are usually not serious pathogens. Other microorganisms (e.g., the fungi *Fusarium* and *Rhizoctonia*) enter feeding lesions and may cause more severe disease, such as bacterial wilt of carnation, caused by *Burkholderia caryophylli,* and bacterial wilt of tomato, caused by *Ralstonia solanacearum.*

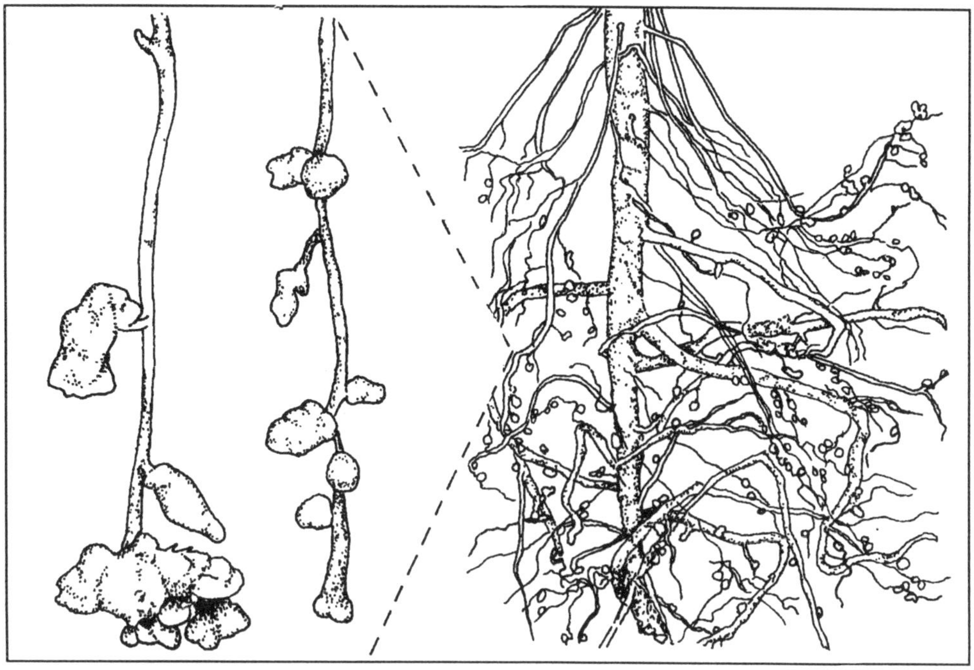

**Figure 3.16.** Citrus roots with irregularly shaped galls produced by *Hemicycliophora arenaria.*

## *Hemicycliophora:* Sheath Nematodes

### Class Secernentea, order Tylenchida, family Criconematidae

#### HOSTS

*Hemicycliophora* species (sheath nematodes) are distributed almost worldwide, especially in wet, sandy soils, and feed on plants including *Alchemilla,* alfalfa, almond, apple, apricot, banana, barley, beans (*Phaseolus* spp.), beet (garden beet, sugar beet), blackberry, blueberry, boxwood, broccoli, cabbage, cacao (cocoa, *Theobroma cacao*), catjang (*Vigna unguiculata* subsp. *cylindrica*), cardamom (*Elettaria cardamomum*), carrot, cassava (*Manihot esculenta*), cauliflower, celery, citrus, coffee, corn (maize), cowpea (*Vigna unguiculata*), cranberry, date palm (*Phoenix dactylifera*), dill, dogwood, eggplant, *Eucalyptus,* various evergreens, ginger (*Zingiber officinale*), grape, grasses (forage grasses, turfgrasses, and wild species), *Hedera,* honeysuckle (*Lonicera*), *Hymenoclea salsola,* kohlrabi, lettuce, mango, maples, melons, mints, oaks, okra, olive, various ornamentals, holly osmanthus (*Osmanthus heterophyllus*), palms, papaya, peach, peanut, pepper (garden or bell pepper and ornamentals, *Capsicum annuum*), periwinkle (*Vinca*), pines, plums, poison ivy, *Polygonum,* poplars, *Portulaca,* potato, privet, pumpkin, radish, ragweed (*Ambrosia*), raspberry, rice, rose, strawberry, squashes, sugarcane, sweet potato, sweetgum (*Liquidambar*), tomato, paperbark tea tree (*Melaleuca quinquenervia*), wheat, yam (*Dioscorea*), and others.

#### PARASITISM AND HABITAT

*Hemicycliophora* species are widely distributed sedentary ectoparasites sometimes associated with distinct, irregularly shaped galls at the tips of young roots on which they feed (Figure 3.16). Shoot and root growth and yields may be reduced. Females, males, and juveniles are all found in soil. Males are unknown in many species.

#### BODY

The sexes are dimorphic, differing greatly in appearance (Figure 3.17). The females are rather stout and 0.6–2.0 mm long (a = 14–30); the males are more slender, cylindrical, and 0.4–1.0 mm long. The outermost cuticle of the female is typically covered by a thin, detached, loosely fitted sheath ("double cuticle") secreted along with the cuticle at each molt and marked by numerous (up to 400) coarse, rounded, nonretrorse annules (Figure 3.17B). The head or the tail or both may be withdrawn into the loose sheath. Males lack a sheath cuticle. The female body is straight or curved slightly ventrally. Juveniles resemble adult females. Nematodes in some species take on a more or less semicircular shape when relaxed by heat.

The lip region is bluntly rounded, not offset, and marked by two or three annules. The cephalic framework is well sclerotized. The lateral fields in females are usually simple, without markings, but the cuticle may be variously ornamented with longitudinal zigzag lines, squares, dots, or scratches. The intestine extends into the tail. The anus and rectum are indistinct. The excretory pore and hemizonid are near the anterior end of the intestine, at the level of the esophago-intestinal junction.

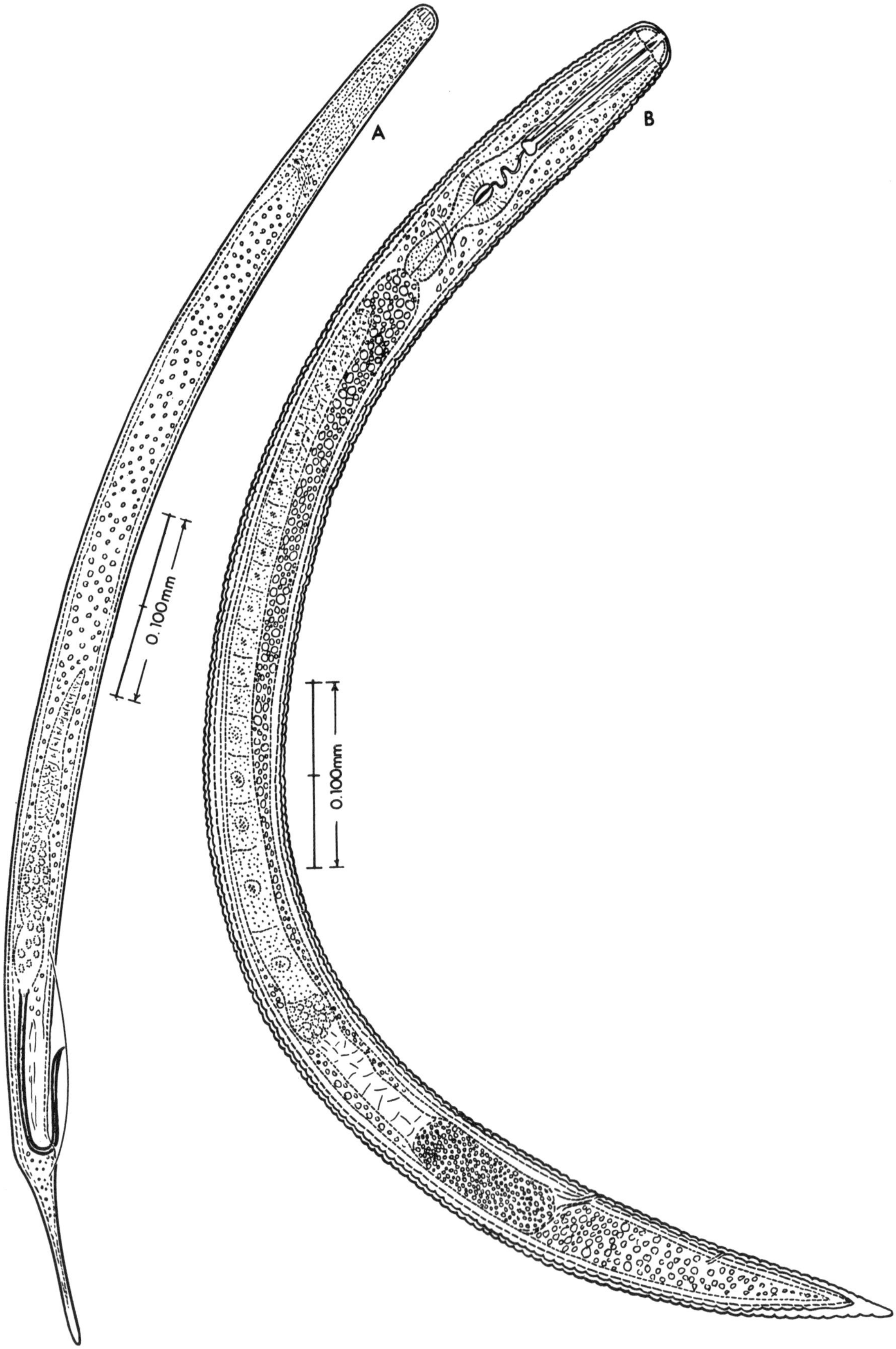


**Figure 3.17.** *Hemicycliophora.* **A,** Adult male. **B,** Adult female. (Reprinted, by permission, from Taylor, 1971)

#### STYLET

The female stylet is slender, strong, long (86–100 μm), and slightly curved, extending through 25 to 40 annules, with a tapered anterior section, longer than the posterior portion, and with large, rounded, posteriorly sloping basal knobs (not anchor-shaped). Males lack a stylet.

#### ESOPHAGUS

The median bulb (metacorpus) and procorpus in females are fused together into a large swelling surrounding the stylet base and in which the esophageal lumen is coiled when the stylet is retracted. The isthmus is short and well defined. The terminal bulb is pear-shaped, with three large gland nuclei usually visible. The esophageal gland lobe does not overlap the intestine. The hemizonid is conspicuous on bisexual species, while on monosexual forms it is missing or is very obscure. The esophagus in males is degenerate, but the nerve ring, excretory pore, and hemizonid are visible.

#### REPRODUCTIVE STRUCTURES

Females have a single ovary, which is anteriorly outstretched and not reflexed. Spermathecae are present in bisexual species but not in monosexual forms. The spermathecae are spheroid.

Males have a single, outstretched testis. The spicules are typically large and curved or sickle- or hook-shaped, with a sheath and mamillate process on the distal posterior edge. The gubernaculum is simple.

#### VULVA

The vulva is a transverse slit with slightly protruding lips and a tube through the sheath cuticle, located in the posterior fourth of the body (V = 75–90).

#### TAIL

The tail varies in different species. It is the region where the cuticular sheath is most prominent. The female tail is tapered to cylindrical and pointed to bluntly rounded to cone-shaped, with a vestigial anus and rectum. The male tail is long (frequently several anal body widths) and narrowly conical and comes to a rounded point, with a small, broad, adanal bursa confined to the region of the cloaca (Figure 3.17A). Juveniles are similar to females but lack the extracuticular scales found in other genera of the family Criconematidae.

#### GENERA RESEMBLING *HEMICYCLIOPHORA*

*Criconemella* and *Hemicriconemoides.* These nematodes also have a closely adpressed double cuticle, but in these genera the females (about 0.5 mm long) are shorter than in *Hemicycliophora,* the stylet knobs are directed anteriorly and are anchor-shaped, the tail is short and conoid, and the cuticle usually has fewer than 200 coarse annules.

#### COMMENTS

There are some 100 or more species of *Hemicycliophora.* The major species is *H. arenaria,* which feeds as an ectoparasite on root tips, forming galls. The galls always develop at root tips, in contrast to those caused by *Meloidogyne,* which form along roots and only occasionally at root tips. Females are attached to the root during feeding by a polysaccharide plug at the lips. A gelatinous substance on the eggs causes them to adhere to soil particles and other objects. Males are rare and not essential for reproduction. These nematodes feed on epidermal and cortical cells and prefer sandy soils.

*Hemicycliophora* species are sluggish, and recovery by the Baermann funnel method is poor. The centrifugal flotation technique should be used for extraction.

## *Heterodera:* Lemon-Shaped Cyst Nematodes

### Class Secernentea, order Tylenchida, family Heteroderidae

#### HOSTS

*Heterodera* species (lemon-shaped cyst nematodes) have numerous hosts worldwide. The more important species and their principal hosts are listed below.

***H. avenae.*** Hosts include barley, beet (garden beet, sugar beet), cabbage, carnation, carrot, celeriac, clover (*Trifolium*), corn (maize), *Cycas,* geranium (*Pelargonium*), *Glycyrrhiza,* many grasses, heather (*Calluna vulgaris*), jasmine, European linden (*Tilia cordata*), mustards, oat, pea, potato, primrose, rutabaga, rye, *Salvia,* tassel-hyacinth (*Muscari comosum*), triticale, tulip, turnip, and wheat.

***H. cacti.*** The most important hosts of this species are cacti growing as ornamentals. It also feeds on asparagus, larkspur, lavender, philodendrons, *Portulaca,* potato, *Sansevieria,* and soybean.

***H. cajani.*** Hosts include black gram (urd, *Vigna mungo*), celeriac, chickpea (*Cicer arietinum*), cluster bean (*Cyamopsis tetragonoloba*), cowpea (*Vigna unguiculata*), horse gram (*Macrotyloma uniflorum*), hyacinth bean (*Lablab purpureus*), jack bean (*Canavalia ensiformis*), moth bean (*Vigna aconitifolia*), mung bean (*Vigna radiata*), pigeon pea (red gram, *Cajanus cajan*), rice bean (*Vigna umbellata*), sesame (*Sesamum*), soybean, and tepary bean (*Phaseolus acutifolius*).

***H. carotae.*** Hosts are cultivated and wild carrots.

***H. ciceri.*** This species is in the *H. trifolii* group but has abundant males, a different host range, and indistinct morphological characteristics, and it is found only in southern Syria. Its hosts include chickpea (*Cicer arietinum*), garden pea, grass pea (*Lathyrus sativus*), and lentil (*Lens culinaris*).

***H. cruciferae.*** This species is commonly found together with *H. schachtii* but has a somewhat smaller host range, including *Alyssum,* beet (garden beet, sugar beet), all species of *Brassica* and most other crucifers and related cole crops, candytuft (*Iberis*), carrot, garden cress (*Lepidium sativum*), pea, potato, sweetsop (*Annona squamosa*), wallflower, and weeds.

***H. fici.*** Hosts include black gram (urd, *Vigna mungo*), figs and other *Ficus* spp., and tuberose.

***H. glycines.*** Hosts include adzuki bean (*Vigna angularis*), beans (dry beans, kidney bean, snap bean, and haricot; *Phaseolus vulgaris*) (Plate 1I), beet (garden beet, sugar beet), black gram (urd, *Vigna mungo*), broad bean (faba bean, *Vicia faba*), cabbage, catjang (*Vigna unguiculata* subsp. *cylindrica*), clovers (*Trifolium*), corn (maize), cowpea (*Vigna unguiculata*), grass pea (*Lathyrus sativus*), hemp (*Cannabis sativa*), hyacinth bean (*Lablab purpureus*), jack bean (*Canavalia ensiformis*), lentil, lespedezas, lima bean, lupines, moth bean (mat bean, *Vigna aconitifolia*), mung bean (*Vigna radiata*), pea (field pea, garden pea), pearl lupine (*Lupinus mutabilis*), penstemon, pigeon pea (red gram, *Cajanus cajan*), rice bean (*Vigna umbellata*), sesame (*Sesamum indicum*), soybean (Figure 1.3F and Plate 1I), sweet potato, tepary bean (*Phaseolus acutifolius*), tomato, and vetches (*Vicia* spp.).

*H. glycines* also feeds on numerous weeds, e.g., chickweeds, cranesbill, hemp sesbania, henbit, hop clover, joint vetch, milkvetch, mullein (common mullein and moth mullein), pokeweed, sicklepod, and toadflax.

***H. goettingiana.*** Hosts include amaryllis, asparagus-fern (*Asparagus plumosus*), *Aspidistra,* bamboo, bay (*Laurus nobilis*), begonia, broad bean (faba bean, *Vicia faba*), cacti, *Campanula,* canna, carnation, carrot, celery, chickpea (*Cicer arietinum*), chrysanthemum, citrus, coleus, *Corylus,* English ivy (*Hedera helix*), ferns, fig, fir, fuchsia, gardenia, geranium (*Pelargonium*), gloxinia, grape, grass pea (*Lathyrus sativus*), grasses (forage grasses, turfgrasses, and wild species), heather, hemlock, hens-and-chickens (*Sempervivum tectorum*), hibiscus, horseradish, hyacinth, impatiens, iris, jasmine, lavender, leek, lentil (*Lens culinaris*), lilies, lupines, magnolia, mints, narcissus, nasturtium, oleander, onions, palms, pea (field pea, garden pea), peony, pepper (garden or bell pepper, *Capsicum annuum*), potato, primrose, oaks, rose, rosemary, rubber plant (*Ficus elastica*), rutabaga, salvia, *Sansevieria,* sedum, soybean, spruce, strawberry, sweet alyssum, sweet pea, sweetgum (*Liquidambar*), turnip, vetches (*Vicia* spp.), and wheat.

***H. humuli.*** Hosts include beans (*Phaseolus* spp.), black mustard, clovers (*Trifolium*), cucumber, European hop, Indian hemp (*Apocynum cannabinum*), pea, rubber plant (*Ficus elastica*), and hairy vetch (*Vicia villosa*).

***H. lespedezae.*** Hosts are clovers (*Trifolium*), lespedezas, soybean, sweet-clover (*Melilotus*), and hairy vetch (winter vetch, *Vicia villosa*).

***H. mani.*** Hosts include barley, grasses (forage grasses, turfgrasses, and wild species), oat, and wheat.

***H. oryzae* and *H. oryzicola.*** Hosts include barley, corn (maize), rice, *Setaria,* sorghums, and teosinte (*Zea mays* subsp. *mexicana*).

***H. punctata.*** Hosts include arborvitae, cacti, carrot, celeriac, clematis, clover (*Trifolium*), dahlia, fir, hyacinth, potato, primrose, rose, and sweet alyssum.

***H. sacchari.*** Hosts are rice and sugarcane.

***H. schachtii.*** This species has over 220 hosts, including adzuki bean (*Vigna angularis*), *Agave, Aloe,* apple, *Arabis* spp., asparagus broccoli (sprouting broccoli, *Brassica oleracea* var. *italica*), asparagus-fern, baby's-breath, basil, beans (*Phaseolus* spp.), beet (garden beet, sugar beet) (Plate 1J), begonia, bermudagrass, *Billbergia,* blackberry, broccoli, Brussels sprouts, cabbage (Plate 1K), cacti, catjang (*Vigna unguiculata* subsp. *cylindrica*), calla, candytuft (*Iberis*), canna, carnation, carrot, catalpa, cauliflower, celeriac, celery, chestnut, chickpea (*Cicer arietinum*), chicory, Chinese cabbage, citrus, clovers (*Trifolium*), coleus, collards, *Commelina,* corn (maize), cowpea (*Vigna unguiculata*), cyclamen, dahlia, dill, English ivy (*Hedera helix*), *Episcia,* ferns, fir, garden cress (*Lepidium sativum*), gardenia, garlic, geranium (*Pelargonium*), grape, grass pea (*Lathyrus sativus*), grasses (forage grasses, turfgrasses, and wild species), heather, *Hebe,* hens-and-chickens (*Sempervivum tectorum*), holly, horseradish, hyacinth, hydrangea, jasmine, kale, kohlrabi, lavender, leek, lentil (*Lens culinaris*), lespedezas, lettuce, lilies, lily-of-the-valley, lupines, mangel (mangold, *Beta vulgaris* subsp. *vulgaris*), marijuana, mignonette (*Reseda odorata*), mints, mustards, nasturtium, oleander, onions, orchids, *Oxalis,* parsnip, pea, peony, pinks, potato, radish, rape (canola), rhubarb, rose, rosemary, rutabaga, *Sansevieria, Saponaria,* scarlet pimpernel, sedum, *Sesbania,* soybean, spinach, spruce, strawberry, sweet alyssum, sweet William, Swiss chard, *Syzygium,* toadflax (*Linaria*), tobacco, tomato, *Tradescantia,* turnip, wallflower, walnut, wheat, witloof (*Cichorium intybus* var. *foliosum*), and *Xanthosoma.*

Weed hosts of *H. schachtii* include charlock (*Sinapis arvensis*), *Chenopodium,* chickweeds (*Stellaria*), dock (*Rumex*), knotweed, lamb's-quarters, mustards, nightshades, pigweeds, pokeweed, purslane, radishes, shepherd's-purse, *Sisymbrium,* smartweed, and stinging nettle.

***H. sorghi.*** Hosts are peanut and sorghums.

***H. trifolii.*** Hosts include apple, barley, beans (*Phaseolus* spp.), beet (garden beet, sugar beet), bird's-foot trefoil, carnation, chickpea (*Cicer arietinum*), clovers (*Trifolium*), corn (maize), cucumber, gherkin, grasses (forage grasses, turfgrasses, and wild species), heart's-ease (lady's-thumb, *Polygonum persicaria*), lespedezas, lupines, *Melilotus* spp., muskmelon, oat, pea, peach, potato, pumpkin, rainbow pink (*Dianthus chinensis*), rhubarb, soybean, spinach, squashes, tobacco, tomato, vetches (*Vicia* spp.), and many weeds.

***H. zeae.*** Hosts are corn (maize) and wheat.

#### PARASITISM AND HABITAT

*Heterodera* species are sedentary semi-endoparasites on roots and tubers of many species in tropical and especially in temperate zones. Adult females have elongated necks and embed themselves in plant roots, with their swollen bodies exposed, to form subspherical, hard-walled cysts, 0.2–1.1 mm in diameter (Figures 1.3F and 3.18 and Plate 1I). Second-stage juveniles, males, and cysts are found in the soil. The contents of some cysts may remain viable for up to 15 years. The characteristic symptom of a high population is a small, oval to elongated area in a field where the shoots may be stunted, chlorotic, flowering is delayed, and weeds are numerous (Plate 1I and J). The infected area commonly enlarges each year until most of the crop is affected. Populations build up quickly with continuous planting of a susceptible host. Host roots with large numbers of cyst nematodes are poorly developed, nutrient balance is disturbed, yields are sharply reduced, and nitrogen fixation of leguminous crops is inhibited by some races of the nematode.

#### BODY

The enlarged females (Figure 3.18C) are 0.3–1.0 mm long when mature. They are white when young, usually becoming pale yellow, then brown, and finally tough, dark brown to black cysts after death. The cysts are generally lemon-shaped (0.3–1.0 mm long and 0.3–0.75 mm wide) to spheroid or pear-shaped (or nearly the same shape as females of *Meloidogyne*) and are found attached to roots and in soil. They vary in size, depending on the species and nutrition. Cysts on a single plant may also be variable. The cuticle is thick, protecting the many eggs retained inside. The cuticle surface has a lacelike or zigzag pattern. The cuticle has two translucent areas, the fenestrae, located on either side of the vulval slit, and distinct lateral fields, usually with four incisures. The lips are annulated and are not offset from the short, distinct neck protruding from the swollen body.

Males are present in some species but absent in others. They may be abundant at certain times of the year but scarce at other times. The male body is elongate (about 1 mm long) and vermiform; the length–width ratio is between 32 and 51 (Figure 3.18A). The head is hemispherical, offset with three to six annules, and equipped with a heavily sclerotized cephalic skeleton. The posterior region of males is usually twisted 90 to 180° when relaxed by heat. The lateral fields have four incisures.

Second-stage juveniles emerge from eggs and invade host roots to establish feeding sites close to the stele. The juveniles are 0.4–0.6 mm long, slender (the length–width ratio is between 21 and 26), and almost straight when killed by heat (Figure 3.18B). The head is hemispherical, offset with four annules, and equipped with a strong cephalic skeleton.

#### STYLET

In adult males, the stylet is long (20–40 µm) and robust, with rounded basal knobs. In juveniles, it is prominent (strong) and 20–30 µm long, with prominent basal knobs (stylets of *Meloidogyne* juveniles are 10–15 µm long). In females, the stylet is 25–30 µm long and strongly developed, with small, rounded basal knobs.

#### ESOPHAGUS

The esophagus is cylindrical, with a well-developed, oval to spherical median bulb. The esophageal glands are in a long basal lobe overlapping the intestine mainly vertically. In young females the excretory pore is at the level of the median bulb or posterior to it.

#### REPRODUCTIVE STRUCTURES

Females have paired, highly convoluted ovaries, containing up to 600 eggs (although normal populations have fewer eggs), which almost fill the body cavity at maturity. In some species the female secretes a gelatinous sac in which some eggs are deposited.

Males have a single, blunt-ended testis, directed anteriorly. The spicules are slender, 30–40 µm long, curved ventrally, and located close to the terminus, with a single or bifurcate tip protruding subterminally. A gubernaculum is present.

#### VULVA

The vulva is a subterminal slit on the terminal vulval cone near the anus. Species are distinguished by differences in the anatomy of this structure.

#### TAIL

The male tail is short and bluntly rounded and lacks a bursa. Incisures in the lateral field of males extend around the tail. The small phasmids are more or less adanal. In adult females, the tail is reduced to a vulval cone containing a minute anus just dorsal to the vulva. In second-stage juveniles, the tail is acutely conical, with a length of 3 to 4 anal body widths, and the terminal portion (constituting 30–60% of the tail) is clear and tapers to a narrow, rounded or pointed tip.

#### GENERA RESEMBLING *HETERODERA*

*Cactodera.* Nematodes in this genus form spheroid to lemon-shaped cysts, with a small posterior cone; the vulva is terminal, becoming a circumfenestrate opening in the cyst.

*Globodera* and *Punctodera.* These genera were established for the classification of species that had previously been assigned to *Heterodera. Globodera* is distinguished by the globular shape of its cysts, in contrast to the lemon shape of *Heterodera. Punctodera* is pyriform (elongate-ovoid), has a protruding neck, and lacks a vulval cone. See the section on *Globodera* and Figures 1.3F and 3.14D.

*Sarisodera.* This genus differs from *Heterodera* in having hypertrophied vulval lips and lacks structures associated with the vulva in *Heterodera* and *Punctodera.* Males of *Sarisodera* have a pointed tail with straight spicules at the very tip. *Sarisodera* species induce a single huge syncytium with a nucleus that is much larger than normal.

#### COMMENTS

There are about 60 species of *Heterodera,* most with narrow host ranges. The major species include *H. avenae, H. cajani, H. carotae, H. cruciferae, H. glycines, H. goetting-*

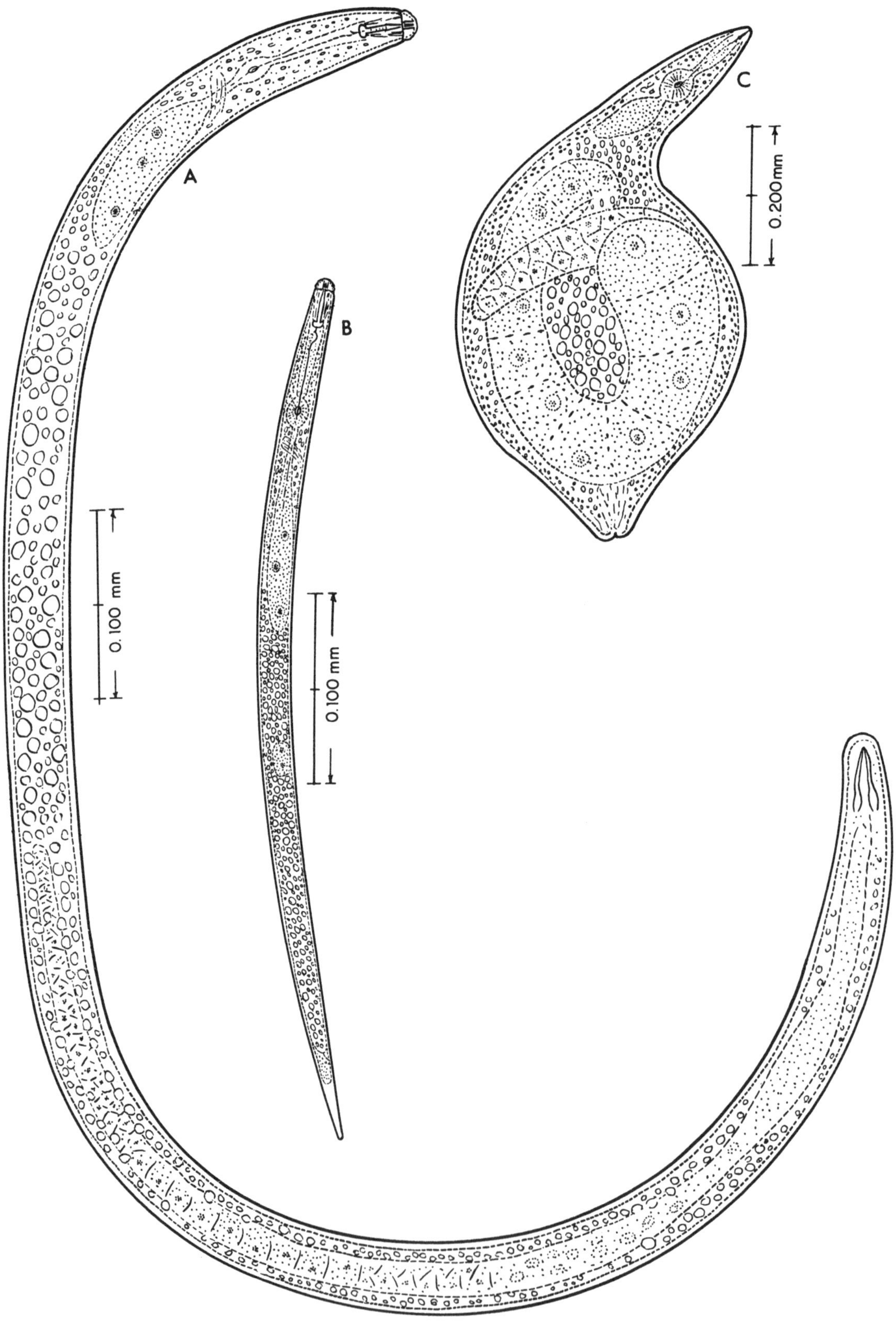


**Figure 3.18.** *Heterodera.* **A,** Adult male. **B,** Second-stage juvenile. **C,** Adult female. (Reprinted, by permission, from Taylor, 1971)

*iana, H. humuli, H. oryzae, H. sacchari, H. schachtii,* and *H. trifolii.* Some species have several to many biotypes (pathotypes or races) which may be important for management based on crop rotation and resistant varieties.

*Heterodera* and other genera of cyst nematodes are specialized semiendoparasites well adapted for long-term survival in the soil without suitable host plants. When the female dies, embryonated eggs stay within the body (the cyst). A few embryonated eggs are known to remain viable in the absence of a host plant for up to 15 years. A portion of the juveniles hatch from time to time over a period of years and invade host roots.

The unhatched juveniles of some species are quiescent within cysts until a stimulus from nearby growing roots reaches them. The juveniles then become active, hatch, leave the cyst through openings in the wall, and migrate to growing roots. Once inside a root, a juvenile moves through the root cortex for a few days before locating close to the stele. It then establishes a feeding site, altering host cells to induce the formation of a syncytium, consisting of enlarged and interconnected cells upon which it feeds for the remainder of its life. The sedentary third- and fourth-stage juveniles are somewhat sausage-shaped. Those that develop into mature females enlarge greatly. The posterior end of the enlarging female eventually ruptures the root cortex, and most of the body protrudes from the root, leaving only the neck portion remaining in the root. The vermiform males leave the root tissue to seek out and mate with young females exposed on the root surface. A gelatinous matrix is sometimes produced, into which some of the eggs are laid. Upon death, the female body becomes a cyst, dislodges from the root, and remains in the soil. Galls are not formed.

The life cycles of these nematodes may be as short as 22 days and as long as a year, depending on the species, the temperature, and other factors. For example, *H. avenae* completes one life cycle per season; *H. cruciferae* completes a life cycle in 30 days; *H. glycines,* 22 days; *H. schachtii,* 30 days; and *H. trifolii,* 28 days (Barker et al., 1998).

If roots are carefully removed from the soil, yellow to dark brown cysts may be seen on feeder roots.

Cysts, eggs, and juveniles within cysts of *Heterodera* species are prey to numerous predators and parasites, including many fungi, other nematodes, collembolans, beetles, centipedes, and amoeboid organisms.

The combined effects of the soybean cyst nematode (*H. glycines*), stem canker fungus (*Diaporthe phaseolorum* var. *caulivora*), and soybean looper (*Pseudoplusia includens*) are additive in reducing soybean growth.

#### DETERMINING RACES OF THE SOYBEAN CYST NEMATODE

The soybean cyst nematode, *H. glycines* (Figure 1.3F and Plate 1J), is a major problem in many soybean-growing areas. Resistant cultivars may be more or less race-specific, and therefore the races of *H. glycines* present in a field should be identified. The races are distinguished by their ability to infect and reproduce on a specified set of soybean cultivars. Morphological, anatomical, and other characteristics of the nematode are not used for race determination.[1]

Presently, five soybean cultivars are used to determine the 16 designated races of the soybean cyst nematode. When more differentials are used, more races can be distinguished. Eighteen differentials have been used to distinguish 36 distinct groups of cyst nematodes among 38 populations. In the future, races will likely be rapidly differentiated with DNA probes and biochemical techniques. Race determination presently requires four to six weeks and is usually done with potted plants in a greenhouse during the off-season. However, race determination can be done in the field during the growing season by planting test strips of differential cultivars or commercial cultivars with resistance to the different races. The general laboratory and greenhouse procedure is as follows:

1. Four to seven days before setting up the test, plant seed of the five differential soybean cultivars (Pickett, Peking, PI 88788, PI 90763, and Lee) in a germinating medium (vermiculite). Proper watering is essential to ensure vigorous seedlings.
2. Obtain infested soil from the rhizosphere of infected soybean plants in the field and place it in five 8-cm (3-inch) clay pots on a greenhouse bench. (It is best to replicate the test at least three times, so 15 pots are usually used for each race determination.)

   If the quantity of infested soil is inadequate for 15 pots, mix it with an equal volume (or less) of sterilized soil prior to placing the soil in the pots. If the sample is very small, it can be mixed with sterilized soil, placed in large containers, and planted with a susceptible cultivar, such as Bragg, Essex, or Lee, to obtain a larger quantity of infested soil.
3. Transplant two seedlings of each differential host to each pot.
4. Grow the plants for four to five weeks.
5. Gently remove the plants and soil from the pots to minimize injury to the root systems. Gently wash the soil from the roots into a container of water.
6. Place the roots on a 20-mesh sieve nested over a 60-mesh sieve, and spray them with high-pressure water to dislodge cysts.

   **NOTE** Clean the sieves well before processing the next sample.
7. Transfer the cysts caught on the 60-mesh sieve by rinsing them into a 100-ml beaker or counting dish.
8. With the aid of a dissecting microscope, count only the white and yellow cysts collected from each root system.

[1]This section is based in part on Riggs and Schmitt (1988) and Riggs and Wrather (1992).

*Do not count brown cysts,* as they are the inoculum. Record the numbers on a chart with a column for each host.

9. Compute the reproductive index (RI):

$$RI = \frac{\text{no. of cysts recovered from a differential host}}{\text{no. of cysts recovered from a susceptible host (Lee)}} \times 100$$

Record the reaction of each host in the chart, with a plus sign if RI ≥ 10 and a minus sign if RI < 10. Determine the race by referring to Table 3.7, which defines the 16 accepted races of *H. glycines.* See the example in Table 3.8.

Although designating races is useful in controlling soybean losses caused by the nematode, it is somewhat arbitrary, because *Heterodera* is genetically variable and capable of race shifts as new cultivars of soybeans are extensively grown.

Some races of the soybean cyst nematode may be suppressed in cultivars bearing resistance to other races. For example, a soybean variety resistant to race 3 will suppress race 2 or race 5 (i.e., reduce the reproduction of these nematodes, thus increasing soybean yield, relative to that of a susceptible variety), but not as effectively as it suppresses race 3.

**Table 3.7.** Reproductive indices of races of *Heterodera glycines* on differential soybean cultivars[a]

| | Differential host | | | | |
|---|---|---|---|---|---|
| Race | Pickett | Peking | PI 88788 | PI 90763 | Lee |
| Original races | | | | | |
| 1 | – | – | + | – | + |
| 2 | + | + | + | – | + |
| 3 | – | – | – | – | + |
| 4 | + | + | + | + | + |
| 5 | + | – | + | – | + |
| New races | | | | | |
| 6 | + | – | – | – | + |
| 7 | – | – | + | + | + |
| 8 | – | – | – | + | + |
| 9 | + | + | – | – | + |
| 10 | + | – | – | + | + |
| 11 | – | + | + | – | + |
| 12 | – | + | – | + | + |
| 13 | – | + | – | – | + |
| 14 | + | + | – | + | + |
| 15 | + | – | + | + | + |
| 16 | – | + | + | + | + |

[a] The reproductive index (RI) is the number of cysts recovered from a differential host divided by the number of cysts recovered from the susceptible cultivar Lee and then multiplied by 100. A plus sign represents RI ≥ 10; a minus sign represents RI < 10.

**Table 3.8.** Example of soybean cyst nematode race determination

| | Differential host | | | | |
|---|---|---|---|---|---|
| | Pickett | Peking | PI 88788 | PI 90763 | Lee |
| Number of cysts in sample | 10 | 12 | 65 | 2 | 85 |
| Reproductive index[a] | 12 | 14 | 76 | 2 | 100 |
| Reaction[a] | + | + | + | – | + |
| Race 2 definition (Table 3.7) | + | + | + | – | + |
| Conclusion: The sample contains members of race 2. | | | | | |

[a] See Table 3.7, note a.

#### REFERENCES

For further reference, see Golden et al. (1970), Inagaki (1979), Riggs and Schmitt (1988), Riggs and Wrather (1992), and Taylor and Brown (1981). Complete citations are presented at the end of the chapter.

## *Hirschmanniella:* Rice Root Nematodes and Related Species

### Class Secernentea, order Tylenchida, family Pratylenchidae

#### HOSTS

*Hirschmanniella* species are distributed worldwide in wet, very wet, and aquatic soils. Their hosts include wild blackberry, bulrushes (*Scirpus*), cattail (*Typha*), clovers (*Trifolium*), corn (maize), cotton, cucumber, aquatic grasses, horse gram (*Macrotyloma uniflorum*), lotus, muskmelon, okra, pearl millet (*Pennisetum glaucum*), rice, rushes (*Juncus*), sedges (*Carex* spp.), squashes, strawberry, sugarcane, taro (*Colocasia esculenta*), tomato, watermelon, wheat, and related weeds.

#### PARASITISM AND HABITAT

*Hirschmanniella* species are migratory endoparasites, mainly of roots in very wet soils. These nematodes are unusual in being perfectly adapted to constant flooding. They may be found in all stages in soil (up to 12 months in drier rice soils) and roots. Weeds and other hosts in paddy fields serve to overseason the population between rice crops. Maximum penetration occurs in 20- to 30-day-old rice plants. Eggs are laid in roots, and development to the adult stage may take place in four to six weeks. One or two generations are produced per year, depending on the species.

These nematodes kill cortical cells and dissolve the cell walls, creating large cavities. The cortex becomes yellowish to reddish brown. Brown, patchy, necrotic lesions form in roots, and whole roots may later turn yellowish brown and rot. Growth is retarded, especially early growth, as tillering and flowering are delayed and fewer shoots are produced; culm height, number of panicles, and weight of rice grains are reduced.

*H. gracilis* is commonly associated with a serious disease of rice called *mantek.* Small discolored areas first appear on the roots. These areas gradually enlarge and merge to involve the whole root cortex. When the damage is severe, the parenchyma tissue is entirely destroyed, and only the epidermis and vascular cylinder are left.

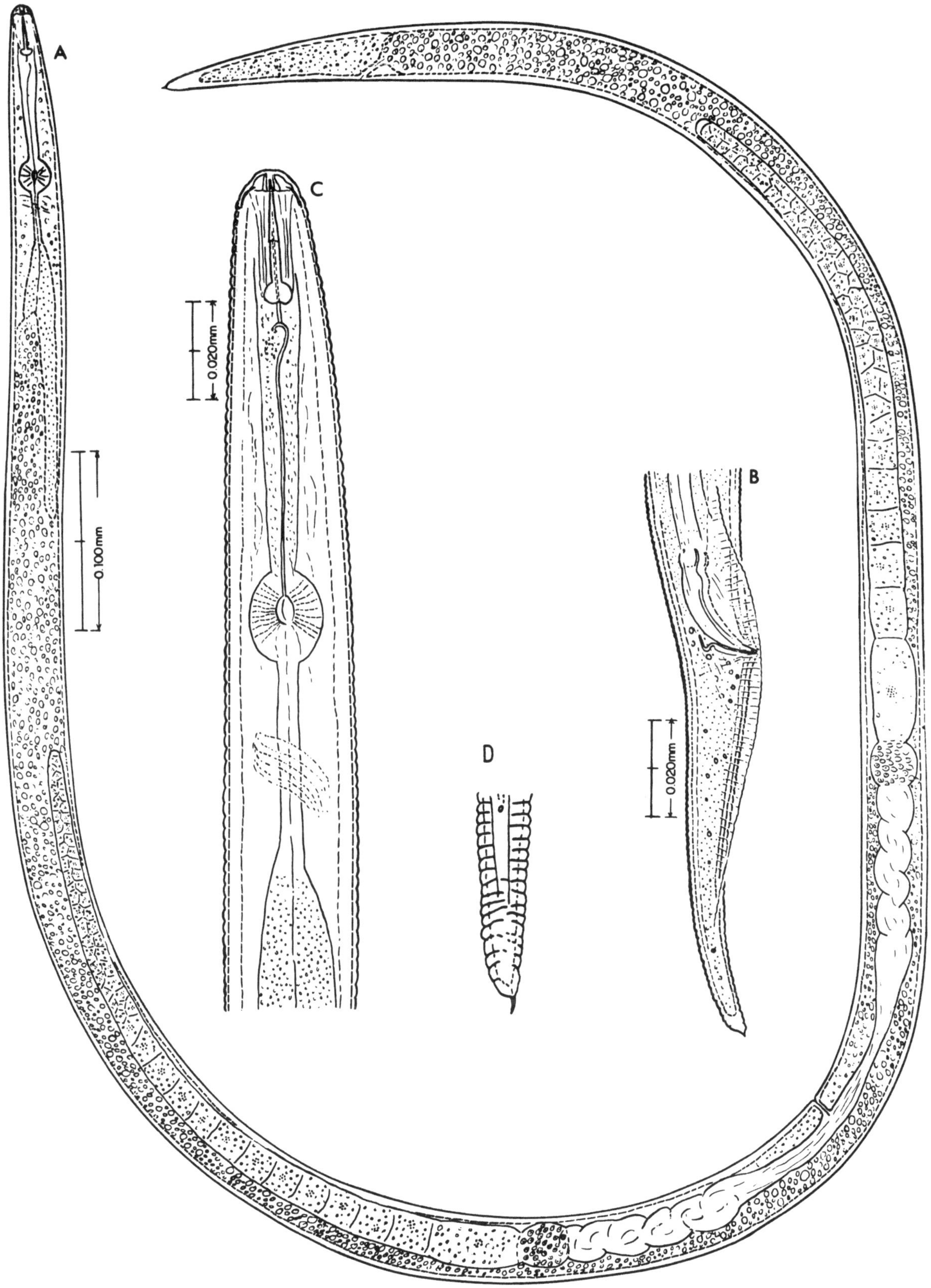


**Figure 3.19.** *Hirschmanniella.* **A,** Adult female. **B,** Male, posterior part. **C,** Female, anterior part. **D,** Female tail with mucro. (**A–C** reprinted, by permission, from Taylor, 1971; **D** courtesy H. Hirschmann)

#### BODY

The sexes are separate, with the males resembling the females, except for secondary sex characteristics. Adults are typically long or moderately long (0.9–4.2 mm), slender (a = 42–72), and straight or slightly curved (Figure 3.19). In some species the lip region is flattened and low, with rounded edges; in other species it is hemispherical and not offset from the body. The cephalic framework is well developed, with the outer margins extending to several body annules. The cuticle is annulated. There are four incisures in the lateral field. A unique characteristic of the genus is the large, elongated, thin-walled, unequal-sized cells "in tandem" located between the body muscle and the intestinal tract. When heat-killed, these nematodes lie more or less straight or ventrally arcuate.

#### STYLET

The stylet is typically strong and short (15–50 µm), with rounded basal knobs.

#### ESOPHAGUS

The median bulb is oval and conspicuous. The esophageal glands are in a long lobe ventrally overlapping the anterior end of the intestine. The excretory pore is at the level of the esophago-intestinal junction or somewhat anterior to it.

#### REPRODUCTIVE STRUCTURES

*Hirschmanniella* species have two equally developed ovaries, which are opposed and outstretched, joining at the vulva. The spicules are slender, distinctly cephalated, arcuate, and flanged; the gubernaculum is simple. Most species are bisexual.

#### VULVA

The vulva is typically near middle of the body (V = 52).

#### TAIL

The tail is elongate (3 or more anal body widths) and rounded or elongate-conoid and may have a short, sharp mucro at the tip. The male tail is partially enveloped by a bursa. The phasmids are in the posterior third of the tail.

#### GENUS RESEMBLING *HIRSCHMANNIELLA*

*Radopholus.* Nematodes in this genus have much shorter bodies (nearly always less than 1 mm long) than members of *Hirschmanniella. Radopholus* is also distinguished by dorsally overlapping esophageal glands and the absence of a mucro on the tail.

*Hirschmanniella* and *Radopholus* resemble *Pratylenchus* in gross respects, but they both have two opposing ovaries, whereas *Pratylenchus* has only one ovary.

---

## Plate 1

**1A.** Wheat seed galls caused by *Anguina tritici.* (Courtesy M. A. McClure)

**1B.** Chrysanthemum infected by a foliar nematode, *Aphelenchoides ritzemabosi.* (Courtesy BASF Corporation)

**1C.** Rice white tip, caused by *Aphelenchoides besseyi.* (Photo by R. D. Riggs, reproduced from "Nematodes and nematode diseases of plants," Slide Set No. 20, American Phytopathological Society, 1978)

**1D.** Putting green damaged by a sting nematode, *Belonolaimus longicaudatus.* (Courtesy D. W. Dickson)

**1E.** Corn roots damaged by a sting nematode, *Belonolaimus longicaudatus.* (Courtesy D. W. Dickson)

**1F.** Scotch pines, *Pinus sylvestris,* dying of wilt caused by the pinewood nematode, *Bursaphelenchus xylophilus* (left), and blue stain fungus in nematode-infested wood (right). (Courtesy R. B. Malek [left], E. Dutky [right])

**1G.** Large onion sets injured by a stem and bulb nematode, *Ditylenchus dipsaci* (left). *Narcissus* bulb (cross section) infested with *D. dipsaci* (right). (Courtesy Department of Plant Pathology, University of Illinois [left], H. M. Darling [right])

**1H.** Golden potato cyst nematode, *Globodera rostochiensis,* on potato roots. (Courtesy Society of Nematologists)

**1I.** Soybeans damaged by the soybean cyst nematode, *Heterodera glycines* (left). Cysts of *H. glycines* on snap bean roots (right). (Courtesy D. I. Edwards [left], G. R. Noel [right])

**1J.** Table beets infested with the beet cyst nematode, *Heterodera schachtii.* (Courtesy W. F. Mai)

**1K.** Cabbage plants severely damaged by *Heterodera schachtii.* (Courtesy D. W. Dickson)

**1L.** Putting green damaged by a lance nematode, *Hoplolaimus* sp. (Courtesy D. W. Dickson)

**1M.** Corn seedling damaged by a needle nematode, *Longidorus* sp. (Courtesy Society of Nematologists)

**1N.** Tobacco severely injured by a root-knot nematode, *Meloidogyne* sp. (Courtesy D. W. Dickson)

**1O.** Potato tuber distorted by *Meloidogyne incognita.* (Courtesy R. B. Malek)

**1P.** Parsnip root with injury due to a root-knot nematode, *Meloidogyne hapla.* (Reprinted, by permission, from G. W. Simone, 1980, cover photo, Plant Dis. 64[3])

# Plate 1

See the full captions on the facing page.

**1A.** Wheat. *Anguina tritici.*

**1B.** Chrysanthemum. *Aphelenchoides ritzemabosi.*

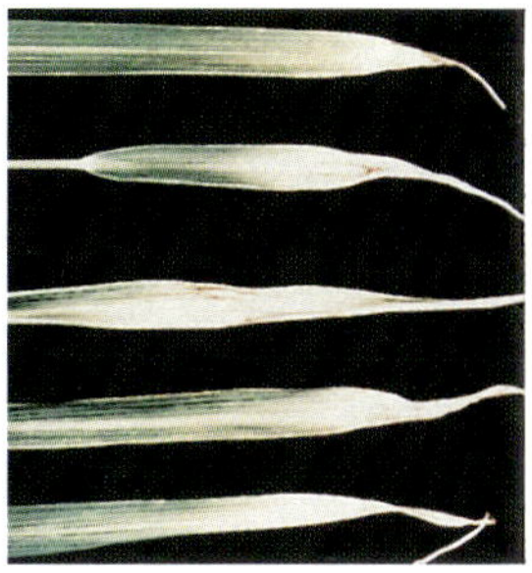

**1C.** Rice. *Aphelenchoides besseyi.*

**1D.** Turfgrass. *Belonolaimus longicaudatus.*

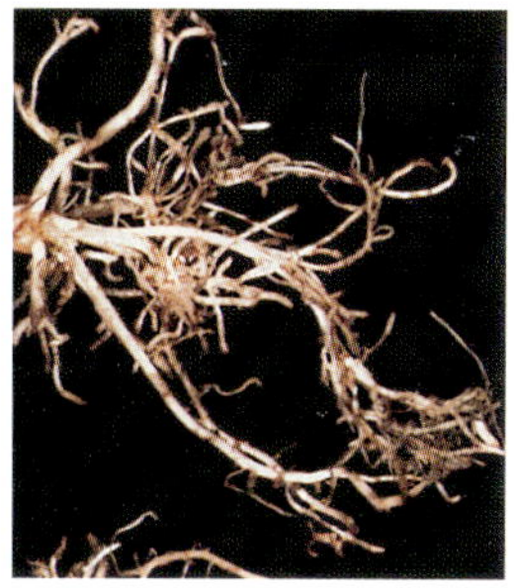

**1E.** Corn. *Belonolaimus longicaudatus.*

**1F.** Scotch pine. *Bursaphelenchus xylophilus.*

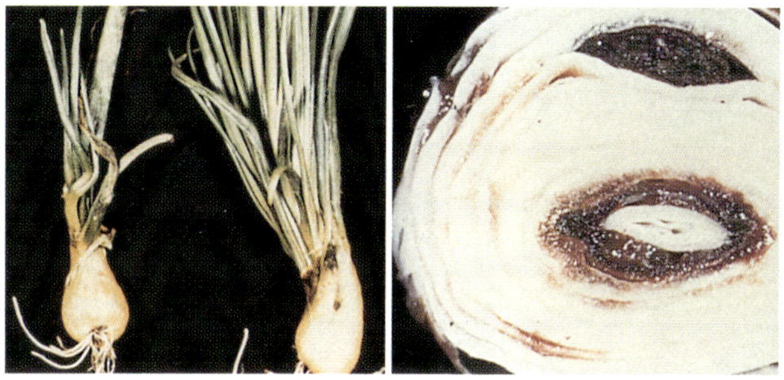

**1G.** Onion (left) and *Narcissus* (right). *Ditylenchus dipsaci.*

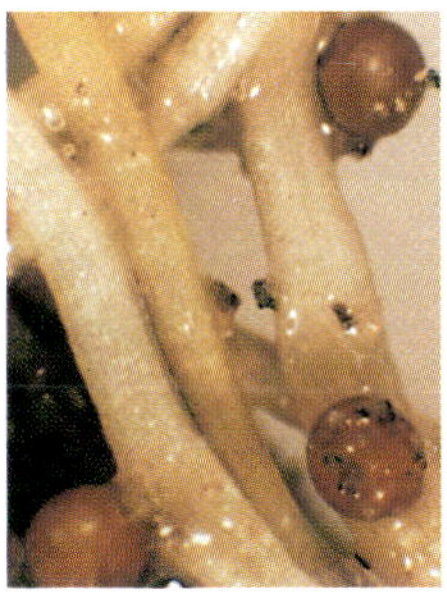

**1H.** Potato. *Globodera rostochiensis.*

**1I.** Soybean (left) and snap bean (right). *Heterodera glycines.*

**1J.** Beets. *Heterodera schachtii.*

**1K.** Cabbage. *Heterodera schachtii.*

**1L.** Turfgrass. *Hoplolaimus* sp.

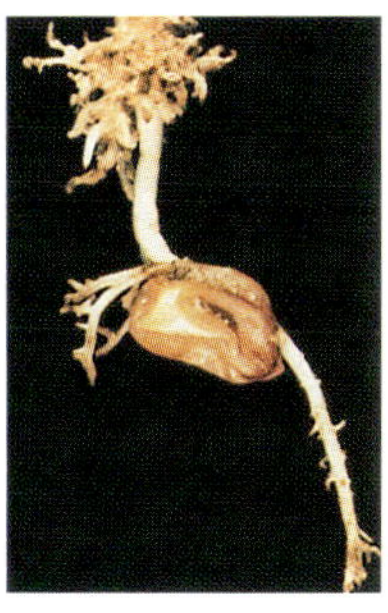

**1M.** Corn. *Longidorus* sp.

**1N.** Tobacco. *Meloidogyne* sp.

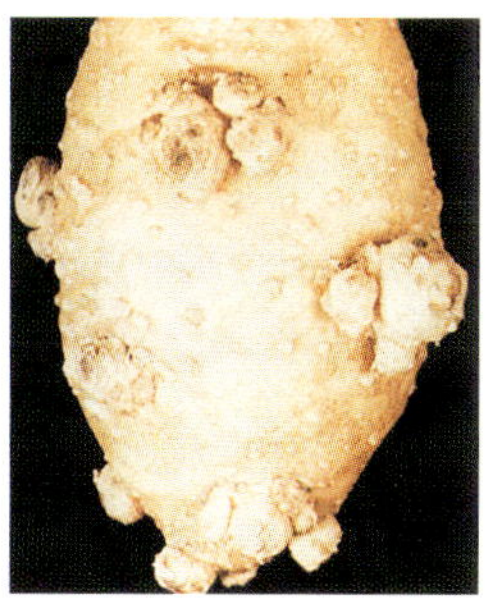

**1O.** Potato. *Meloidogyne incognita.*

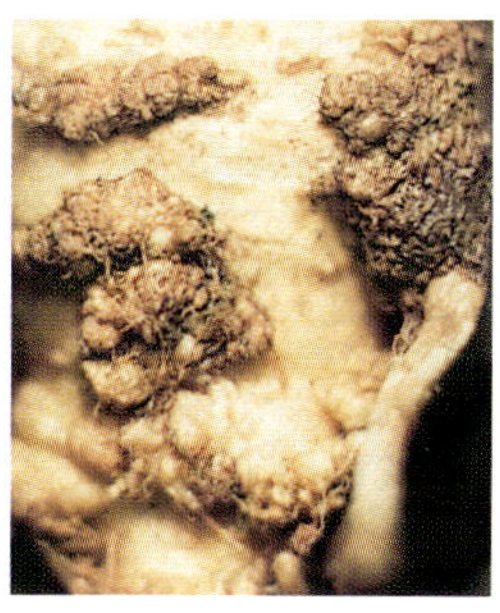

**1P.** Parsnip. *Meloidogyne hapla.*

# Plate 2

See the full captions on the facing page.

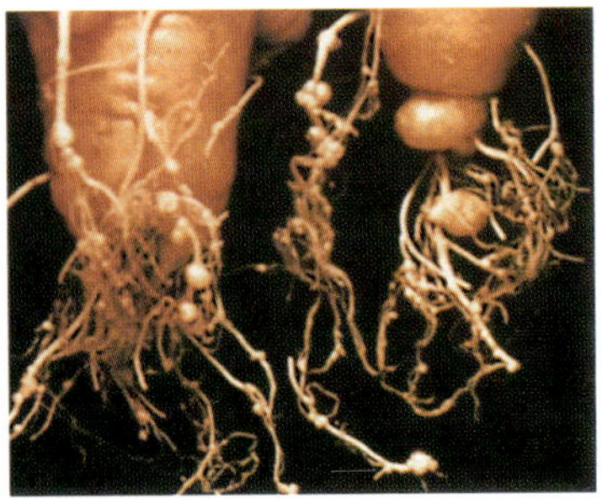

**2A.** Carrot. *Meloidogyne hapla.*

**2B.** Soybean. *Meloidogyne javanica.*

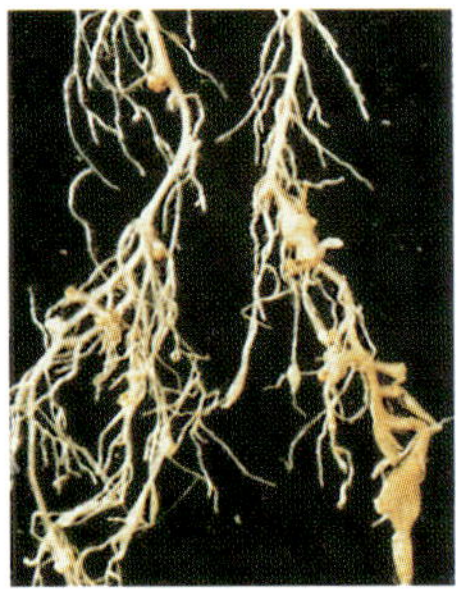

**2C.** African violet. *Meloidogyne hapla.*

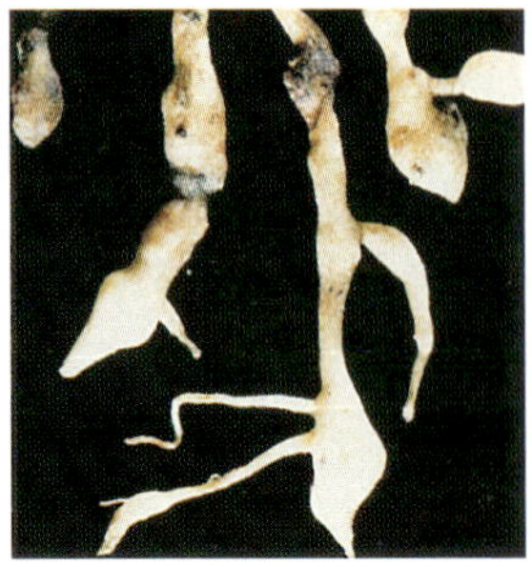

**2D.** Peony. *Meloidogyne hapla.*

**2E.** Sweet potato. *Meloidogyne* sp.

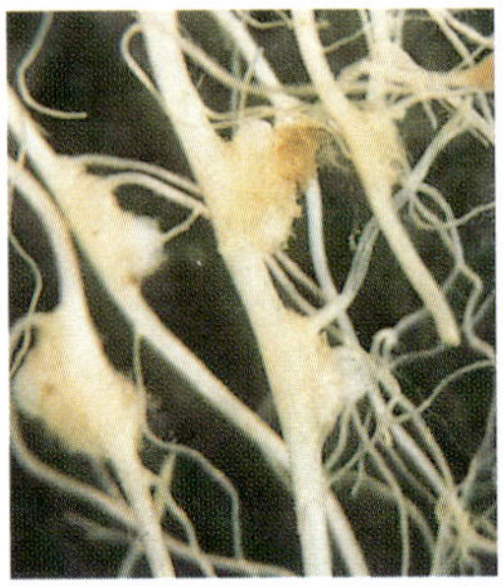

**2F.** Bermudagrass. *Meloidogyne javanica.*

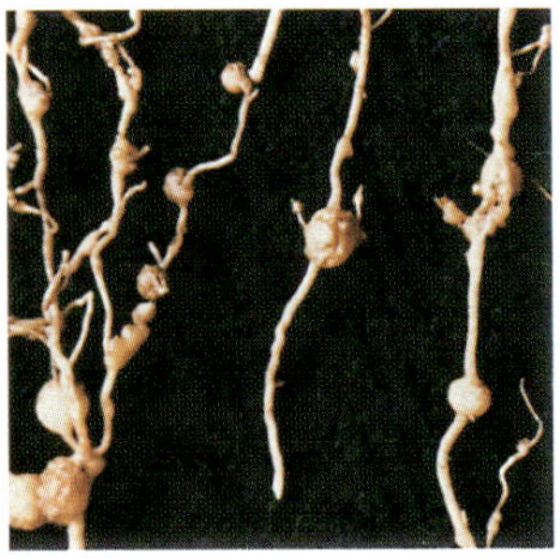

**2G.** Potato. *Nacobbus* sp.

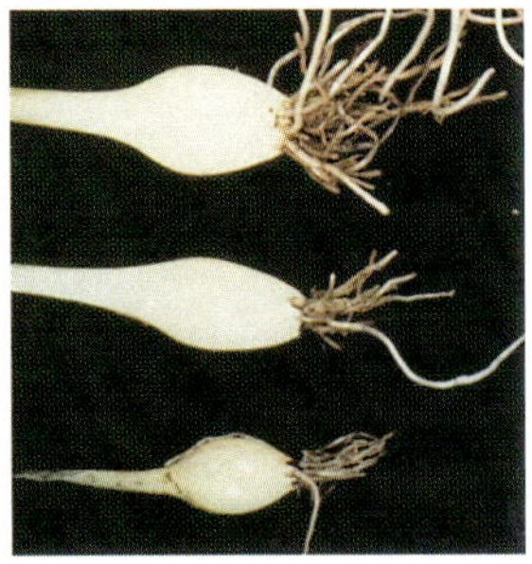

**2H.** Onion. *Paratrichodorus christiei.*

**2I.** Corn. *Paratrichodorus* sp. and *Belonolaimus* sp.

**2J.** Wheat. *Paratrichodorus* sp.

**2K.** Boxwood. *Pratylenchus vulnus.*

**2L.** Peanut. *Meloidogyne arenaria.*

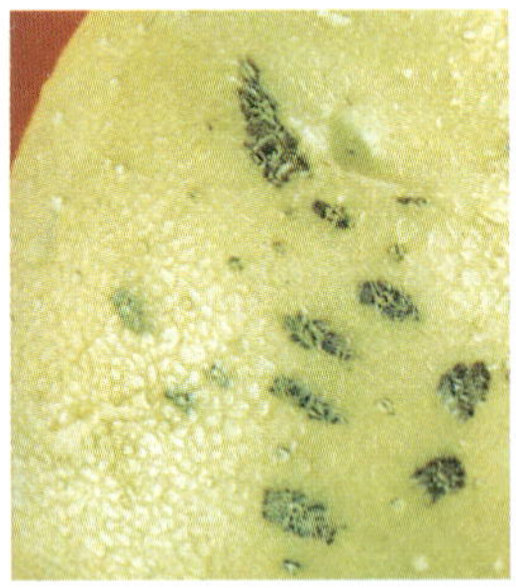

**2M.** Potato. *Pratylenchus crenatus.*

**2N.** Peanut. *Pratylenchus* sp.

**2O.** Corn. *Pratylenchus* sp.

**2P.** Citrus. *Radopholus similis.*

**2Q.** Coconut palm. *Rhadinaphelenchus cocophilus.*

#### COMMENTS

There are 29 or so valid species of *Hirschmanniella.* They are widespread in rice plantings throughout the world, causing yield losses as high as 25%. *H. oryzae* is of major economic importance. Six other species also infect rice. These nematodes may be associated with *Rhizoctonia solani, Pythium,* and *Fusarium* spp.

### *Hoplolaimus:* Lance Nematodes

**Class Secernentea, order Tylenchida, family Hoplolaimidae**

#### HOSTS

*Hoplolaimus* species (lance nematodes) have numerous hosts worldwide, including abacá (*Musa textilis*), *Acacia, Acalypha,* alfalfa, amaryllis, apple, bamboo, banana, barley, beans (*Phaseolus* spp.), beet (garden beet, sugar beet), begonia, bird's-foot trefoil, black gram (urd, *Vigna mungo*), black pepper (*Piper nigrum*), blackberry, blueberry, broad bean (faba bean, *Vicia faba*), cabbage, cacao (cocoa, *Theobroma cacao*), catjang (*Vigna unguiculata* subsp. *cylindrica*), cardamom (*Elettaria cardamomum*), carnation, Carolina geranium (*Geranium carolinianum*), cassava (*Manihot esculenta*), cauliflower, chestnut, chickpea (*Cicer arietinum*), Chinese cabbage, cinnamon, citrus, clove (*Syzygium aromaticum*), clovers (*Trifolium*), cluster bean (*Cyamopsis tetragonoloba*), coconut palm (*Cocos nucifera*), coffee, corn (maize), cotton, cowpea (*Vigna unguiculata*), cranberry, cucumber, date palm (*Phoenix dactylifera*), eggplant, elms, *Eucalyptus,* fenugreek (*Trigonella foenum-graecum*), flax, ginger (*Zingiber officinale*), grape, grass pea (*Lathyrus sativus*), grasses (forage grasses, turfgrasses [Plate 1L], and wild species), guava (*Psidium*), holly, hyacinth bean (*Lablab purpureus*), jute (*Corchorus capsularis* and *C. olitorius*), lespedezas, lima bean, *Lycoris,* maples, mango, melons, millets (pearl millet and proso millet), morning-glory, moth bean (*Vigna aconitifolia*), mulberry, mung bean (*Vigna radiata*), muskmelon, mustards, various nursery crops, nutmeg (*Monodora*), oaks, okra, onions, *Opuntia,* pagoda-tree (*Plumeria rubra*), palms, papaya, pawpaw, pea, peach, peanut, pear, pecan, pepper (garden or bell pepper, *Capsicum annuum*), pigeon pea (red gram, *Cajanus cajan*), pineapple, pines, plantain (*Musa* × *paradisiaca*), white poplar (*Populus alba*), potato, ragi (African millet, *Eleusine coracana*), rice, rose, rye, sesame (*Sesamum*), sorghums, soybean, spruces, strawberry, sugarcane, sweet potato, sweetgum (*Liquidambar*), sycamore, tea, tobacco, tomato, tree-of-heaven (*Ailanthus altissima*), turmeric (*Curcuma domestica*), watermelon, wheat, white willow (*Salix alba*), yams (*Dioscorea* spp.), and many others.

Weed hosts include ball nightshade (*Solanum carolinense*), crabgrass, hairy vetch (*Vicia villosa*), milkweed, nutgrass (*Cyperus*), plantain (*Plantago*), and ragweed (*Ambrosia*).

---

## Plate 2

**2A.** Carrot roots damaged by a root-knot nematode, *Meloidogyne hapla.* (Courtesy K. R. Barker)

**2B.** Soybean roots severely distorted by *Meloidogyne javanica.* (Courtesy K. R. Barker)

**2C.** African violet roots infested with *Meloidogyne hapla.* (Courtesy Department of Plant Pathology, University of Minnesota)

**2D.** Peony roots damaged by *Meloidogyne hapla.* (Courtesy R. B. Malek)

**2E.** Sweet potato severely injured by a root-knot nematode, *Meloidogyne* sp. (Courtesy R. D. Riggs)

**2F.** Bermudagrass roots attacked by *Meloidogyne javanica.* (Courtesy R. D. Riggs)

**2G.** Potato roots damaged by a false root-knot nematode, *Nacobbus* sp. (Courtesy P. Jatala)

**2H.** Onion damaged by a stubby-root nematode, *Paratrichodorus christiei.* (Courtesy W. F. Mai)

**2I.** Corn with injury due to a stubby-root nematode (*Paratrichodorus* sp.) and a sting nematode (*Belonolaimus* sp.). (Courtesy K. R. Barker)

**2J.** Wheat roots damaged by a stubby-root nematode, *Paratrichodorus* sp. (Courtesy R. D. Riggs)

**2K.** Boxwood, *Buxus sempervirens,* damaged by a lesion nematode, *Pratylenchus vulnus.* (Courtesy K. R. Barker)

**2L.** Peanut plants (left) and pods (right) infested with a root-knot nematode, *Meloidogyne arenaria.* (Courtesy D. W. Dickson)

**2M.** Potato damaged by a lesion nematode, *Pratylenchus crenatus.* (Courtesy W. F. Mai)

**2N.** Peanut pods damaged by a lesion nematode, *Pratylenchus* sp. (Courtesy D. W. Dickson)

**2O.** Corn roots damaged by a lesion nematode, *Pratylenchus* sp. (Courtesy Department of Plant Pathology, University of Illinois)

**2P.** Spreading decline of citrus, caused by *Radopholus similis.* (Courtesy Society of Nematologists)

**2Q.** Red ring of coconut palm, caused by *Rhadinaphelenchus cocophilus* (left), and cross section of the trunk of a diseased tree (right). (Courtesy D. W. Dickson)

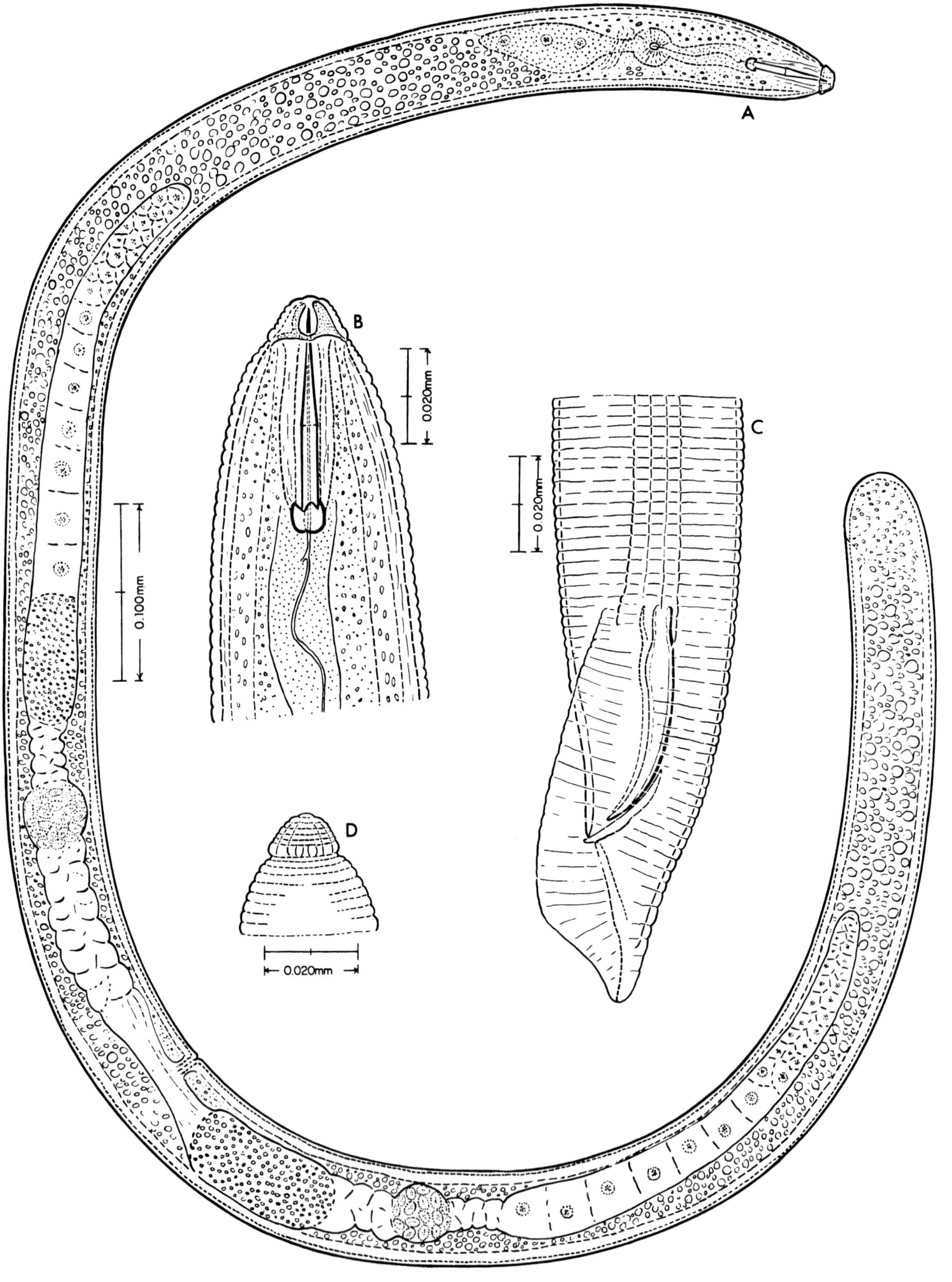


**Figure 3.20.** *Hoplolaimus.* **A,** Adult female. **B,** Female, anterior part. **C,** Male, posterior part. **D,** Female, lip region. (Reprinted, by permission, from Taylor, 1971)

#### PARASITISM AND HABITAT

*Hoplolaimus* species are migratory ecto- and endoparasites. They are very commonly found in all stages in soil and roots. These nematodes usually feed some distance from the root tip, often on roots of considerable size. Cortex cells are destroyed during feeding, and extensive necrotic lesions may form. The nematodes sometimes cross the endodermis and damage xylem and phloem tissues in the stele.

#### BODY

Adult males and females are 0.9 to 2.0 mm long (a = 22–40). In general, they are rather stout, vermiform, and rounded at both ends (Figure 3.20). The sexes are separate in most species. The males resemble the females, except for secondary sex characters. The lip region is typically high and offset by a constriction (caplike), and some or all of its annules are subdivided into small, tilelike segments by longitudinal and lateral striations, visible under 100× magnification. The cephalic framework is massive and yellowish in some species. The enlarged, shieldlike phasmids (scutellae) are not opposite one another on each side of the body; usually one is located anterior to the vulva and the other in the lateral field between the vulva and the anus on the other side. The lateral field has four or fewer incisures and is areolated. When relaxed by gentle heat, these nematodes assume an almost straight position or are slightly curved (arcuate) ventrally.

#### STYLET

The stylet is strongly developed and 30–61 µm long, with typically elongated, forward-pointing (or tulip-shaped), massive basal knobs.

#### ESOPHAGUS

The median bulb is well developed and oval to rounded. The esophageal glands are in a short lobe overlapping the anterior end of the intestine dorsally and laterally for 1 or 2 body widths with three to six nuclei. The dorsal gland orifice is a short distance (less than one-fourth the stylet length) from the stylet. The location of the excretory pore varies in different species, ranging from the level of the midmedian bulb to a level behind the esophago-intestinal junction (above or below the hemizonid).

#### REPRODUCTIVE STRUCTURES

Females have two opposed, outstretched ovaries.

Males have a single, anteriorly outstretched testis. They have small, well-developed, arcuate spicules and a distinct, often troughlike gubernaculum, which protrudes through the cloaca.

#### VULVA

The vulva is typically just posterior to the midpoint of the body (V = 50–62).

#### TAIL

The female and juvenile tails are blunt, with a length of usually less than 1 anal body width; they are hemispheroid to convex-conoid and have annulations extending to the terminus. The male tail is short, conical, pointed, and enveloped by a small bursa.

#### GENERA RESEMBLING *HOPLOLAIMUS*

*Aorolaimus.* In this genus, the spear knobs are rounded or lack distinct anterior projections, and there are four incisures areolated at the phasmids and anteriorly.

*Helicotylenchus.* The dorsal gland orifice is at least one-half the stylet length posterior to the stylet in this genus.

*Peltamigratus.* The cephalic framework is not as massive in this genus as in *Hoplolaimus.*

*Rotylenchus* and *Scutellonema.* Nematodes in these genera have broader stylet knobs and, in males, shorter tails than members of *Hoplolaimus.*

#### COMMENTS

There are 17 or more species of *Hoplolaimus.* The major species include *H. columbus, H. galeatus, H. indicus, H. magnistylus, H. pararobustus,* and *H. seinhorsti.*

*H. columbus* occurs mainly in the eastern United States while *H. galeatus* is more widespread in the Western Hemisphere and India. Juveniles and adults frequently occur together in abundance.

*H. columbus* reproduces parthenogenically. Females repeatedly lay single eggs in and along the surface of roots.

*Hoplolaimus* species are associated with species of *Fusarium* (including *F. moniliforme* and *F. oxysporum*) and numerous other microorganisms that enter lesions created by nematode feeding.

### *Longidorus:* Needle Nematodes

**Class Adenophorea, order Dorylaimida, family Longidoridae**

#### HOSTS

*Longidorus* species (needle nematodes) have numerous hosts worldwide, including alders, alfalfa, apple, arborvitae, artichoke, *Arundinaria,* ashes, asparagus, avocado, azaleas, banana, barley, beans (*Phaseolus* spp.), beeches, beet (garden beet, sugar beet), black pepper (*Piper nigrum*), blackberry, broad bean (faba bean, *Vicia faba*), cabbage, cacao (cocoa, *Theobroma cacao*), caraway (*Carum carvi*), carnation, carrot, cauliflower, celery, cherries, chives, chrysanthemum, citrus, clovers (*Trifolium*), coconut palm (*Cocos nucifera*), coffee, corn (maize), cotton, cucumber, currants, dill, *Euonymus,* fig, firs, gherkin, gladiolus, grape, grasses (forage grasses, turfgrasses, and wild species), hawthorns, hazelnut (*Corylus*), holly, hop, hornbeam, jack bean (*Canavalia ensiformis*), larch, leek, lettuce, maples, marigold, millets, mints, mulberry, narcissus, oaks, oat, onions, parsley, pea, peanut, pear, pecan, pepper (garden or bell pepper, *Capsicum annuum*), peppermint, petunia, phlox, pineapple, pines, plums, pomegranate (*Punica granatum*), poplars, potato, privet, pumpkin, raspberry, rice, *Robinia* spp., roses, sage, savory, sorghums, spinach, spruces, strawberry, sugar beet, sugarcane, sweet potato, tassel-flower (*Amaranthus caudatus*), tobacco, tomato, tree-of-heaven (*Ailanthus altissima*), tulips, turmeric (*Curcuma domestica*), turnip, walnut, wheat, willows, yams (*Dioscorea*), and many weeds.

#### PARASITISM AND HABITAT

*Longidorus* species are cosmopolitan migratory ectoparasites. They feed near the root cap, inhibiting the development of feeder roots and inducing the formation of short, thickened, brown to black, round to hook-shaped galls on the tips of roots (Plate 1M). Seedlings die if parasitism is heavy. Shoots of older plants are severely stunted, foliage becomes chlorotic, and yield is reduced.

The feeding habits of needle nematodes are much like those of dagger nematodes (see the section on *Xiphinema*). *Longidorus* species cause cycles of cell enlargement and the formation of galls, upon which the nematodes feed. The galls soon collapse and are invaded by bacteria. The galling is less severe than that due to *Xiphinema.*

In some species, two years may elapse during development from egg to adult. The life span may be three years or more, depending on the species, but a life cycle can be completed in 10 to 12 days at 18°C (64°F). Most species live in temperate and tropical regions, but *L. elongatus* favors cooler climates. Needle nematodes are usually confined to sandy and sandy loam soils.

#### BODY

Both sexes are very long (2–11 mm) and slender (a = 84–158) (Figure 3.21). The lips are flattened anteriorly and formed in the same contour as the neck and body, and they bear the usual two circlets of six to 10 papillae. The head is usually not offset. The body at the lips is two-thirds the body width at the level of the single guide ring, located around the anterior half of the odontostyle. The cuticle generally appears smooth when viewed with a light microscope. The lateral chords are broad. The lateral fields are wide and marked with a series of pores in two lines. The amphids are abnormally large and pouchlike, almost encircling the head, and open by a minute (and very difficult to see), slitlike pore at the base of the lip region. The amphids contain coiled fibrillar terminals and refractive bodies believed to be nervous in function.

#### STYLET

The stylet is very long (more than 180 μm) and formed in two parts: the odontostyle, a very long, anterior, lightly sclerotized, needle-like spear (44–88 μm long or longer); and the odontophore, a rather weak posterior stylet extension (30–90 μm long or longer), which is not sclerotized and lacks the flanges possessed by *Xiphinema.* The odontophore is not forked at the base, and the odontostyle is broader at its base. The basal knobs are inconspicuous. Both parts of the stylet are less strongly cuticularized than in *Xiphinema.* A prominent, cuticular guiding sheath surrounds the anterior half or more of the odontostyle. The odontostyle penetrates root cells, and the odontophore contains nerve tissues adjacent to the food canal, which probably enable the nematode to discriminate between feeding sites within plant roots. Juveniles have two odontostyles, a functional one and a replacement in line with and just posterior to it.

#### ESOPHAGUS

The esophagus is reduced to a slender, cylindrical, flexible tube with an elongated gland or bulb. The esophagus is usually looped at the junction between the narrow anterior portion and the wider, muscular posterior gland that abuts the intestine. The dorsal and the anterior pair of submedian gland nuclei are easily visible, whereas the posterior submedian pair is rather small and obscure.

#### REPRODUCTIVE STRUCTURES

Females have paired ovaries, which are short, opposed, and reflexed. Each genital tract consists of a uterus, oviduct, and ovary, in which germ cell formation is localized at the distal end.

Males have two opposed testes. The spicules are paired, massive (50–105 μm long), and banana-shaped, with small, lateral accessory pieces but no gubernaculum. Prominent, ventral supplementary papillae are also present.

#### VULVA

The vulva is a transverse slit close to the midsection (V = 43–59).

#### TAIL

The male and female tails are similar, being dorsally convex and bluntly rounded to conoid. The male terminus curls ventrally when the nematode is killed by heat, because of the prominent copulatory muscles in the posterior region. The anus is a transverse slit. There is no bursa.

#### GENERA RESEMBLING *LONGIDORUS*

*Paralongidorus* and *Xiphinema.* Nematodes in *Paralongidorus* have funnel- to stirrup-shaped amphid pouches and wide amphidial apertures at the base of the lip region, which resemble those of *Xiphinema.*

#### COMMENTS

There are some 90 or more species of *Longidorus.* The major species include *L. attenuatus, L. breviannulatus, L. elongatus,* and *L. macrosoma.*

Males are very rare or absent in most populations of some species, and reproduction is parthenogenetic.

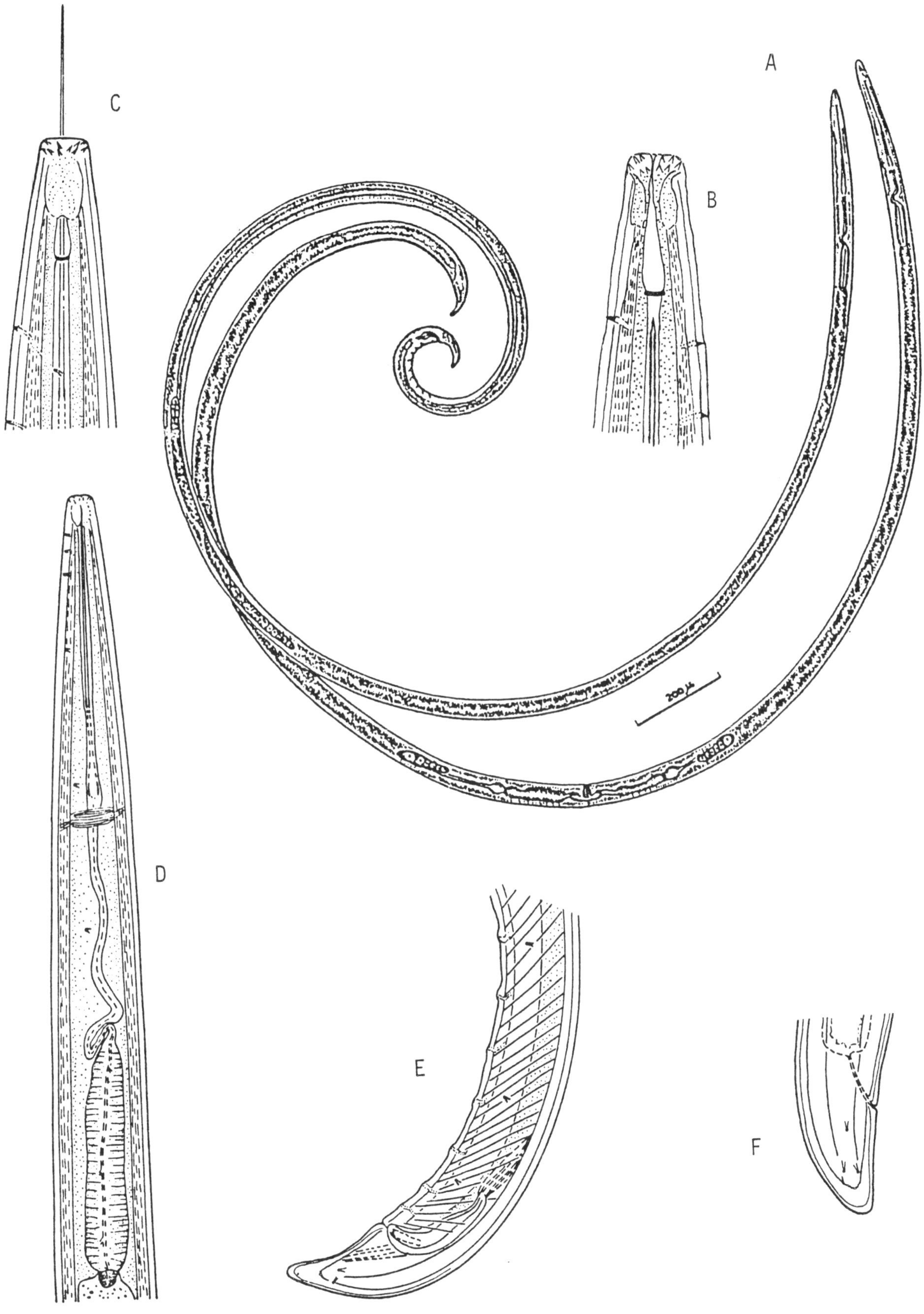


**Figure 3.21.** *Longidorus elongatus.* **A,** Adult male (left) and female (right). **B,** Head region, with the stylet not extended. **C,** Head region, with the stylet extended. **D,** Anterior part. **E,** Male tail. **F,** Female tail. (**A** reprinted, by permission, from Smart and Khuong, 1983; **B–F** reprinted, by permission, from Hooper, 1961)

In some soils most needle nematodes are 60 to 90 cm deep. In other soils they are closer to the surface. Individuals in a few species may live for three to five years. The length of the life cycle depends on the soil structure, temperature, and moisture and growth of the host plant, and it varies from several weeks to several years, depending on the species. Needle nematodes may feed in the same position for as long as four days. Irrigation favors damage by these nematodes.

Eight species of *Longidorus* transmit several troublesome nepoviruses, including artichoke Italian latent, cherry rosette, mulberry ringspot, peach rosette mosaic, raspberry ringspot, and tomato black ring (see Appendix, Table A.1). Virus transmission by *Longidorus* is specific for both the nematode and the virus. The nematodes retain the ability to transmit these viruses for usually less than two months. Many weeds are hosts of both the nematodes and the viruses, and thus transmission to crop plants continues, as the viruses are readily reacquired. *Longidorus* species can transmit these viruses in all stages of development.

## *Macrotrophurus*

### Class Secernentea, order Tylenchida, family Tylenchorhynchidae

#### HOSTS

Hosts of *Macrotrophurus arbusticola,* the only species of this genus, include apple, pear, and poplars.

#### PARASITISM AND HABITAT

This nematode is a migratory ectoparasite on plant roots in the Netherlands and Switzerland. It is found in all stages in the soil.

#### BODY

The body is of moderate length (1.0–2.1 mm), rather slender (a = 38–58), and cylindrical, tapering rapidly in the same body contour to a very narrow, smooth, rounded lip region (Figure 3.22). The head skeleton is poorly developed. The amphid apertures are elongate slits with pocketlike inner connections. The lateral fields are two-fifths as wide as the body and are marked by four faint incisures. The phasmids are just anterior to the middle of the tail.

#### STYLET

The stylet is very long (90–120 µm) and slender. The anterior part is 2.5 times as long as the shaft, which has minute, rounded, basal knobs.

#### ESOPHAGUS

The esophageal lumen is convoluted in the anterior portion of the esophagus. The procorpus is enlarged. The median bulb is large and rounded and has a conspicuous valve. The isthmus is rather long and narrow, with the nerve ring just anterior to the middle of it. The terminal bulb is pear-shaped and elongate and does not overlap the intestine. The hemizonid is adjacent to the excretory pore. Cardia are present. The opening of the dorsal esophageal gland is a short distance posterior to the stylet base.

#### REPRODUCTIVE STRUCTURES

Females have two opposed, outstretched ovaries.

Males have spicules with proximal knobs and a gubernaculum.

#### VULVA

The vulva is near middle of the body (V = 45–50).

#### TAIL

The female tail is short (c = 28–42) and rounded and has a very thick terminal cuticle packed with coarse granular material. The male tail is short (c = 20–28), archlike, and enveloped by a large bursa.

#### GENERA RESEMBLING *MACROTROPHURUS*

*Dolichodorus.* The lips are set off by a distinct constriction in this genus.

*Trophurus*

*Tylenchorhynchus*

#### COMMENTS

*M. arbusticola* is a problem in commercial plantings of apple, pear, and poplars in clay soil in the Netherlands and Switzerland.

## *Meloidodera:* Cystoid Nematodes

### Class Secernentea, order Tylenchida, family Heteroderidae

#### HOSTS

Hosts of *Meloidodera* species include blueberry, camellia, corn (maize), Douglas fir (*Pseudotsuga menziesii*), American dunegrass (*Leymus mollis,* syn. *Elymus mollis*), elms, lily-of-the-valley, mesquite (*Prosopis*), pepper (garden or bell pepper, *Capsicum annuum*), pines (species listed below), sage (*Salvia* spp.), sea chickweed (*Honckenya peploides*), Pennsylvania smartweed (*Polygonum pensylvanicum*), tea-plant (*Camellia sinensis*), and turfgrasses (including centipedegrass, St. Augustine grass, and zoysia grass).

Pine hosts include Austrian (black, Corsican; *Pinus nigra*), Caribbean (Cuban, *P. caribaea*), loblolly (frankincense, old-field; *P. taeda*), Georgia (longleaf, southern yellow; *P. palustris*), long-tag (shortleaf, yellow; *P. echinata*), pitch (*P. rigida*), sand (*P. clausa*), and slash (*P. elliottii*).

#### PARASITISM AND HABITAT

*Meloidodera* species are sedentary endoparasites in roots. They do not form galls on roots.

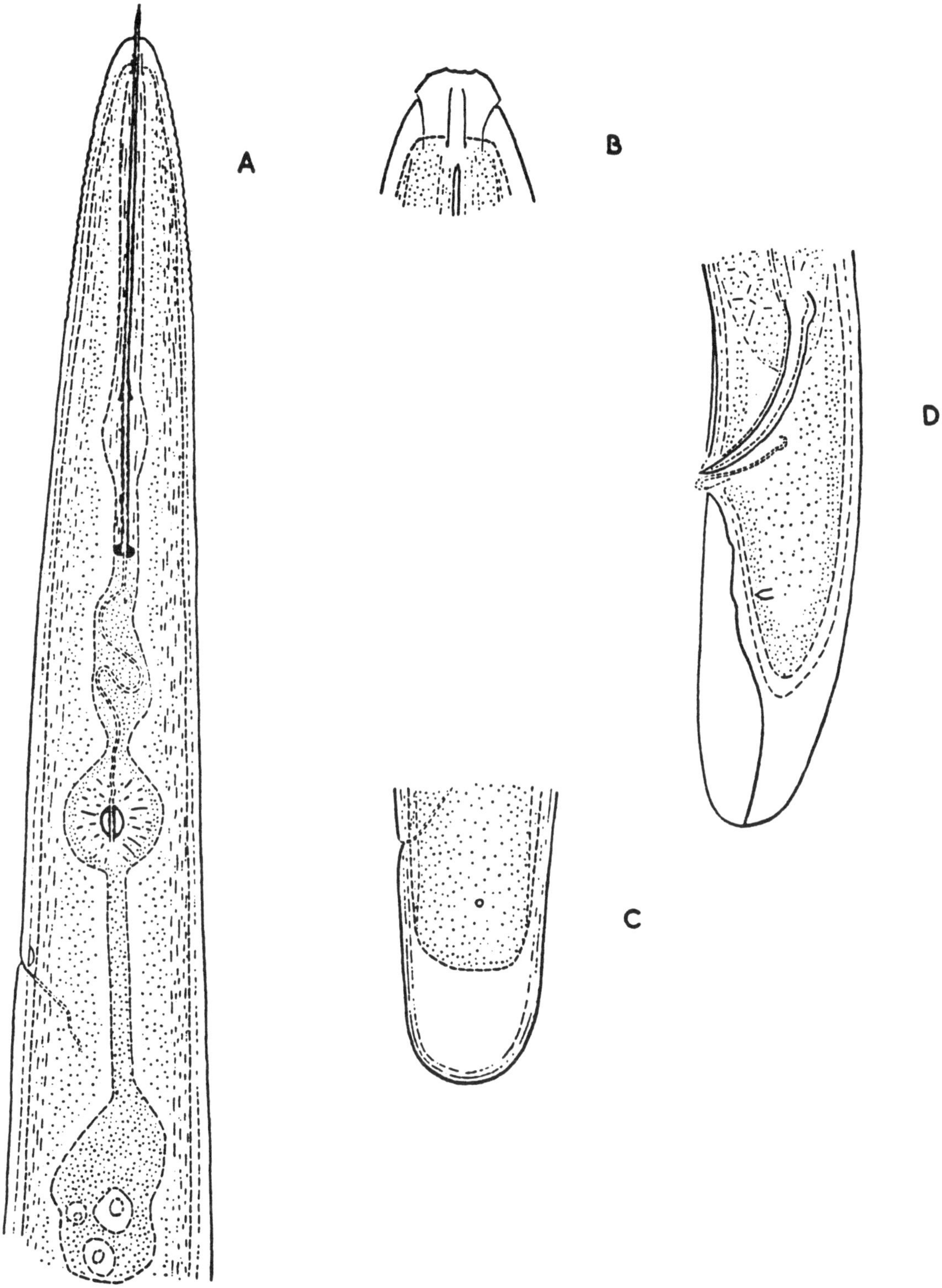


**Figure 3.22.** *Macrotrophurus arbusticola.* **A,** Female, anterior part. **B,** Female lip region. **C,** Female tail. **D,** Male, posterior part. (Reprinted from Mai and Mullin, 1996, © 1960 by W. F. Mai and H. H. Lyon, by permission of the publisher, Cornell University Press)

### BODY

Adult females are pear-shaped to spherical or ovoid (0.5–0.8 mm long and 0.22–0.4 mm wide) with a white, thick, tough cuticle, which does not turn yellow or brown at maturity (Figure 3.23C). The cuticle has well-marked annulations and rows of subcuticular dots. The head is annulated and lacks the large amphidial pouches present in *Meloidogyne.* The neck constitutes 20% of the body length. The body and internal structures are similar to those of *Globodera* and *Meloidogyne.* The phasmids are terminal.

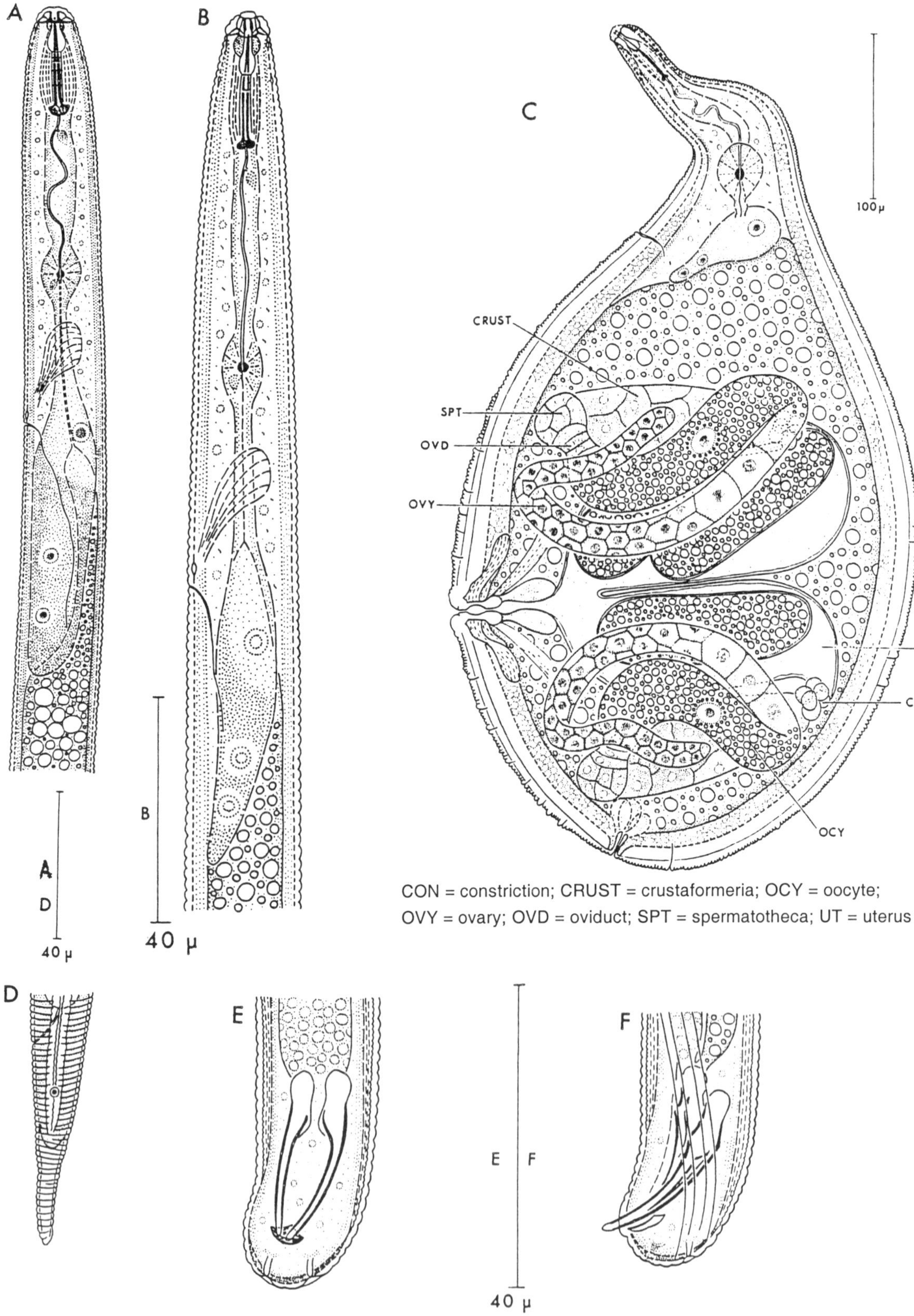


**Figure 3.23.** *Meloidodera floridensis.* **A,** Second-stage juvenile, anterior part. **B,** Adult male, anterior part. **C,** Adult female. **D,** Second-stage juvenile tail. **E,** Male tail, ventral view. **F,** Male tail, lateral view. (Courtesy H. Hirschmann)

Males (Figure 3.23B, E, and F) may be absent or difficult to find in some populations. They are smaller (less than 0.6 mm long) than males of *Heterodera* and *Meloidogyne.* Males have longitudinal striations, which are absent on the basal lip annules. The lateral field has four incisures.

Juveniles resemble those of cyst nematodes. *Meloidodera* species hatch as second-stage juveniles. Their amphids are occasionally quite large, and they have four lateral lines. The clear part of the juvenile tail is less than 25 μm long. Juveniles are 0.5–0.6 mm long (a = 26–29, b = 6.4–7.1, and c = 8.4–10.0) (Figure 3.23A and D).

#### STYLET

The female stylet is strong (35–37 μm long), with distinct, rounded knobs. The juvenile spear is 27–29 μm long, with strong basal knobs. The male stylet is 20–24 μm long.

#### ESOPHAGUS

The median bulb is very large and ovate and has a conspicuous valve. The excretory pore is prominent and slightly posterior to the median bulb. The dorsal gland orifice is about 5 μm posterior to the stylet base. The esophageal glands extend in lobes over the anterior end of the intestine. The phasmids are conspicuous, as wide as one of the caudal alae, and less than one-third the distance from the anus to the terminus.

#### REPRODUCTIVE STRUCTURES

Females have two opposed, outstretched ovaries, similar to those of *Globodera* (see the section on *Globodera*). Eggs (115–118 × 55 μm) are retained in the body of the female, but a persisting cyst is not formed.

Males have spicules 20–22 μm long and a gubernaculum about 4 μm long.

#### VULVA

The vulva is similar to that of *Meloidogyne,* but in *Meloidodera* the anus is near the tail tip (terminus), not protruding out of the body contour, and the subequatorial vulva is widely separated from it.

#### TAIL

The tail is similar to that of *Globodera.*

#### GENERA RESEMBLING *MELOIDODERA*

*Cryphodera* and *Zelandodera.* These genera appear to be closely related to *Meloidodera.* In *Cryphodera* the vulva is subterminal. In *Zelandodera* the anus is dorsal to the vulva in the dorsal body curvature of the female.

*Globodera* and *Heterodera.* Females of *Meloidodera* differ from females of these genera in having well-developed annulations and a subequatorial vulva and in producing fewer eggs contained in the body. Juveniles of *Meloidodera* closely resemble those of *Heterodera* but are distinguished by their large, easily observed phasmids.

*Meloidogyne.* Females in this genus form root galls and egg sacs, whereas those of *Meloidodera* do not. *Meloidodera* females also differ in having a thickened cuticle, a subequatorial vulva, and subcuticular dots in the cuticle.

#### COMMENTS

There are six species of *Meloidodera.* They are intermediate between *Meloidogyne* (in which the female is white) and *Globodera* (in which the cuticle becomes leathery, and eggs are retained in the body). See the section on *Globodera* and Figure 3.14D.

## *Meloidogyne:* Root-Knot Nematodes

### Class Secernentea, order Tylenchida, family Meloidogynidae

#### HOSTS

*Meloidogyne* species (root-knot nematodes) feed on more than 3,000 plant species worldwide, including vegetables, legumes, cereals and grasses, bush and tree fruits, and herbaceous and woody ornamentals (Plates 1N–2F). Some less common species are very host-specific (Table 3.9). Two or more species of *Meloidogyne* are sometimes found in the same host.

#### PARASITISM AND HABITAT

Females of *Meloidogyne* species are sedentary endoparasites in roots of many plants. They induce root enlargements (galls) of various sizes and shapes—irregular and spindle-shaped or spherical (Figure 3.24, Plate 1O and P, and Plate 2A–F and L). The size and form of the galls depends on the nematode species, number of nematodes in the host tissue, host, and plant age. Initially there is little root necrosis. These nematodes cause general unthriftiness, severe stunting, chlorosis, and nutrient deficiencies. Infected plants tend to wilt on hot days (Plates 1N and 2L) and may die in severe infections. Males and preparasitic, second-stage juveniles are migratory and found in soil.

Infection by *Meloidogyne* alters the entire physiology of the infected plant. Cortex and pericycle tissues close to infective juveniles enlarge and divide, and root galls are formed as a result. Second-generation juveniles emerge from eggs at the surface of a gall and commonly reinvade nearby roots. Each juvenile stimulates enlargement of the gall, which in certain hosts may grow to 2.5 cm in diameter in heavy infections. Galls occur singly or coalesce to form typically knotted roots (massive galls).

Infective juveniles also penetrate intercellularly through the endodermis and reach the stele. Up to about seven cells around the nematode's head enlarge and become specialized, thick-walled, multinucleate giant cells, which provide a source of nutrition throughout the nematode's development. The giant cells disintegrate when the nematode dies.

Normal differentiation of xylem and phloem is disturbed in roots infected with *Meloidogyne* spp., and root functions decline. Transport of water and nutrients to the shoot is di-

**Table 3.9.** Characteristics of important species of *Meloidogyne*

| Species[a] | Climate | Geographic distribution | Type host and other important hosts | Length of second-stage juveniles (mm ± 10%) |
|---|---|---|---|---|
| *M. ardenensis* | Cold | England | Myrtle (*Vinca minor*) | 0.412 |
| *M. arenaria** | Warm to tropical | Worldwide | Peanut, wheat, watermelon, many others | 0.470 |
| *M. brevicauda* | Warm | Sri Lanka | Tea-plant (*Camellia sinensis*), a few other species | 0.525 |
| *M. chitwoodi* | Cool | Oregon, Washington | Potato, cereals | 0.376 |
| *M. coffeicola* | Warm | Brazil | Common coffee, a few others | 0.425 |
| *M. exigua* | Warm | Brazil | Common coffee | 0.346 |
| *M. graminicola* | Warm to tropical | Louisiana, southeast Asia, India | Jungle ricegrass (*Echinochloa colona*), rice, sugarcane, oats | 0.449 |
| *M. graminis* | Warm | Louisiana | St. Augustine grass (*Stenotaphrum secundatum*), other grasses | 0.465 |
| *M. hapla** | Cool, temperate | Worldwide | Potato, tomato, carrot, celery, peanut, many others | 0.430 |
| *M. incognita** | Warm to tropical | Worldwide | Carrot, cotton, onion, pepper, tomato, cucurbits, many others | 0.376 |
| *M. indica* | Warm | India | Lime (*Citrus aurantifolia*) | 0.414 |
| *M. javanica** | Warm to tropical | Worldwide | Sugarcane, many others | 0.370 |
| *M. mali* | Cool | Japan | Plum-leaved apple (*Malus prunifolia*) | 0.420 |
| *M. microtyla* | Cool | Ontario | Red fescue, tomato, a few others | 0.375 |
| *M. naasi* | Cold, temperate | Europe, United States | Barley, grasses, other cereals | 0.441 |

[a] Species marked with an asterisk are by far the most widespread.

minished, and plant growth and yield are reduced as a result. *Meloidogyne* spp. cause more damage in sandy soils than in heavier soils.

#### BODY

Adult females are typically soft, pearl white, pear-shaped to saccate or globose, with a short, slender, protruding neck region (0.4–1.3 mm long and 0.6 mm wide), which is usually embedded in root tissue and does not form a cyst (Figure 3.25C and F). The lip region has a weakly developed framework, six minute lips, and three annules. The amphid openings are slitlike. The female cuticle is marked by fine, transverse striae, which in some species are interrupted on the lateral fields by narrow incisures, most often visible near the posterior end. Females have six large, unicellular rectal glands, in the posterior part of the body, which excrete a gelatinous matrix to form an egg sac, in which many eggs are deposited.

The male body is vermiform (1.0–2.0 mm long) and cylindrical, with the neck tapering to a strongly sclerotized head region (Figure 3.25A, B, and E). The lateral and medial lips are prominent, and there is a distinct head cap, which contains the amphidial pouches (Figure 3.26). The lateral fields are marked by four incisures extending from near the middle of the neck to the blunt, rounded terminus. The excretory pore and hemizonid are near the latitude of the nerve ring and anterior to the median bulb, often near the stylet base.

Infective second-stage juveniles are common in soil. They are short (0.3–0.5 mm) and have a weak cephalic framework (Figure 3.25D).

#### STYLET

The male stylet is moderately strong (10–24 μm long), with rounded basal knobs. The female stylet is mostly 9–18 μm long, with three small, prominent, dorsally curved basal knobs (Figure 3.27); it is more slender than the stylet in *Heterodera*. The second-stage juvenile stylet is weak and 10–15 μm long, with small, indistinct basal knobs.

#### ESOPHAGUS

In females, the median bulb is very large and spherical and leads to a short isthmus. The esophageal glands are large, compact, and close to the median bulb; they overlap the anterior end of the intestine and contain three large nuclei. The dorsal esophageal gland duct enlarges into an ampulla just posterior to its junction with the lumen of the esophagus. The excretory canal opens in a pore far forward, sometimes just behind the stylet base. The intestine loses its form in the adult female and is not connected to the rectum.

In males, the median bulb is reduced and is followed by a slender isthmus. The esophageal gland lobe overlaps the intestine ventrally in males and second-stage juveniles.

#### REPRODUCTIVE STRUCTURES

Females have two greatly convoluted, ovaries, which fill most of the swollen body cavity. The uteri join just anterior to the vulva. The spermatheca are pouchlike and have cellular walls. In most species the eggs are deposited in an external gelatinous sac secreted by the rectal glands. The gelatinous matrix of the egg mass turns brown on aging.

Males have one or two testes. The spicules and gubernaculum are near the tail terminus. Intersex males, having a rudimentary or well-developed vulva, are common in some species and may be absent in others.

#### VULVA

The vulva is typically terminal with the anus, flush with or slightly raised from the body contour, and surrounded by cuticular striae, which form a pattern of fine lines resembling human fingerprints, called the perineal pattern (Figures 3.28 and 3.34). Perineal patterns are used for species identification, as described below (pp. 108–109). Certain variations, however, occur among specimens of the same species.

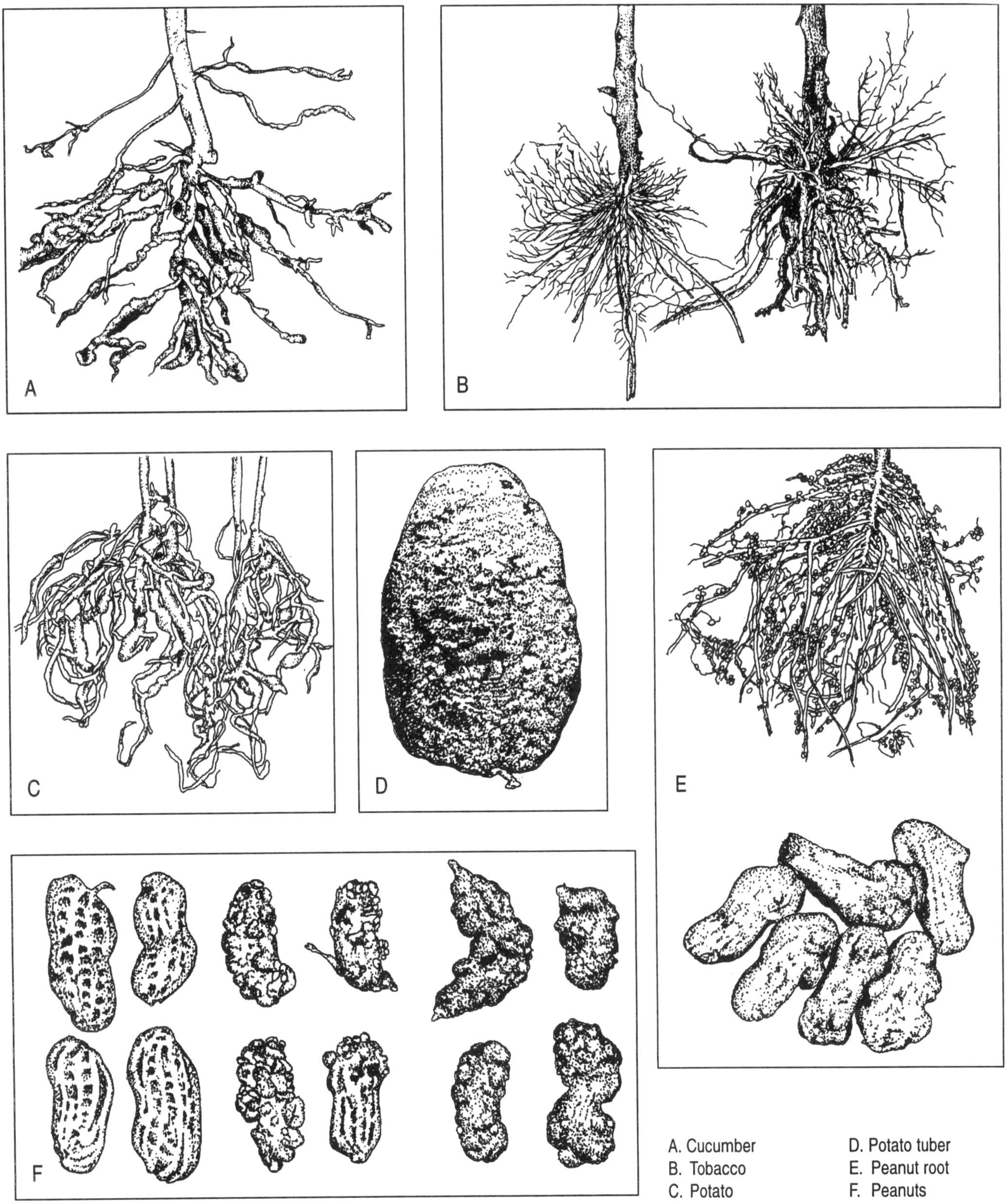


**Figure 3.24.** Damage caused by *Meloidogyne.* Irregular and spindle-shaped galls form on plant roots and other underground organs. **A,** Roots of a young cucumber plant with galls caused by *M. incognita.* **B,** Tobacco roots damaged by a root-lesion nematode (*Pratylenchus*) (left) and roots damaged by *Meloidogyne* (right). **C,** Potato roots with root-knot galls. **D,** Potato tuber with root-knot damage. **E,** Peanut roots with root-knot galls (top) and peanut pods infected with *M. incognita* (bottom). **F,** Healthy peanut pods (the two on the left in each row) and pods with distortion caused by *M. arenaria.*

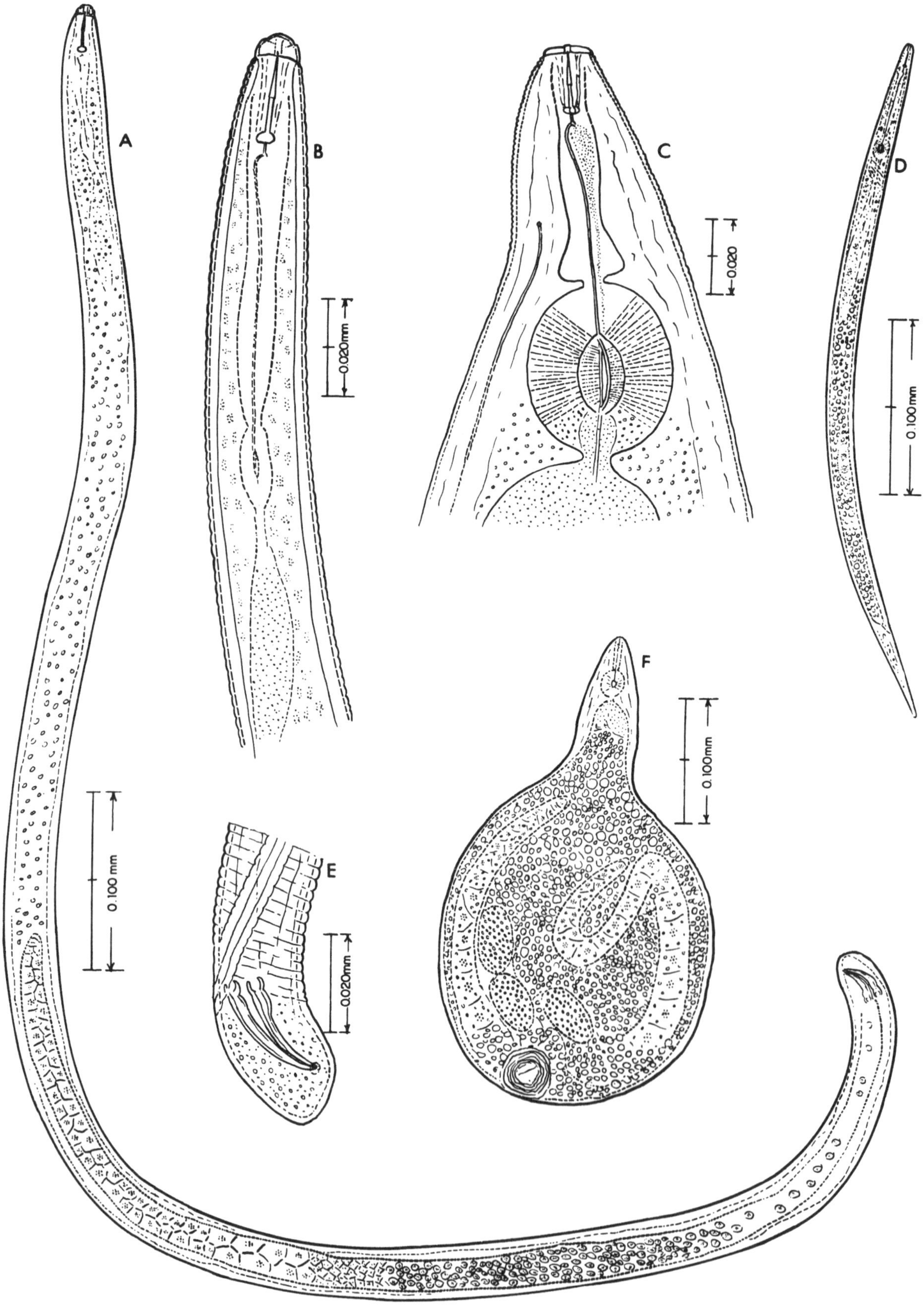


**Figure 3.25.** *Meloidogyne.* **A,** Adult male. **B,** Male, anterior part. **C,** Female, anterior part. **D,** Second-stage juvenile. **E,** Male, posterior part. **F,** Adult female. (Reprinted, by permission, from Taylor, 1971)

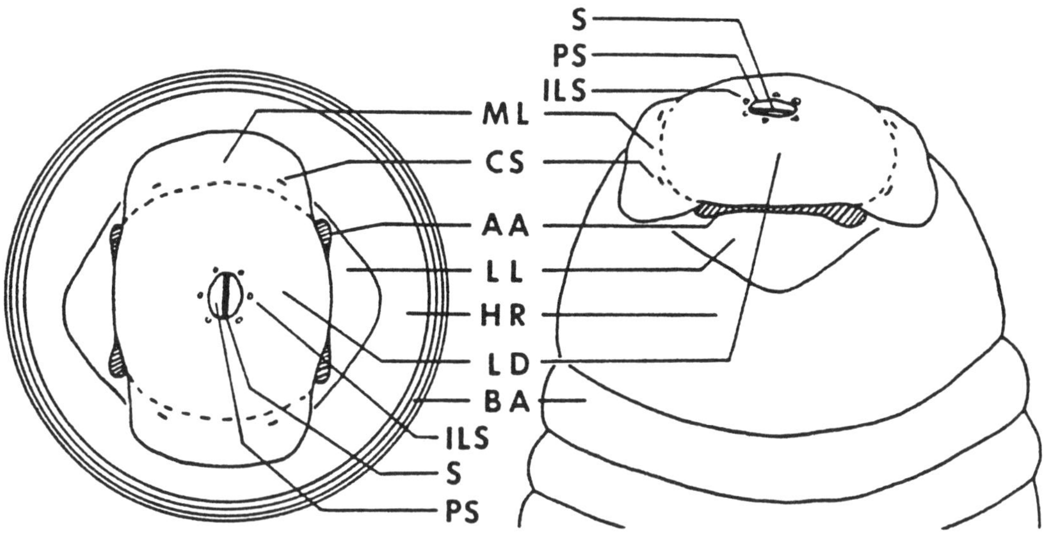


**Figure 3.26.** Generalized morphology of the male head in *Meloidogyne* species (970×), face view (left) and oblique face view (right). AA = amphidial aperture, BA = body annule, CS = cephalic sensillum, HR = head region, ILS = inner labial sensillum, LD = labial disk, LL = lateral lip, ML = medial lip, PS = prestoma, S = stoma. (Courtesy J. D. Eisenback and H. Hirschmann)

#### TAIL

The male tail is very short and bluntly rounded and lacks a bursa. Males twist 180° or more from head to tail. The juvenile tail is conoid, variable in length (5 to 7 anal body widths), often ending in a narrow, rounded terminus in which there is commonly a clear region.

#### GENERA RESEMBLING *MELOIDOGYNE*

*Globodera* and *Heterodera.* Second-stage juveniles of *Meloidogyne* are less robust than *Globodera* and *Heterodera* juveniles but are sometimes confused with them.

*Hypsoperine*
*Meloidodera*
*Nacobbus*
*Punctodera*

#### COMMENTS

Root-knot nematodes are frequently encountered in plant clinics, because of their extensive host range and wide geographic distribution and the conspicuous galls they form on roots. Identification of *Meloidogyne* species is described in the next section.

Infection with root-knot nematodes may reduce the level of bacterial nodulation on legumes, and in some plants it may increase the severity of infection by pathogenic fungi, including *Alternaria, Fusarium, Macrophomina, Pythium, Phytophthora, Rhizoctonia, Sclerotium,* and *Verticillium,* and some pathogenic bacteria, including *Clavibacter michiganensis* subsp. *insidiosus, Pseudomonas marginalis, Ralstonia solanacearum,* and *Agrobacterium tumefaciens.* Infection by *Meloidogyne* may predispose a host to attack by fungi not regarded as major pathogens or not generally associated with it, such as species of *Alternaria, Aspergillus, Botrytis, Curvularia, Penicillium,* and *Trichoderma.*

During migration through soil, juveniles use their stored lipid reserves and eventually die after several months if a suitable host is not located. Soil free of host plants may contain infective juveniles for more than a year.

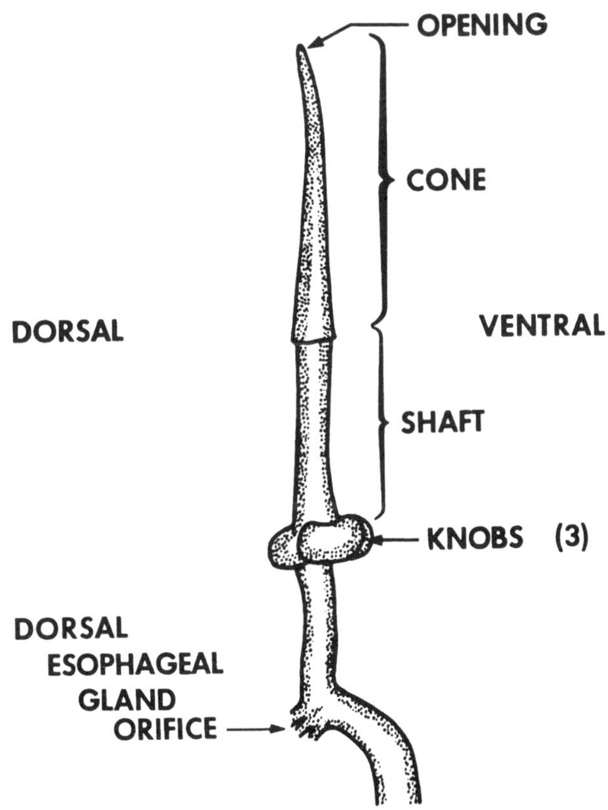


**Figure 3.27.** Morphology of the female stylet in *Meloidogyne* species (970×). (Reprinted, by permission, from Eisenback et al., 1981)

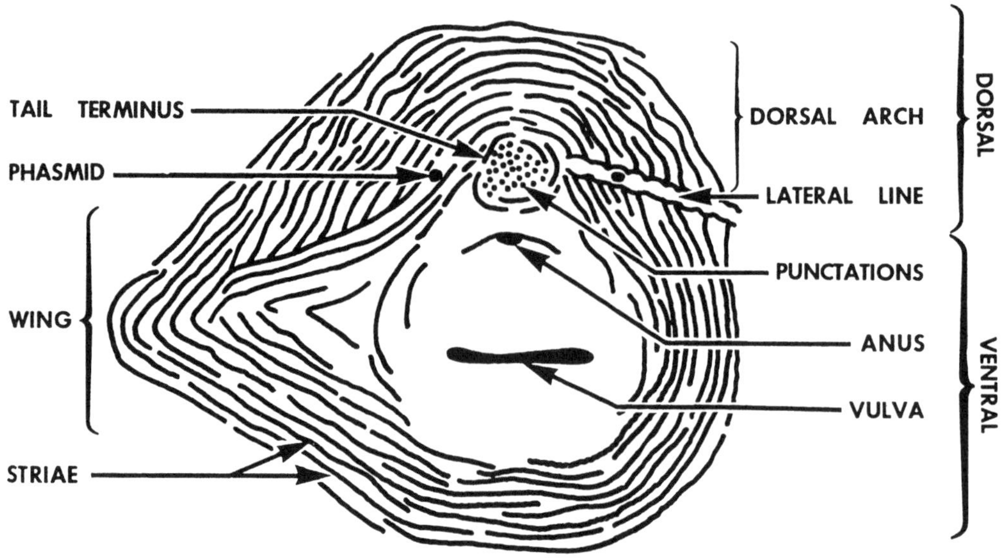


**Figure 3.28.** Generalized morphology of a perineal pattern in *Meloidogyne* (400×). (Reprinted, by permission, from Eisenback et al., 1981)

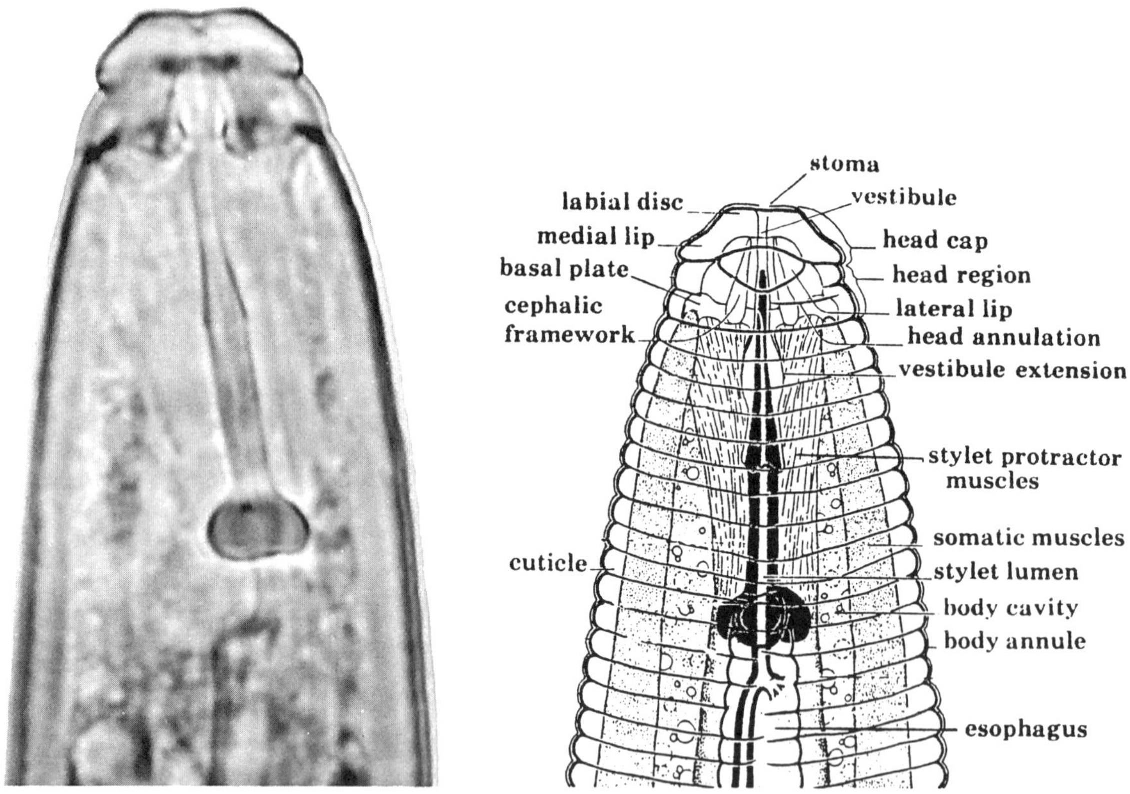


**Figure 3.29.** Basic morphology of the head and stylet of a male *Meloidogyne incognita* (400×). (Reprinted, by permission, from Eisenback et al., 1981)

Penetration usually occurs in the region of root elongation. Once within a root, the juvenile moves intercellularly until it is close to the stele. Here it becomes sedentary, altering the cells on which it feeds.

The life cycle varies from three weeks to several months, depending on the temperature and host. Males are more numerous in heavily infected roots or roots stressed from inadequate nutrition.

Females are very prolific, often producing up to 1,000 or more eggs. Most eggs are deposited in a gelatinous matrix, which usually protrudes from galled root tissue.

#### IDENTIFICATION OF *MELOIDOGYNE* SPECIES

Identification of *Meloidogyne* species, and in some cases even host races (pathotypes), is critical to growers for implementing control strategies based on crop rotation and the

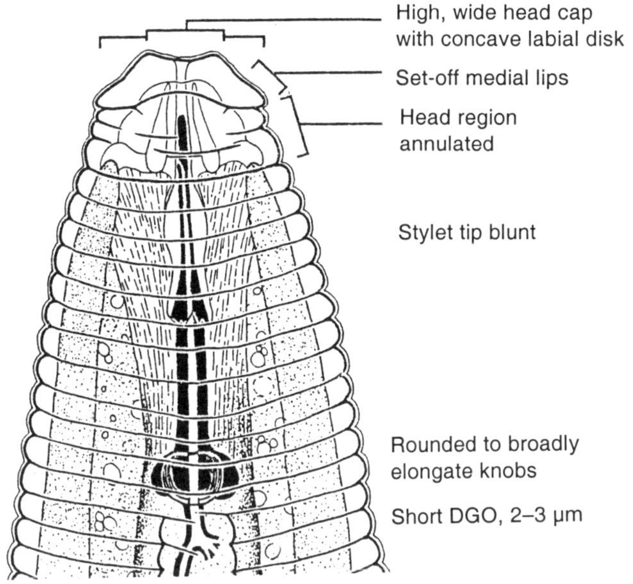


**Figure 3.30.** Anterior region of a male *Meloidogyne incognita.* DGO = dorsal esophageal gland outlet. (Reprinted, by permission, from Eisenback et al., 1981)

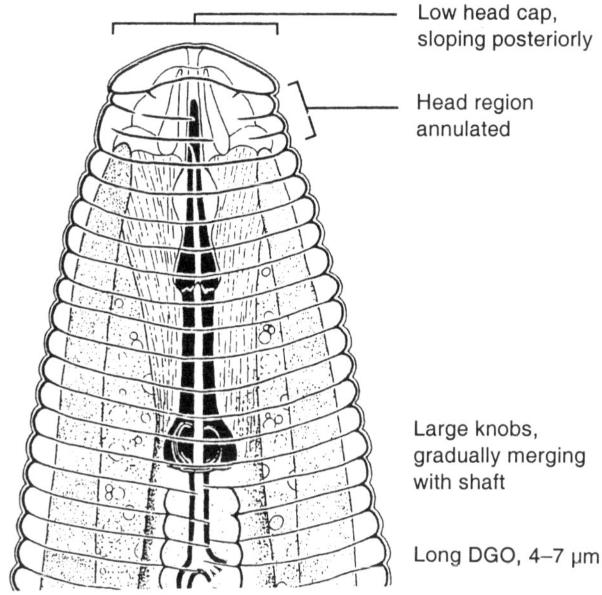


**Figure 3.32.** Anterior region of a male *Meloidogyne arenaria.* DGO = dorsal esophageal gland outlet. (Reprinted, by permission, from Eisenback et al., 1981)

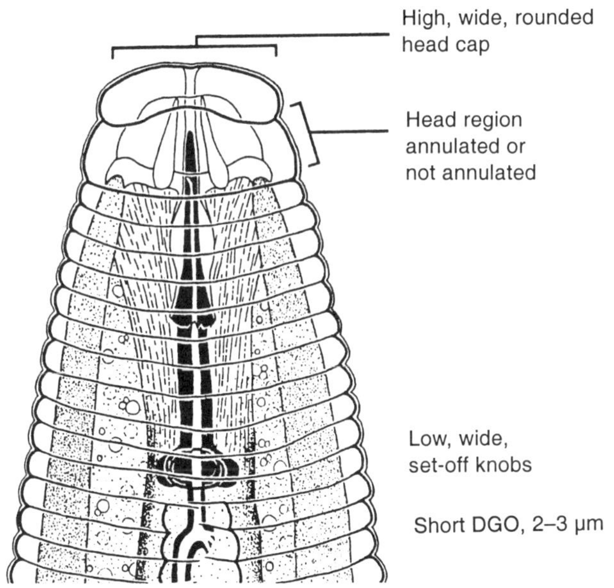


**Figure 3.31.** Anterior region of a male *Meloidogyne javanica.* DGO = dorsal esophageal gland outlet. (Reprinted, by permission, from Eisenback et al., 1981)

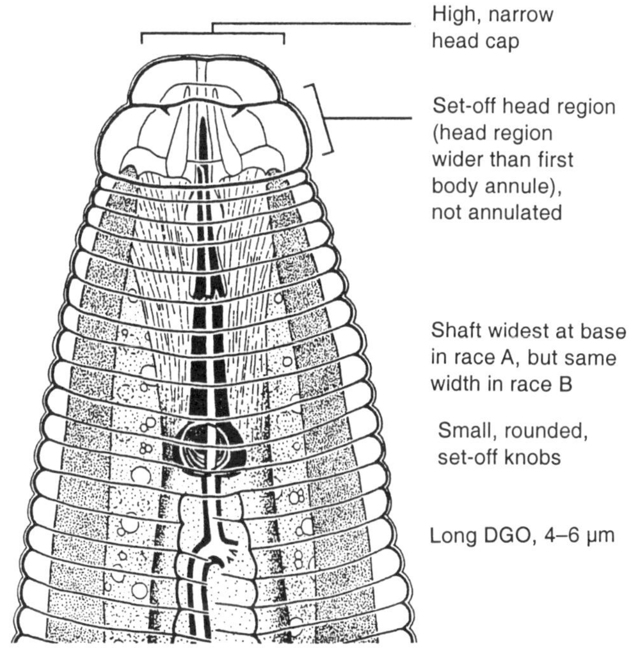


**Figure 3.33.** Anterior region of a male *Meloidogyne hapla.* DGO = dorsal esophageal gland outlet. (Reprinted, by permission, from Eisenback et al., 1981)

use of resistant varieties. Isozyme (esterase and malate dehydrogenase) analysis is commonly used in some clinics to rapidly identify species of root-knot nematodes.

About 60 species of *Meloidogyne* have been named, and new species are still being added. The most important species are listed in Table 3.9. Some described species are rare, and some may be invalid or may have been combined with other species. The less common species occur in limited geographic areas and usually have narrow host ranges; some are restricted to a single host. However, four species are distributed worldwide and have extensive host ranges: *M. arenaria; M. hapla,* the northern root-knot nematode; *M. incognita,* the southern root-knot nematode; and *M. javanica,* the Javanese root-knot nematode. These four species are, by

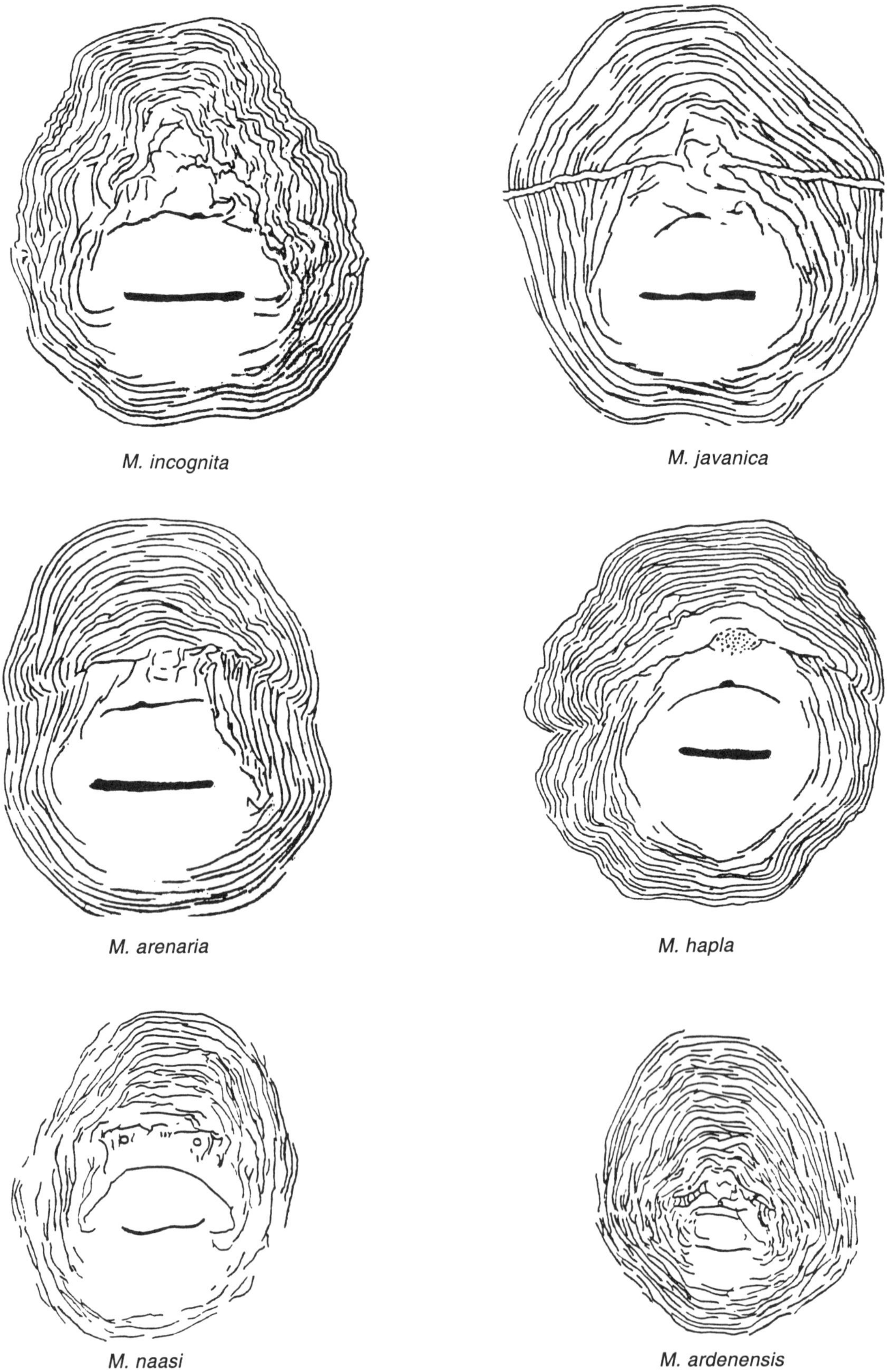


**Figure 3.34.** Perineal patterns of six species of *Meloidogyne.* (Top and middle rows reprinted, by permission, from Eisenback et al., 1981; bottom row reprinted from Franklin, 1965 [left], and from Santos, 1967 [right])

**Table 3.10.** Typical responses[a] of differential hosts to species and races of *Meloidogyne*

| | Differential host[b] | | | | | | |
|---|---|---|---|---|---|---|---|
| | Cotton | Tobacco | Pepper | Watermelon | Peanut | Corn | Tomato |
| *M. arenaria* | | | | | | | |
| Race 1 | – | + | + | + | + | + | + |
| Race 2 | – | + | – | + | – | + | + |
| *M. chitwoodi* | – | – | – | – | – | + | + |
| *M. hapla* | – | + | + | – | + | – | + |
| *M. incognita* | | | | | | | |
| Race 1 | – | – | + | + | – | + | + |
| Race 2 | – | + | + | + | – | + | + |
| Race 3 | + | – | + | + | – | + | + |
| Race 4 | + | + | + | + | – | + | + |
| *M. javanica* | – | + | – | + | – | + | + |

[a] + = Susceptible response; – = resistant response.
[b] Cotton cultivar Deltapine 61, tobacco cultivar NC95, pepper cultivar Early California Wonder, watermelon cultivar Charleston Gray, peanut cultivar Florunner, hybrid field corn, and tomato cultivar Rutgers.

far, the most important and the most common. Approximately 95% of all specimens submitted from around the world to the International Meloidogyne Project (Department of Plant Pathology, North Carolina State University, Raleigh) from 1975 to 1985 were identified as members of these four species.

Identification of *Meloidogyne* species is complicated by their overlapping morphological and physiological characteristics, the diversity of host species in a field, and the ability of more than one root-knot species to infect the roots of a single plant. Nevertheless, the clinician with practice can identify the four important species by using several taxonomically useful criteria. These include geographic location, body measurements, stylet and head morphology (Figures 3.29–3.33), perineal patterns (Figure 3.34), host range, host symptoms, and cytogenetic and biochemical properties (e.g., as determined by isoenzyme analysis). Host range tests and protein separation by electrophoresis are also used to differentiate populations.

Procedures for identifying the four common species of *Meloidogyne* (*M. arenaria, M. hapla, M. incognita,* and *M. javanica*) in the clinic may vary depending on the degree of confidence required, time constraints, and knowledge of species present in the area. Body morphology determination is the quickest. A differential host test requires a month or more to complete. The host, geographic area, symptoms, and type of galls produced give an indication of the species (Table 3.9). After assessing morphological characteristics and effects on differential hosts (Table 3.10), the diagnostician can work through the keys to identify the species (Table 3.11).

The diagnostician must minimize the risk of incorrect species identification in examining samples containing mixed populations of species, individual nematode variability, and overlapping taxonomic characters. This is done by 1) obtaining three or more root samples from individual plants in different areas of the field and keeping them separate, 2) establishing morphological norms for each sample based on averages of five or more individual nematodes, and 3) using three or more taxonomic criteria.

### Morphological Characteristics in the Identification of *Meloidogyne* Species

All species of *Meloidogyne* have similar morphologies (Figure 3.25).

The length of second-stage juveniles (Table 3.9) is often reported but is a poor diagnostic characteristic, because of the variability of individuals; the range of lengths of different individuals in a sample may overlap the average lengths of different species. The best approach is to measure at least 20 juveniles from at least four egg masses in a soil sample (other stages are found in roots) and determine their average length. Relaxed juveniles (suspended in a drop of water on a microscope slide and gently heated with a flame) and fixed juveniles (see pp. 43–44) usually lie in a curve approximating one-sixth of a circle. The true length is estimated by measuring the straight-line distance from the head to the tail terminus and adding 5%.

The head and stylet morphology of males is distinctive. The shape and size of the head and the stylet and the distance from the base of the stylet to the dorsal esophageal gland orifice (DGO) are characteristic features and should be observed at magnifications of 400 and 970× (Figure 3.29). Figures 3.30–3.33 illustrate the head morphology of the four important species of *Meloidogyne.*

***M. incognita.*** The male head is distinctive (Figure 3.30) and is not easily confused with the heads of the other three important species. The labial disk is large and round, often centrally concave, and raised above the medial lips. The medial lips are set off and as wide as the head region, which is usually marked by two or three incomplete annulations.

The stylet tip is blunt and wider than the medial portion of the cone. A projection on the ventral side of the cone marks the opening of the stylet lumen. The distance from the tip of the stylet to the opening is one-fourth the cone length. The shaft is generally cylindrical and may narrow near the basal knobs. The knobs are set off from the shaft, broadly elongate to rounded, and anteriorly indented. The DGO is a short distance (2–3 µm) from the base of the stylet. Males produced from eggs of a single female have the same number of testes, either one or two.

***M. javanica.*** The male head cap is high and almost as wide as the head region (Figure 3.31). The large, smooth labial disk and medial lips are fused. The head region is often not annulated, but some populations may have two or three annulations. Males have three postlabial annules in the lip region and two testes.

The stylet cone is narrow anteriorly and very wide posteriorly. The shaft is cylindrical, often narrowing near the junction with the basal knobs. The stylet knobs are low, wide, and set off from the shaft. The DGO is a short distance (2–3 μm) from the base of stylet.

**Table 3.11.** Key to the four common species of *Meloidogyne*

A. Perineal pattern with high dorsal arch, and without distinct lateral lines (Figure 3.34). Male head with centrally concave labial disk raised above medial lips (head cap) (Figure 3.30). Anterior portion of male stylet blunt and paddle-shaped. Stylet knobs rounded to broadly elongate. Dorsal esophageal gland orifice (DGO) a short distance (2–3 μm) from base of knobs. Reproduction on pepper and watermelon but not on peanut. Root galls usually coalesce to form massive galls. ....... *M. incognita*
   1. No reproduction on cotton:
      - No reproduction on resistant tobacco ....... Race 1
      - Reproduction on resistant tobacco ....... Race 2
   2. Reproduction on cotton:
      - No reproduction on resistant tobacco ....... Race 3
      - Reproduction on resistant tobacco ....... Race 4

B. Perineal pattern of low, rounded arches, with or without distinct lateral line. Male head with fused labial disk and medial lips (head cap) in the same body contour. Male stylet pointed at tip. Reproduction on resistant tobacco but not on cotton:
   1. Perineal pattern with distinct lateral lines (Figure 3.34). Male head high and nearly as wide as head region. Male stylet knobs low and wide (Figure 3.31). DGO a short distance (2–3 μm) from base of knobs. Reproduction on watermelon but not on strawberry, pepper, cotton, or peanut. Root galls similar to those caused by *M. incognita*. ....... *M. javanica*
   2. Perineal pattern with rounded to flattened arches and indistinct lateral lines (Figure 3.34). Male stylet knobs not low and wide. DGO a long distance (4–7 μm) from base of knobs. Reproduction on pepper and resistant tobacco:
      a. Male head cap low, sloping posteriorly, and nearly as wide as head region (Figure 3.32). Perineal pattern with forked dorsal and ventral striae meeting at an angle in lateral field. Dorsal striae slightly indented, forming a shoulder on the arch. Stylet knobs gradually merge with shaft. DGO a long distance (4–7 μm) from base of knobs. Reproduction on pepper, watermelon, and resistant tobacco but not on cotton or strawberry. Root galls often small, numerous, beadlike, and lacking rootlets. Root galls cannot be differentiated with certainty from those produced by *M. incognita* and *M. javanica*. ....... *M. arenaria*
         - Reproduction on peanut ....... Race 1
         - No reproduction on peanut and frequently none on pepper ....... Race 2
   3. Male head cap high, not sloping posteriorly, and not as wide as head region. Stylet knobs rounded and set off from shaft:
      a. Perineal pattern nearly rounded, hexagon to flattened oval, often with punctations in tail terminus (Figure 3.34). Male head usually set off from body annules (Figure 3.33). Stylet narrow and short. Stylet knobs rounded and set off from shaft. DGO a long distance (4–6 μm) from base of knobs. Reproduction on pepper, peanut, and resistant tobacco but not on cotton, okra, grasses, grains, or watermelon. Root galls often small, with many branch rootlets, and root system thick and matted. ....... *M. hapla*
      b. Characteristics different from those described in 3a ....... Other species

Males of *M. javanica* are rare, but intersex males are numerous.

***M. arenaria.*** The male head cap is low, sloping posteriorly, and forming a continuous structure almost as wide as the head region (Figure 3.32). The head region is usually smooth, rarely with one or two incomplete annulations in some populations.

The stylet cone is pointed, and the lumen is marked on the ventral side by a slight protuberance. The posterior part of the cone is much wider than the anterior part of the shaft. The shaft is generally cylindrical, but its diameter may increase somewhat medially. The basal knobs are large and merge gradually with the shaft. The DGO is a long distance (4–7 μm) from the stylet.

***M. hapla.*** The male head is high, and the head cap is much narrower than the head region (Figure 3.33). The head region is not annulated and, with its larger diameter, is generally set off from the first body annule. The body annules near the head region are narrower and less protuberant than those more distal.

The stylet is much shorter and thinner than those of the other three common species. The DGO is a long distance (4–6 μm) from the base of the stylet.

Races 1 and 2 of *M. hapla* can be distinguished by stylet morphology. In race 1, the cone gradually widens posteriorly, and the basal knobs are round and set off from the shaft. In race 2, the stylet is longer (in proportion to the width of the shaft) and the basal knobs are larger than in race 1. The shaft is often indented at the junction with the basal knobs.

No males are present in many populations of *M. hapla,* while in others males are numerous.

## Perineal Patterns

The perineal patterns of *Meloidogyne* species (Figures 3.28 and 3.34) are fingerprint-like patterns of grooves and ridges around the tail terminus of mature females. Perineal patterns can be prepared for observation in less than an hour and are useful in species identification.

***M. incognita.*** The perineal pattern is somewhat oval, with a high, irregular arch made up of closely spaced wavy lines. The interior part of the arch above the anus is marked by numerous zigzag and broken striae, which sometimes form a whorl. Lateral lines are not evident, but the dorsal and ventral striae tend to become forked along the areas where they meet. There are often short transverse striae that extend from the inner striae toward the vulva.

***M. javanica.*** The lateral lines are distinct incisures cutting through the striae of the perineal pattern, with few or no striae extending unbroken from the dorsal to the ventral section. The arch is usually low and rounded. The ventral striae are smooth and usually unbroken.

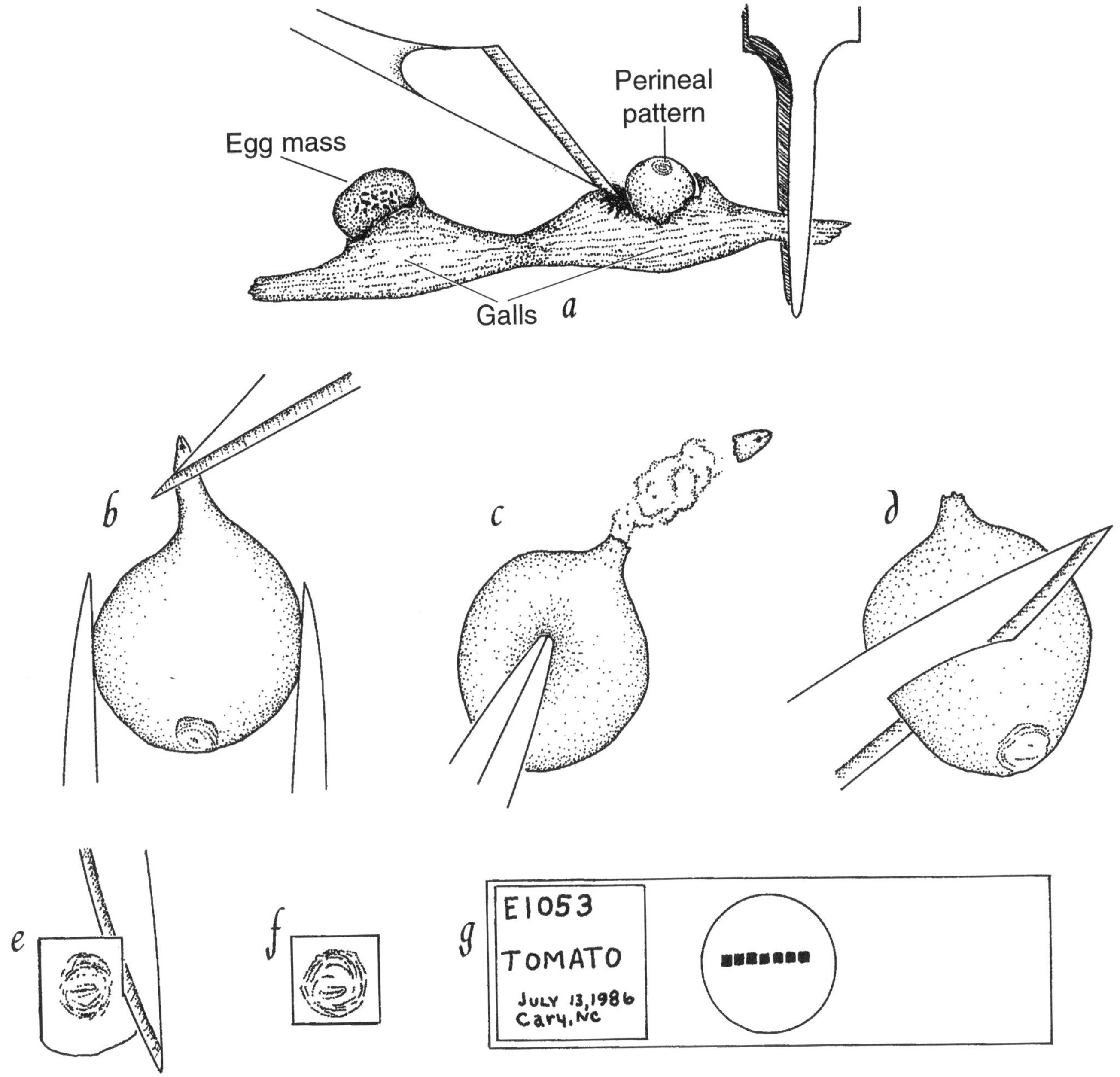


**Figure 3.35.** Preparation of cuticle for examination of perineal patterns in *Meloidogyne.* **A,** An adult female is removed from a root. In this example, two galls are present on the root, and the heads of the nematodes are embedded in the stele of the root. An egg mass is shown on the gall on the left. Note the perineal pattern of the nematode on the right. **B,** The head is removed. **C,** The body contents are pushed out. **D,** The body cuticle is cut in half. **E,** The cuticle is trimmed around the perineal pattern. **F,** Prepared cuticle. **G,** Finished slide on which several cuticles are mounted with the anus oriented downward. A coverslip is sealed to the slide, and the slide is properly labeled.

***M. arenaria.*** Numerous short, disorderly striae are present near the lateral lines, where the dorsal and ventral striae meet. The arch is rather low and rounded, except where the ventral striae extend to form wings. The perineal patterns closely resemble those of *M. hapla* but lack punctations above the anus.

***M. hapla.*** Distinct punctations are usually present just above the anus. The lateral lines may be marked by only slight irregularities in the striae, or the striae of the dorsal and ventral sections may meet at a slight angle along the lines. The arch is low and somewhat rounded. The ventral striae frequently extend laterally to form wings on one or both sides, breaking the symmetry of the perineal pattern. Striae in both sectors are smooth or slightly wavy.

### Preparing Root-Knot Nematodes for Mounting and Observation

Wash roots in running tap water, and select single, mature galls. Place them in a plastic (disposable) petri dish partially filled with tap water. With the aid of a stereoscopic dissecting microscope at a magnification of 20–30×, carefully remove about 20 females from root galls with sharp-pointed tweezers and a small scalpel (Figure 3.35A). Cut off the heads of the females with a sharp scalpel (Figure 3.35B), and gently push out the body contents by pressing with tweezers (Figure 3.35C).

NOTE The nematodes may be placed in a drop of warm 0.5% acid fuchsin in lactophenol to stain the cuticle. Females stained in fresh root material are preferable, as their body contents are more easily removed.

Place the body cuticle (shell) in a drop of 45% lactic acid in a small petri dish, and cover the dish for 30 min to several hours. Cut the cuticle in half equatorially (Figure 3.35D). Trim the posterior cuticle to a square, with the anus near the middle (Figure 3.35E and F). Remove debris from the square cuticle with a small needle (dental pulp canal files are often used), and use the needle to transfer the trimmed and cleaned cuticle to a fresh drop of 45% acetic acid. Similarly, transfer up to 10 or more perineal patterns to a drop of glycerin on a clean glass microscope slide (Figure 3.35G), making sure that the outer surface of the cuticle is facing the objective lens. Arrange the perineal patterns in one or two neat rows, with the anus oriented down. Gently press the cuticles to the slide, and cover them with a coverslip. Supports are optional. For semipermanent mounts, seal the coverslip to the slide with Glyceel, Permount, paraffin, or clear fingernail polish, and label the slide (Figure 3.35G). Finally, examine the perineal patterns under a compound microscope at magnifications of 400 to 970×.

Perineal patterns of six species of *Meloidogyne* are shown in Figure 3.34. Notice that the lateral lines point to the tail terminus and that the anus is between the vulva and the terminus. When viewing a perineal pattern, slowly focus up and down to be sure that the cuticle is not folded.

The shape of the lip region and the position of the excretory pore in mature females are also an aid in identifying species. See Gerber and Taylor (1988) for details of preparation and mounting to show both the anterior end and the perineal pattern of the same specimen.

### Differential Host Response to Infection by *Meloidogyne*

The response of differential hosts to infection by *Meloidogyne* (Table 3.10) is an aid in identifying species and races of root-knot nematodes. An appreciable deviation from the usual reaction of a differential host indicates that the nematode population is different from the norm. For instance, a few populations of *M. javanica* infect and reproduce on pepper and peanut, although most populations of that species do not. Differential host tests are useful for identification of the four common species of *Meloidogyne,* but the species *cannot* be mixed. However, these tests are not entirely reliable, because the nematode population may contain more than one species or race, or the population may be composed of members of a species or race for which there are limited data on host responses. The final identification of a root-knot population should also be based on morphological, cytological, and biochemical characteristics.

To conduct such a test, grow vigorous seedlings of the differential hosts (Table 3.10, note b) in a 1:1 mixture of fertile, sandy loam and coarse sand pasteurized with moist (steam) heat. (To pasteurize soil, moisten it and heat it to 82°C, or 180°F, for 30 min to 1 h.) Collect root samples with well-developed galls, wash them in running tap water, chop them into pieces, and mix the pieces together. Uniformly incorporate 15 to 50 pieces into the soil, and place similar amounts of soil in each pot. If the amount of root gall inoculum is insufficient, it can be increased in the pots and grown for six to eight weeks at 24–32°C (75–90°F). Establish checks (controls) by transplanting seedlings into nematode-free soil. The test should be replicated four or five times. Observe plant growth and symptom development for four to eight weeks, and provide adequate water to maintain soil moisture near field capacity. *Do not overwater the plants,* or root rots may become a problem. Apply a water-soluble complete fertilizer once a week.

Nematode reproduction on a host is indicated by the presence of egg masses. Gently remove the root system of each plant from the wet soil, and wash the roots in a gentle stream of tap water. Float the roots in a shallow layer of water against a dark background, and spread them out so that accurate observations can be made. Viewing is facilitated by the use of a large magnifying glass. Egg masses are enhanced by staining them for 15 min in a 0.0015% solution of phloxine B.

Assign a rating index number to each root system according to the number of galls or egg masses (sacs) present:

0 = no galls or egg masses (sacs)
1 = 1 or 2
2 = 3–10
3 = 11–30
4 = 31–100
5 = more than 100

Record the score for galls and egg masses or sacs in each replicate, and then calculate the average score for each differential host. A host with an average rating index of 2 or less is designated resistant, provided that the tomato rating index is 4 or more.

The precision of the differential host test is enhanced when a known quantity of eggs is used as inoculum, as in the following procedure:

1. Chop the roots into 2- to 5-cm lengths, placing them in 0.5% sodium hypochlorite (a 10% solution of Clorox bleach) in a stainless steel container for 4 min, stirring the solution while they soak. (Stainless steel containers should be used because the eggs tend to adhere to glass and plastic surfaces.)
2. Pour the contents of the container over a 40-mesh sieve to remove the roots and debris. Then pour the solution

through a 500-mesh sieve, and immediately rinse the sieve with water to remove residual bleach.

3. Collect the eggs by rinsing them off the sieve into a beaker.
4. To determine the inoculum density, stir the sample, remove a 1.0-ml aliquot to a counting dish, and count the eggs under a stereoscopic microscope.
5. Inoculate each differential host by uniformly incorporating 5,000 eggs into the soil in each pot.

## *Mesocriconema*

See *Criconemella, Criconemoides,* and *Mesocriconema:* Ring Nematodes.

## *Nacobbus:* False Root-Knot Nematodes

**Class Secernentea, order Tylenchida, family Pratylenchidae**

### HOSTS

*Nacobbus* species have an extensive host range in the Western Hemisphere, including beans (*Phaseolus* spp.), beet (garden beet, sugar beet), broccoli, Brussels sprouts, cabbage, cacti (*Coryphantha, Opuntia*), carrot, cauliflower, Chinese cabbage, cucumber, eggplant, gherkin, kale, kohlrabi, lettuce, mangel, mashua (añu, *Tropaeolum tuberosum*), mustards, oca (*Oxalis tuberosa*), olluco (*Ullucus tuberosus*), pea, pepper (garden or bell pepper, *Capsicum annuum*), potato, pumpkin, radish, rutabaga, salsify, squashes, spinach, sweet potato, Swiss chard, tomato, turnip, and English walnut (Persian walnut).

These nematodes also feed on many weeds, including lamb's-quarters, purslane, and Russian thistle (*Salsola iberica*).

### PARASITISM AND HABITAT

Juveniles and young females are migratory, sedentary endoparasites in roots. Infectious second-stage juveniles enter roots and move intracellularly. They may induce swelling, but this symptom does not always occur. Third- and fourth-stage (preadult) females are mobile, migrate from soil to roots, and reenter large roots, in which they become completely embedded and stimulate the formation of root galls. Adult females induce the formation of spindle-shaped syncytia, up to 3 mm long and 2 mm wide, close to the vascular system, upon which they feed. The body of the adult female swells greatly and develops within a root gall; the tail remains elongate and reaches the root surface at maturity. Eggs are deposited in a gelatinous egg sac. Juveniles and males are found in soil. The life cycle can be completed in 36 days at 25°C (77°F) and 43 days at 20°C (68°F) or 30°C (86°F).

Galled roots bear numerous small rootlets. The galls are characteristically produced in strands in beadlike fashion along the root (Figure 3.36 and Plate 2G). They are formed only in the presence of adult females. Large galls are typical, with four to six females inhabiting each gall. Old galls are masses of dead tissue and usually contain no living females. Shoot growth may be stunted, plants may be chlorotic and tend to wilt at midday, and yield may be reduced

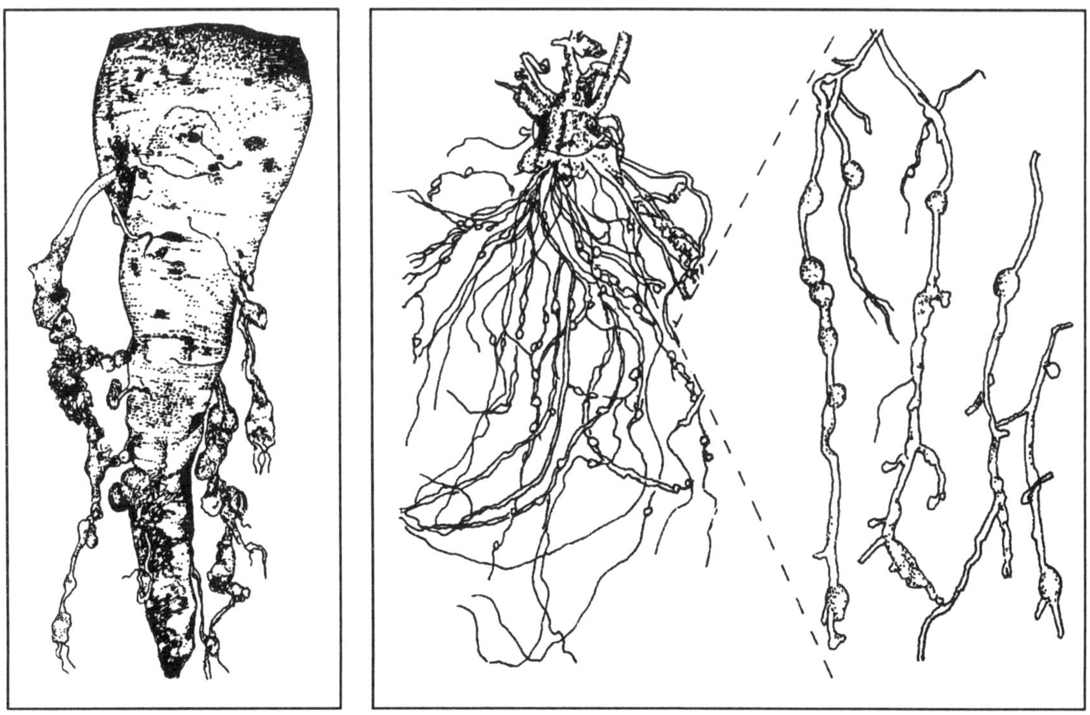

**Figure 3.36.** Sugar beet root (left) and potato roots (right) infected with *Nacobbus aberrans.*

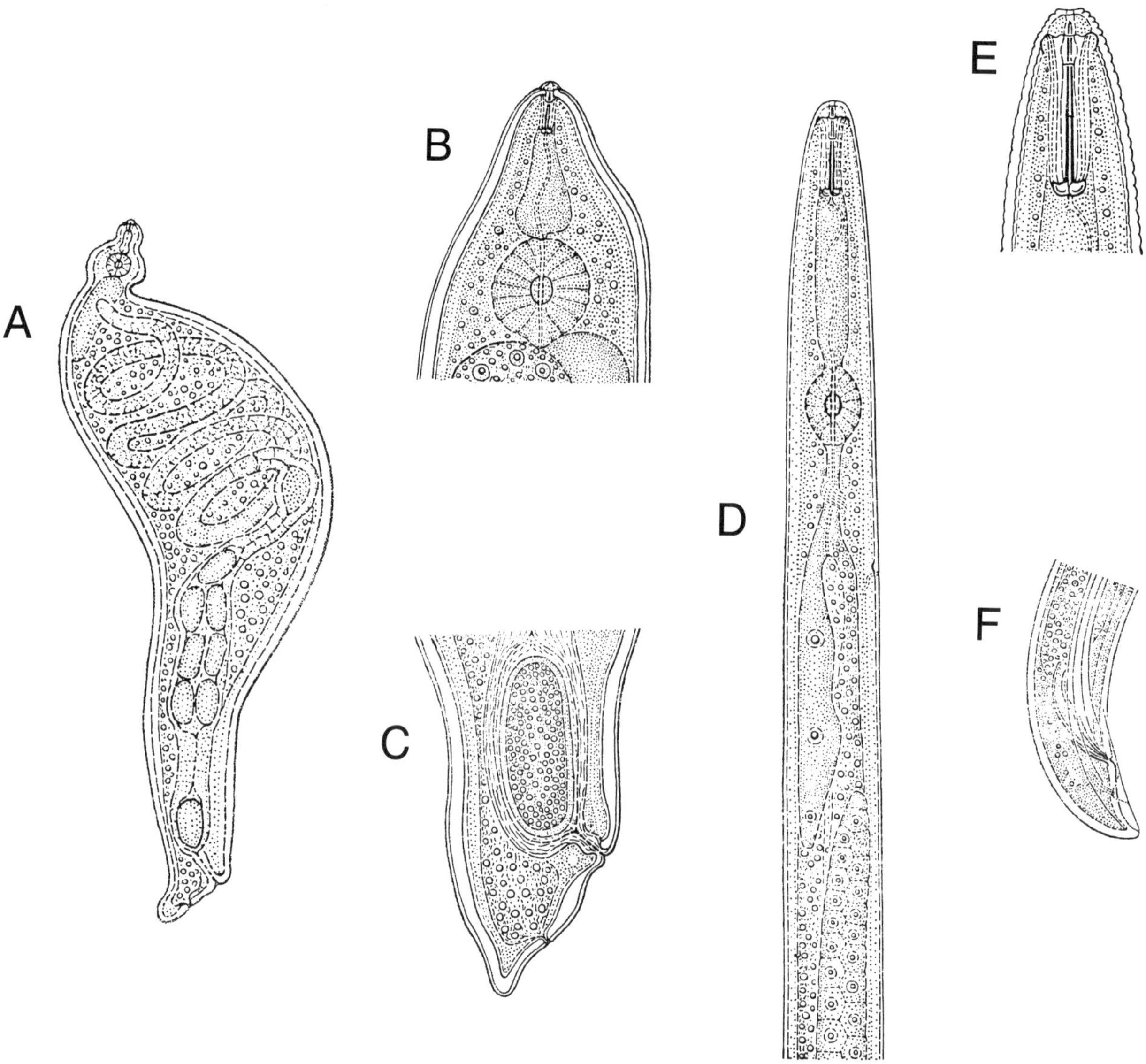


**Figure 3.37.** *Nacobbus.* **A,** Adult female. **B,** Female, anterior part. **C,** Female, posterior part. **D,** Adult male, anterior part. **E,** Male head. **F,** Adult male, posterior part. (Reprinted from Thorne and Schuster, 1956, by permission from *Comparative Parasitology,* formerly *Journal of the Helminthological Society of Washington*)

in some hosts. Infected plants have a lowered ability to withstand drought and other stresses.

The galls resemble those induced by *Meloidogyne* but tend to be more discrete and rounded and often bear numerous secondary rootlets. The starch in galls induced by *Nacobbus,* which stains dark in iodine, readily distinguishes them from those induced by *Meloidogyne.*

### BODY

Adult males and females are markedly different (Figure 3.37). Immature females are 0.6–1.0 mm long. They have a low, rounded lip region with three or four body annules. The lip region is continuous with the body contour. The head skeleton is moderate. The lateral fields have four areolated lines. The adult female is up to 1.6 mm long, much swollen in the middle, with a short, protruding, narrow neck; a narrow, outstretched tail; and offset lips. The female head and neck are sometimes pressed into the anterior end of the body until they are almost indistinguishable. Juveniles, males, and young females are mobile, and all are slender and vermiform. Males are 0.8–1.2 mm long (a = 28–32), their lips are not offset, and they have small caudal alae. The elongate males emerge from roots, searching for females. Several males may be produced in a single egg sac. Second-stage juveniles are about 0.5 mm long (a = 18); their anterior region is similar to that of immature females.

### STYLET

In adult females, the stylet is robust and 15–18 µm long, with small, rounded, basal knobs. In immature females and adult males, the stylet is 20–25 µm long, with strong, rounded basal knobs, which are slightly cupped anteriorly.

#### ESOPHAGUS

In adults, the median esophageal bulb is spheroid, attached to conspicuous radial muscles, and set off by a narrow constriction. The esophageal glands are elongate, and the basal bulb overlaps the intestine dorsally. The intestinal cells are filled with coarse, refractive granules. The excretory pore is located near the basal region of the esophagus, posterior to the nerve ring.

#### REPRODUCTIVE STRUCTURES

Females have a single, anterior ovary. In a mature female, the ovary is highly convoluted and packed with hundreds of embryonic eggs. The thin-shelled eggs (49 × 83 μm) may be retained in the female body but generally are extruded in a gelatinous matrix.

Males have a single, anteriorly outstretched testis. The spicules are curved and 26 μm long. The gubernaculum is a simple trough, about one-fourth as long as the spicules.

#### VULVA

The vulva is a deep transverse slit close to the anus, and both the anus and vulva are near the terminus (V = 90–95). The uterus ends in an ovate or pyriform spermatheca, which is axial and generally filled with sperm.

#### TAIL

The male tail is short (about 1 anal body width), tapering, and enveloped by a small, narrow bursa. The tails of juveniles and immature and mature females are bluntly rounded, with a length of 1 to 1.5 anal body widths.

#### GENERA RESEMBLING *NACOBBUS*

*Meloidogyne.* Adult females of this genus somewhat resemble those of *Nacobbus.*

*Pratylenchus.* Nematodes in this genus resemble *Nacobbus* juveniles.

#### COMMENTS

*Nacobbus* contains only two closely related species, *N. aberrans* (*N. batatiformis*) and *N. dorsalis*, and possibly several races. The species can easily be distinguished by the number of annules between the vulva and the anus of immature, vermiform females. The two species are typically found in light, coarse, sandy soils. Populations decline very rapidly in the absence of a host. Several generations can be produced in a year, if temperature and soil moisture conditions are favorable and a suitable host is present.

Infective second-stage juveniles enter small rootlets and migrate through the tissue, where they molt twice and may or may not induce swelling. They move intracellularly, causing extensive necrosis of the roots, and affected rootlets are severely injured or killed. This life cycle distinguishes *Nacobbus* from other sedentary endoparasites.

*N. aberrans* is a serious problem in sugar beets in the western United States. In South America it causes a serious disease of potato, called *rosario.*

## *Paratrichodorus* and *Trichodorus:* Stubby-Root Nematodes

### Class Adenophorea, order Dorylaimida, family Trichodoridae

#### HOSTS

Species of *Trichodorus* and the closely related genus *Paratrichodorus* (stubby-root nematodes) are polyphagous, feeding on a very large range of plants worldwide, including abacá (*Musa textilis*), *Aeonium,* African violet, alder (*Alnus*), alfalfa, almond, anemone, *Anthurium,* apple, artichoke (*Cynara cardunculus*), ashes, asters, avocado, azaleas, bamboo, banana, barley, beans (*Phaseolus* spp.), beeches, beet (garden beet, sugar beet), bird-of-paradise (*Strelitzia*), bird's-foot trefoil, black pepper (*Piper nigrum*), blackberry, blueberry, boxwood, boysenberry, broccoli, Brussels sprouts, buckwheat, butternut, cabbage, cacao (cocoa, *Theobroma cacao*), camellia, cardamom (*Elettaria cardamomum*), carrot, cauliflower, celery, *Chamaecyparis obtusa,* chayote (*Sechium edule*), cherries, chestnut (*Castanea*), Chinese cabbage, citrus, clove (*Syzygium aromaticum*), clovers (*Trifolium*), corn (maize) (Plate 2I), cotton, cowpea (*Vigna unguiculata*), cranberry, *Cryptomeria,* cucumber, cypress (*Cupressus*), dahlia, date palm (*Phoenix dactylifera*), dill, Douglas fir, eggplant, elder (*Sambucus*), elms, endive, fenugreek (*Trigonella foenum-graecum*), fig, firs, gardenia, garlic, *Gerbera,* gherkin, gladiolus, grape, grasses (forage grasses, turfgrasses, and wild species), hawthorn, hemlock, hop, hyacinth, jasmine, kale, kenaf (*Hibiscus cannabinus*), kohlrabi, larches, lespedezas, lettuce, lilac, lilies, litchi (lychee, *Litchi chinensis*), manzanita, maples, mimosa, mints, mulberry, muskmelon, mustards, *Nandina,* narcissus, nasturtium, oaks, oat, okra, olive, onions (Plate 2H), palms, parsley, pea, peach, peanut, pear, pecan, pepper (garden or bell pepper and ornamentals, *Capsicum annuum*), persimmon, pigeon pea (red gram, *Cajanus cajan*), pineapple, pines, plantain (*Musa* × *paradisiaca*), plums, poplars, potato, privet, quince, radish, raspberry, rhododendrons, rice, rose, rubber-tree (*Hevea brasiliensis*), rutabaga, rye, sorghums, soybean, spinach, spruces, strawberry, sugarcane, sunflower, sweet potato, sweetgum (*Liquidambar*), sycamore, tobacco, tomato, tulips, turnip, hairy vetch (*Vicia villosa*), walnut, wheat (Plate 2J), willows, witloof (*Cichorium intybus* var. *foliosum*), yam (*Dioscorea*), yew (*Taxus*), and many others, including weeds.

#### PARASITISM AND HABITAT

Stubby-root nematodes are migratory ectoparasites on the roots of many plants. They are found in all stages in soil. Both juveniles and adults invade and feed on plants. These nematodes usually feed on epidermal cells just behind the zone of root elongation, moving from cell to cell, feeding for a period rarely exceeding 6 min at a single cell.

Feeding induces a proliferation of short, stunted, stubby, lateral and coarse rootlets and excessive development of

**Figure 3.38.** Damage caused by *Paratrichodorus* and *Trichodorus:* a healthy onion plant and infected plants (left), infected sugar beet roots (center), and corn (maize) roots infected with *P. minor* (right).

upper roots, all of which are parasitized and darkened in turn, resulting in a distinctive appearance called stubby root (Figure 3.38 and Plate 1J). Some species cause root swelling and galling. Large brown lesions may form on roots, mostly near the tips (Figure 3.38, right). Heavily damaged plants may be severely stunted, chlorotic, and wilted and may produce lower yields as a result of the lack of feeder roots (Figure 3.38, left and center), but the disease rarely kills the plants.

Both *Paratrichodorus* and *Trichodorus* occur mostly in sandy or sandy loam soils, often fairly deep, down to 60 cm below the surface. Individuals in both genera are more sensitive to desiccation than most plant-parasitic nematodes. When the soil dries slowly, however, they can survive low moisture.

#### BODY

Adult females are 0.35–1.8 mm long, largely cylindrical, and plump (cigar-shaped) (a = 15–33), tapering at the anterior end, with a thick cuticle that appears smooth when viewed by light microscopy (Figure 3.39). In some specimens, the outer layer of cuticle may appear loose and wrinkled. The more or less rounded head is not offset. The anterior parts of males and juveniles closely resemble those of adult females. In males, ventral papillae are prominent in the posterior region. The amphids are elongate and pocketlike (cup-shaped), with large, postlabial, transverse slit apertures. When killed or moribund, these nematodes are usually straight and cigar-shaped.

In *Trichodorus,* one pair of lateral body pores is located within a distance of 1 body width from the vulva, while *Paratrichodorus* lacks lateral body pores within that distance from the vulva. The excretory pore is usually located just posterior to the lateral cervical pores and the nerve ring. Males are usually abundant in *Trichodorus,* while in *Paratrichodorus* they are present in some species but rare in others. When killed by heat, males of *Trichodorus* curve ventrally in the posterior part of the body, while specimens of *Paratrichodorus* species usually remain straight.

Table 3.12 summarizes the primary diagnostic features distinguishing *Trichodorus* and *Paratrichodorus.*

#### STYLET

The stylet is typically a slim, ventrally curved, solid, elongate "tooth" (onchiostyle), 22–52 μm long, attached to the dorsal esophageal wall. It consists of two parts: an anterior outer spear, which penetrates root hairs and epidermal and subepidermal root cells adjacent to the zone of root elongation, and a posterior inner spear, without basal knobs, to which the protractor muscles are attached. The vestibule is elongated and tubular, serving as a simple guide for the slender spear. A heavily sclerotized guiding ring surrounds the spear about 2 to 3 head widths from the anterior end. Most males have a pair of lateral cervical pores opposite or just posterior to the base of the onchiostyle.

#### ESOPHAGUS

The anterior portion of the esophagus is a narrow cylinder, gradually expanding into typically pear- or spoon-shaped posterior glands, which abut the intestine in *Trichodorus* species or overlap it ventrally in *Paratrichodorus* species. Males of *Trichodorus* have one to four prominent ventral papillae in the onchiostyle region, in addition to the excretory pore.

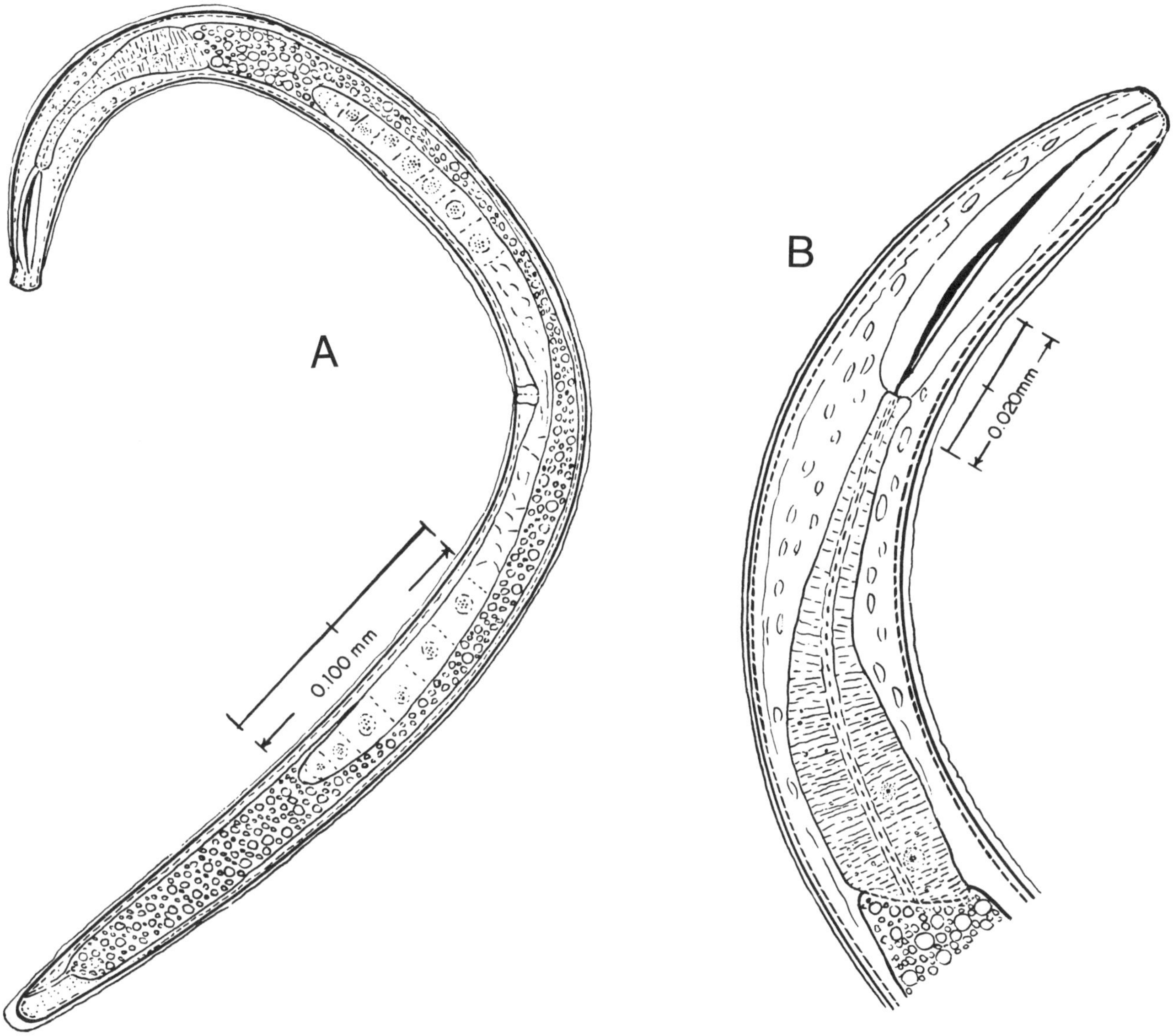


**Figure 3.39.** *Paratrichodorus.* **A,** Adult female. **B,** Female, anterior part. (Reprinted, by permission, from Taylor, 1971)

#### REPRODUCTIVE STRUCTURES

Females are usually amphidelphic, with two reflexed ovaries. The uterus in *Paratrichodorus* is a relatively simple, moderately long tube. *Trichodorus* species have a uterine sac as well as a uterus proper. Most species are bisexual, possessing a spermatheca near the junction of the oviduct and the uterus. The spermatheca is clearly visible when filled with sperm.

Males have a single, anteriorly outstretched testis; massive, paired spicules, which are straight to somewhat ventrally curved; and a long, thin gubernaculum. The copulatory protractor and retractor muscles, which encapsulate the spicule shafts, are well developed in *Trichodorus* but inconspicuous in *Paratrichodorus.* In *Trichodorus* (except *T. cylindricus*), males usually lack a bursa and have three prominent ventromedian papillae opposite and anterior to the spicules. In *Paratrichodorus,* males have a small bursa.

#### VULVA

The vulva is an inconspicuous, narrow, transverse slit near the midpoint of the body (V = 50–59), with strong vaginal sclerotization and musculature in *Trichodorus* and generally weak vaginal sclerotization in *Paratrichodorus.*

#### TAIL

The tail is usually short and bluntly rounded in both females and males. The anus is close to the terminus. The cuticle of the tail has several caudal pores, usually with a number of lateral body pores within the range of the oblique copulatory muscles.

#### GENERA RESEMBLING *TRICHODORUS* AND *PARATRICHODORUS*

*Allotrichodorus* and *Monotrichodorus.* These genera have been described only in South America.

**Table 3.12.** Primary diagnostic features distinguishing *Trichodorus* species from *Paratrichodorus* species[a]

| | *Trichodorus* | *Paratrichodorus* |
|---|---|---|
| Males and females | Body cuticle not swollen in acid fixatives<br>Offset esophageal bulb (in most species) | Body cuticle swollen upon heat-relaxation in acid fixatives<br>Ventral esophageal or dorsal intestinal overlap (in most species) |
| Females | Vagina usually about half the corresponding body width<br>Vaginal sclerotization well developed<br>Lateral pores near vulva (advulvar pores) present<br>Uterine sac present in addition to the uterus proper | Vagina less than half the corresponding body width<br>Vaginal sclerotization poorly developed<br>Lateral pores rare or absent<br>Uterus a uniform sac |
| Males | Tail curved ventrally when fixed<br>Bursa absent, except in *T. cylindricus*<br>Copulatory muscles well developed, extending anteriorly<br>Spicules usually curved (arcuate)<br>One to three ventromedian cervical pores<br>Males usually abundant | Tail straight when fixed<br>Bursa present, usually narrow and weakly developed<br>Copulatory muscles poorly developed, restricted to bursa region<br>Spicules usually straight, with transverse striations<br>One or no cervical pores<br>Males present in some species and rare in others |

[a] Adapted from Decraemer (1995).

#### COMMENTS

There are about 30 species of *Paratrichodorus* and about 35 species of *Trichodorus*. The two genera are distinguished by several differentiating characters (Table 3.12). Major species of *Paratrichodorus* include *P. minor* and *P. pachydermus*. Major species of *Trichodorus* are *T. primitivus, T. similis, T. sparsus,* and *T. viruliferus*. Many of the important plant parasites previously classified in *Trichodorus* have been transferred to *Paratrichodorus*.

To feed, a stubby-root nematode turns its head at right angles to the surface of an epidermal cell and rapidly thrusts its stylet (10 times per second). Some species secrete a feeding tube, which coagulates around the stylet. It is left in position when the nematode withdraws. These nematodes can distinguish between dead and living root cells and can feed on epidermal cells without killing them.

Salivary secretions from *Paratrichodorus* are extremely phytotoxic, and plant injury is much greater than expected from relatively small populations parasitizing the roots.

Eleven species of *Paratrichodorus* and *Trichodorus* transmit tobacco rattle, pea early-browning, and pepper ringspot tobraviruses (see Appendix, Table A.1). These viruses may be retained up to three years by nematodes in fallow soil lacking host roots.

Stubby-root nematodes prefer sandy or sandy loam soils but can occur in large numbers in organic soils. Their vertical distribution depends on the soil pore size, temperature, and host; some stubby-root nematodes may be more than 45 cm deep in soil. The life cycle from egg to egg generally takes about 20 to 45 days.

## *Paratylenchus:* Pin Nematodes

### Class Secernentea, order Tylenchida, family Paratylenchidae

#### HOSTS

Hosts of *Paratylenchus* species (pin nematodes) include *Aconitum,* African violet, alfalfa, *Amaranthus, Annona cherimola* (cherimalla, cherimoya), apple, apricot, arborvitae (*Thuja*), *Artemisia,* aspen, azaleas, bamboo, barley, beans (*Phaseolus* spp.), beeches, beet (garden beet, sugar beet), birches, bird's-foot trefoil, black pepper (*Piper nigrum*), blueberry, boxwood, broccoli, Brussels sprouts, cabbage, cacao (cocoa, *Theobroma cacao*), California buckwheat (*Eriogonum fasciculatum*), camellia, cardamom (*Elettaria cardamomum*), carnation, carrot, cauliflower, celeriac, celery, cherries, chicory (witloof, *Cichorium intybus*), Chinagrass (ramie, *Boehmeria nivea*), chrysanthemum, cinquefoil, cinnamon (*Cinnamomum verum*), citrus, clove (*Syzygium aromaticum*), clovers (*Trifolium*), coffee, coriander (parsley), corn (maize), cotton, cowpea (*Vigna unguiculata*), crabapple, cranberry, cucumber, date palm (*Phoenix dactylifera*), dill, Douglas fir, *Eucalyptus,* fig, ginger (*Zingiber officinale*), grape, grasses (forage grasses, turfgrasses, and wild species), heath (*Erica alba plena*), heather (*Calluna vulgaris*), hemlock, holly, European hop (*Humulus lupulus*), horsechestnut, iris, kale, kohlrabi, larch, laurel, lespedezas, lettuce, linden, loganberry, mango, marsh blue violet (*Viola cucullata*), mints, mulberry, nectarine, nutmeg (*Monodora*), oaks, oat, okra, onions, palms, pansy, parsley, pawpaw, pea, peach, pear, pineapple, pines, pistachio, plums, poplars, potato, radish, rambutan (*Nephelium lappaceum*), raspberry, rhubarb, rice, rose, rutabaga, sedum, sorghums, soybean, spinach, spruces, squash, strawberry, sugarcane, sweet pea, sweet potato, sweetgum (*Liquidambar*), tea-plant (*Camellia sinensis*), tobacco, tomato, turmeric (*Curcuma domestica*), turnip, hairy vetch (*Vicia villosa*), walnut, wheat, willows, yam (*Dioscorea*), and many others.

Weed hosts include *Equisetum* and shadscale (*Atriplex confertifolia*).

#### PARASITISM AND HABITAT

Pin nematodes are very small, cosmopolitan, migratory ectoparasites on roots. Males, females, and juveniles are all found in the soil. Large populations are needed to have much effect on crop production; such populations sometimes develop, especially in fine-textured soils. Young female adults insert their stylets into epidermal cells or the bases of root hairs. Roots may be shriveled, short, and sparse. Severe infection causes a slow, progressive decline in vigor.

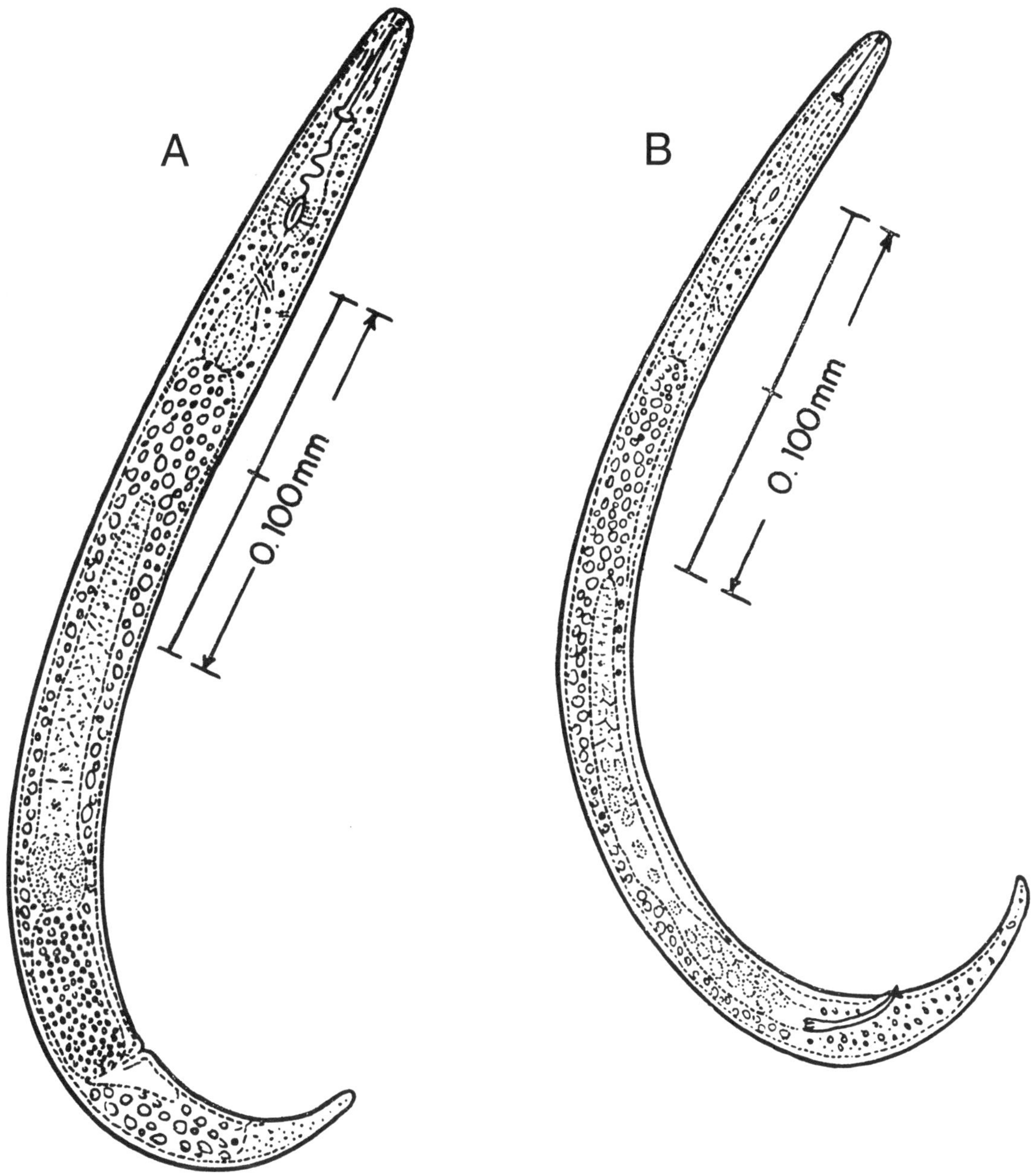


**Figure 3.40.** *Paratylenchus.* **A,** Adult female. **B,** Adult male. (Reprinted, by permission, from Taylor, 1971)

#### BODY

The sexes are separate. The body is short to very short; both males and females are usually less than 0.5 mm long and, in some species, only about 0.2 mm long (a = 11–46, typically 16–24) (Figure 3.40). Females may swell somewhat, are ventrally arcuate, and become sessile. In some species the female body narrows just behind the vulva. The lip region is continuous or offset and shaped like a truncated cone, with a weak cephalic skeleton. The head is conical, rounded, or truncate, usually without annules, and rarely offset. The body is finely and distinctly annulated. The transverse annulations are narrow, with delicate divisions. Many species have lateral fields, usually with four minute incisures. Males are more slender than females, and the spear and esophagus in males are usually faint, poorly defined, or nonexistent. Juveniles resemble adult females in general body shape. However, fourth-stage juveniles may lack a stylet and have a degenerate esophagus. When relaxed by heat or dead, these nematodes commonly assume a C shape.

#### STYLET

In females, the stylet is strong to slender, variable in length (12–40 µm), and not flexible, with well-developed, medially elongated to rounded basal knobs. In males, the stylet is reduced, rudimentary, or absent altogether.

#### ESOPHAGUS

The narrow procorpus merges into a large, rounded or spatulate median bulb with prominent crescentic plates. The isthmus is long and slender, and the esophageal glands are small, but the basal gland is well developed. The esophagus is degenerate in males.

#### REPRODUCTIVE STRUCTURES

Females have a single, anteriorly outstretched ovary. A spermatheca, packed with sperms, is apparently present as a pouchlike branch (diverticulum) of the upper end of the uterus.

Males have a single, anteriorly outstretched testis; long, slender spicules, which are straight to ventrally curved and pointed; and a short gubernaculum.

#### VULVA

The vulva is a prominent transverse slit located in the posterior part of the body (V = 72–90). The vagina stretches halfway or more into the body. The uterus is composed of a large number of cells, sometimes with an open chamber adjoining the vagina.

#### TAIL

The tail curves ventrally and is one-sixth to one-eighth the body length. It is conical to conoid, tapering to a variously shaped terminus. Males lack a bursa.

#### GENERA RESEMBLING *PARATYLENCHUS*

*Cacopaurus.* The cuticle of females in this genus has narrow annules ornamented with minute tubercles. The females are very small (0.2–0.3 µm long) and cylindroid to obese, and they have a very long, flexible stylet (92–102 µm). The vulva is within a body length of the posterior end. *Cacopaurus* males are similar to those of *Paratylenchus* but have a small adanal bursa.

*Gracilacus.* Females in this genus have a flexible stylet, 41–120 µm long, and are much more swollen in the prevulval region than *Paratylenchus* females. *Gracilacus* males and juveniles are similar to those of *Paratylenchus.*

#### COMMENTS

About 100 species of *Paratylenchus* have been described. The major species include *P. bukowinensis, P. dianthus, P. hamatus, P. microdorus,* and *P. projectus.* All are among the smallest plant-parasitic nematodes, with none exceeding 0.5 mm in length. Females of *P. projectus* are 0.3–0.4 mm long (a = 21.5–28.0). Second- and third-stage juveniles feed like adult females. Fourth-stage juveniles lack a complete stylet and can survive in soil without feeding for long periods, requiring exudates from host roots to stimulate the final molt. They may predominate in soil at certain times. The life cycle takes three to four weeks, enabling enormous numbers to build up on host plants.

## *Pratylenchus:* Lesion, Root-Lesion, or Meadow Nematodes

### Class Secernentea, order Tylenchida, family Pratylenchidae

#### HOSTS

*Pratylenchus* species feed on well over 500 hosts worldwide, especially in the tropics and subtropics and in the warmer parts of the temperate zone. Their hosts include abacá (*Musa textilis*), *Aconitum,* adzuki bean (*Vigna angularis*), African violet, *Albizia falcata,* alfalfa, almond, *Alstroemeria,* amaryllis, *Anchusa,* anemone, *Annona cherimola* (cherimalla, cherimoya), *Anthurium,* apple, apricot, arborvitae, arracacha (*Arracacia xanthorrhiza*), arrowroot (*Maranta arundinacea*), ashes, asparagus, *Astilbe,* avocado, bamboo, banana, barberry, barley, beans (*Phaseolus* spp.), beet (garden beet, sugar beet), begonia, birches, bird's-foot trefoil, black gram (urd, *Vigna mungo*), black pepper (*Piper nigrum*), blackberry, blueberry, boxwood (Plate 2K), broad bean (faba bean, *Vicia faba*), broccoli, Brussels sprouts, cabbage, cacao (cocoa, *Theobroma cacao*), cacti, *Caladium,* camellia, camphor-tree (*Cinnamomum camphora*), candytuft (*Iberis*), caraway (*Carum carvi*), cardamom (*Elettaria cardamomum*), carnation, carrot, cassava (*Manihot esculenta*), castor bean (*Ricinus communis*), cauliflower, *Cedrus,* celery, *Chamaecyparis, Chamaedorea,* cherries, chickpea (*Cicer arietinum*), chicory (*Cichorium intybus* var. *sativum*), Chinese evergreen (*Aglaonema*), Christmas-rose (*Helleborus niger*), chrysanthemum, cinquefoil, citrus, *Clematis,* clove (*Syzygium aromaticum*), clovers (*Trifolium*), coconut palm (*Cocos nucifera*), coffee, coleus, *Cordyline,* coriander (parsley), corn (maize) (Plate 2O), cotoneaster, cotton, cowpea (*Vigna unguiculata*), cranberry, crocus, *Crotalaria,* crotons, *Cryptomeria, Ctenanthe oppenheimiana,* cucumber, currants, *Cyclamen,* cypress (*Cupressus*), dahlia, date palm (*Phoenix dactylifera*), *Delphinium, Dicentra, Digitalis, Doronicum,* Douglas fir, *Dracaena,* eggplant, endive, *Eucalyptus,* fennel (*Foeniculum vulgare*), fenugreek (*Trigonella foenum-graecum*), ferns, figs (*Ficus* spp.), European filbert (*Corylus avellana*), firs, flax, forsythia, gardenia, garlic, gentian, geranium (*Pelargonium*), gherkin, ginger (*Zingiber officinale*), ginkgo, gladiolus, goldenrods (*Solidago*), gourds, grape, grasses (forage grasses, turfgrasses, and wild species), guava (*Psidium*), hawthorns, heath (*Erica alba plena*), *Hebe, Hippeastrum,* European hop (*Humulus lupulus*), horsechestnut, hyacinth, hyacinth bean (*Lablab purpureus*), *Hydrangea,* iris, Irish shamrock (*Oxalis acetosella*), jack bean (*Canavalia ensiformis*), junipers, kale, kava (yaqona, *Piper methysticum*), kenaf (*Hibiscus cannabinus*), kohlrabi, *Lantana,* larches, larkspur, leek, lespedezas, lettuce, *Leucaena, Ligularia wilsoniana,* lilac, lilies, lily-of-the-valley, lupines, magnolia, mahogany, mango, manzanita, maples, marguerite (*Chrysanthemum frutescens*), marigold, millets, mints, *Monstera,* mung bean (*Vigna radiata*), muskmelon, narcissus, nectarine, oaks, oat,

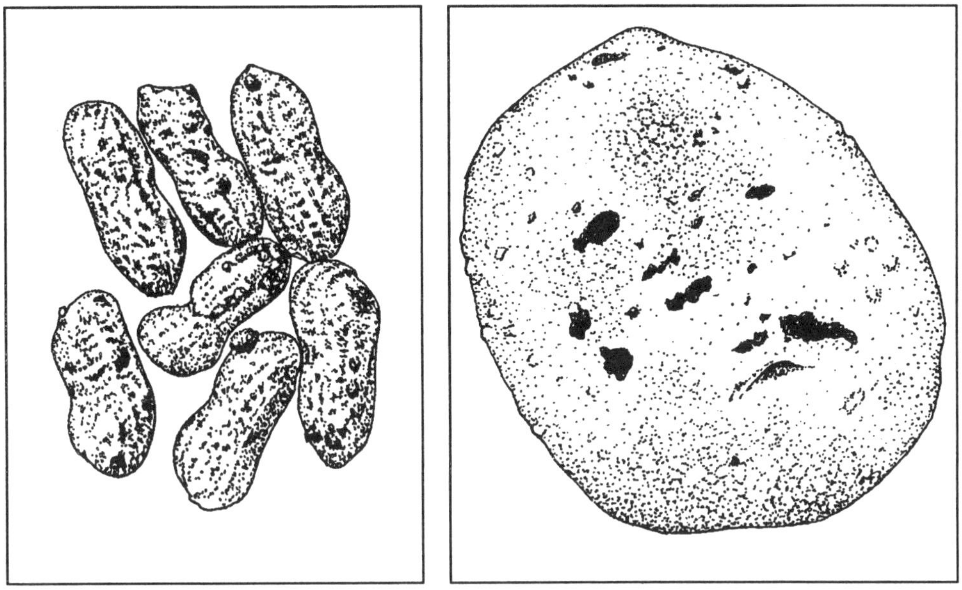

**Figure 3.41.** Peanut pods with lesions caused by *Pratylenchus brachyurus* (left) and a potato tuber infected with *P. crenatus* (right).

okra, olive, onions, orchids, pak choi (*Brassica chinensis*), palms, papaya, parsley, parsnip, pawpaw, pea, peach, peanut (Figure 3.41 and Plate 2N), pear, pecan, peperomia, pepper (garden or bell pepper and ornamentals, *Capsicum*), *Philodendron,* phlox, pigeon pea (red gram, *Cajanus cajan*), pineapple, pines, pistachio, plantain (*Musa* × *paradisiaca*), plums, poppy, potato (Figure 3.41 and Plate 2M), *Potentilla,* primrose, privet, pumpkin, quince, radish, *Ranunculus,* rape (colza, *Brassica napus*), raspberry, rice, jungle ricegrass (*Echinochloa colona*), roses, rubber-tree (*Hevea brasiliensis*), *Rudbeckia,* rutabaga, rye, sapodilla (*Manilkara zapota*), *Schefflera,* shallot, smokebush (smoke-tree, *Cotinus*), snapdragon (*Antirrhinum*), *Sorbus,* sorghums, soybean, spinach, spruces, strawberry, Sudangrass, sugarcane, sunflower, swamp taro (*Cyrtosperma chamissonis*), sweet pea, sweet potato, sweet William, sweetgum (*Liquidambar*), sycamore, *Syngonium,* taro (*Colocasia esculenta*), tarragon (*Artemisia dracunculus*), tea-plant (*Camellia sinensis*), *Thalictrum,* tobacco, tomato, tree-of-heaven (*Ailanthus altissima*), *Trollius,* tulips, tuliptree (yellow-poplar, *Liriodendron*), turmeric (*Curcuma domestica*), turnip, vetches (*Vicia* spp., including hairy vetch, *V. villosa*), *Viola,* violets, walnut, watermelon, *Weigela,* wheat, willows, witloof (*Cichorium intybus* var. *foliosum*), yams (*Dioscorea*), yew (*Taxus*), and many others.

Weed hosts include ball nightshade (*Solanum carolinense*), burdock (*Arctium lappa*), dandelion, and plantain (*Plantago*).

#### PARASITISM AND HABITAT

*Pratylenchus* species are migratory endoparasites that feed in the root cortex (and sometimes in tubers, peanut shells, etc.), migrating intracellularly and killing cells as they feed and deposit eggs. Some damage to the stele may occur, and delicate, succulent roots may be entirely killed. Severely affected plants may produce shoots with chlorotic leaves, and the plants are stunted (Figure 1.3G) and gradually decline or lack vigor. Twig dieback, wilting, and a reduction in flower and fruit production can occur. Such plants may be predisposed to winter injury or other infectious diseases. Severely infected plants are easily pulled from the soil. Plants may be killed in heavy infestations.

Synergistic interaction with root-rotting plant pathogens in soil causes narrow, elongated, light buff to dark or reddish brown to black lesions (Figure 1.3G, center, and Plate 2O), which may girdle roots, resulting in severe root rot. Lesions are also formed on other underground organs (Figure 3.41 and Plate 2M and N).

These nematodes are found in all stages in roots and soil. Males are common in some species and rare or absent in others. Eggs, which give rise to second-stage juveniles, are deposited in clusters within roots or in soil. When conditions are unfavorable, the nematodes move to new locations within the root or leave the plant. Soil populations decline as nematodes invade new roots in late spring or early summer in temperate regions. They return to the soil in late summer and early autumn, when roots decline. On some hosts there is a substantial buildup of nematode numbers without obvious damage to roots; in others, the buildup is slight, with no apparent damage; in still others, there is marked root damage. Cell death always follows extended feeding by certain *Pratylenchus* species (e.g., *P. penetrans*).

#### BODY

The body is small (0.3–1.0 mm long), vermiform, and fairly robust (a = 19–40). The head region is low and broadly flattened. The lip region is slightly offset and bears one to four annules. The cephalic framework is strong, re-

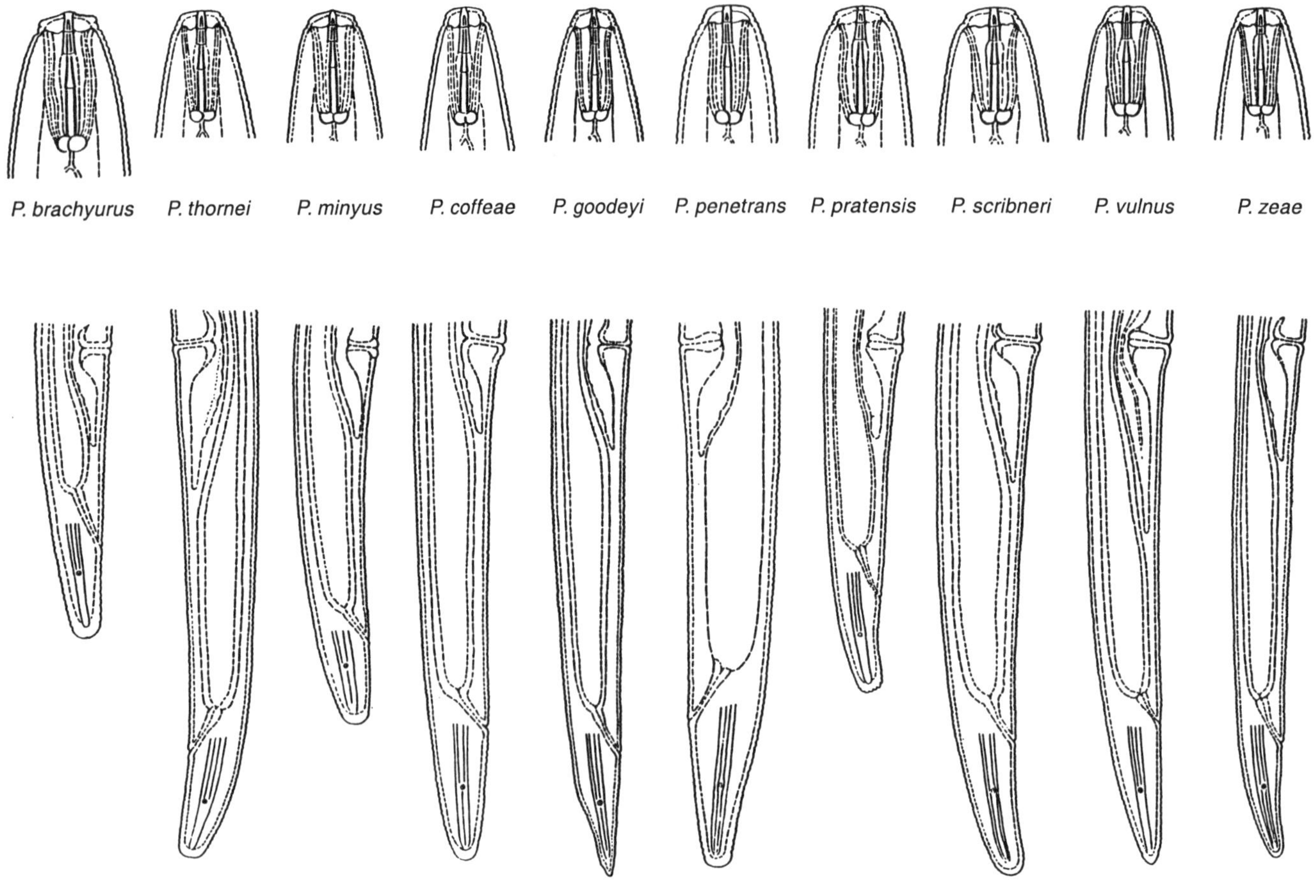


**Figure 3.42.** Head and tail regions of 10 species of *Pratylenchus.* (Reprinted, by permission, from Thorne, 1961)

fractive, and continuous with the body contour (Figures 3.42 and 3.43). The amphids open in the middle of the lateral lips. The cuticular annulations are fine. The lateral fields contain four to six incisures and extend from the level of the median bulb to the tail. In adult females, incisures are frequently absent, because of the stretching of the cuticle. The deirids are located near the latitude of the excretory pore; they are rarely visible. The intestine is packed with numerous dark granules. Males, where present, are smaller than females; they are rare in some species. These nematodes are slightly curved ventrally when relaxed by heat.

#### STYLET

The stylet is typically short (13–20 μm) and thick, with massive, rounded or anteriorly concave basal knobs (Figures 3.42 and 3.43).

#### ESOPHAGUS

The median bulb is spherical to ovoid, more than half as wide as the neck, and heavily muscular, with a conspicuous valve (Figure 3.43). The esophageal lumen and intestine are joined by an obscure muscular valve. The esophageal gland lobes overlap the intestine, usually in a lateroventral position. There are three prominent esophageal gland nuclei. The excretory pore is prominent and located about opposite the nerve ring or opens close to the level of the esophago-intestinal junction. The hemizonid is slightly anterior to the excretory pore.

#### REPRODUCTIVE STRUCTURES

Females have a single, anteriorly outstretched ovary, with oocytes usually in a single or partial double row. The spermatheca are oval or round, usually filled with sperms, and conspicuous in bisexual species. The postuterine sac is small, with a length of typically about 1 body width.

Males have a single, outstretched, anteriorly directed testis, with spermatocytes irregularly arranged, especially in the region of multiplication. The spicules are paired, heavily sclerotized, slender, and slightly to moderately curved (arcuate). They rest on a thin, troughlike gubernaculum.

#### VULVA

The vulva is an adpressed transverse slit in the posterior fourth of the body (V = 65–89). The vagina extends inward and slightly forward.

#### TAIL

The female tail is tapered, with a length of 2 to 3 anal body widths, and varies from broadly rounded to pointed (Figure 3.42). The rectum is slender and muscular and ends in a transverse, slitlike anus. The male tail is short and dorsally convex-conoid, with a distinct, bluntly rounded bursa,

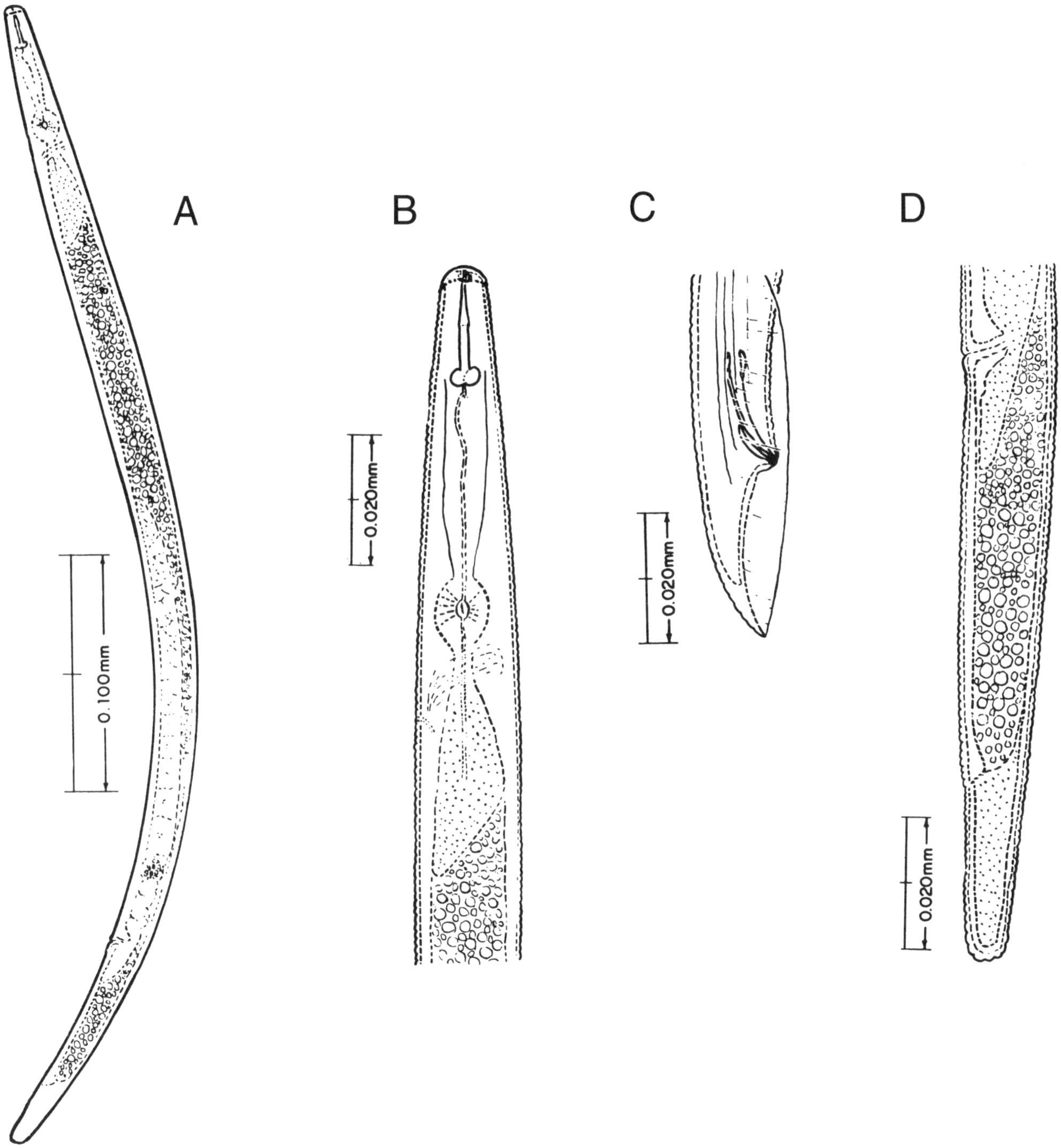


**Figure 3.43.** *Pratylenchus.* **A,** Female, full length. **B,** Male, anterior part. **C,** Male, posterior part. **D,** Female, posterior part. (Reprinted, by permission, from Taylor, 1971)

which extends anteriorly from the tip of the tail to the distal end of the spicules. The female phasmids are located about one-third of the tail length or more behind the anus. The male phasmids are located near the base of the bursa.

#### GENUS RESEMBLING *PRATYLENCHUS*

*Radopholus.* Nematodes in this genus have two ovaries and dorsally overlapping glands.

#### COMMENTS

There are more than 66 species of *Pratylenchus,* most of which have a wide host range within certain groups of plants. The major species include *P. brachyurus, P. coffeae, P. crenatus, P. fallax, P. goodeyi, P. hexincisus, P. loosi, P. neglectus, P. penetrans* (which has over 400 hosts), *P. scribneri, P. thornei, P. vulnus,* and *P. zeae.* These nematodes are active over a wide temperature range (5–30°C, or 41–86°F).

Polyspecific populations are common in the field. Geographic variants also occur among species.

These nematodes prefer coarse-textured, sandy soils. In temperate climates, they overwinter in roots or soil as eggs, juveniles, or adults. During periods of drought, they are quiescent and susceptible to drying if the soil is plowed and fallowed through the hot months. Where soil moisture is low, some species can survive more than a year without a host crop. Species of *Pratylenchus* are unusual in reproducing in larger numbers in stressed plants. In all motile stages, these nematodes can invade and leave roots, so there is no infective stage.

To estimate the size of populations of *Pratylenchus* species, both roots and soil should be sampled. Staining of roots suspected of harboring these nematodes or extraction from such roots may be necessary to confirm their presence.

Fungi and bacteria of many different genera invade root lesions caused by *Pratylenchus* species. The nematodes commonly enhance the growth of these fungi and bacteria, so that damage to the plant is more severe, as in the wilting of peppermint and potato by the action of *Pratylenchus* and *Verticillium;* the wilting of tomato, pepper, and eggplant caused by the nematode and *Ralstonia solanacearum, Pseudomonas fluorescens, Sclerotium rolfsii, Rhizoctonia solani, Verticillium dahliae* (*V. albo-atrum*), and *Fusarium* species; hairy root of roses, caused by *Agrobacterium rhizogenes;* and rot of peanuts caused by *S. rolfsii.* In potato, Verticillium wilt and *Rhizoctonia* disease increase when large numbers of these nematodes are present in the roots. Many *Pratylenchus* species also reduce populations of vesicular-arbuscular mycorrhizal fungi (e.g., *Glomus fasciculatum*).

*P. penetrans* and other plant-parasitic nematodes are a major part of the pome and stone fruit tree replant problem, along with various pathogenic fungi and possibly bacteria and nonparasitic microorganisms, allelopathic compounds from old tree roots, unbalanced soil nutrition, disturbed soil structure, cold injury, and other factors.

## *Radopholus:* Burrowing Nematodes

### Class Secernentea, order Tylenchida, family Pratylenchidae

#### HOSTS

*Radopholus* species (burrowing nematodes) have over 600 hosts in many parts of the world, especially in the subtropics and tropics, including abacá (*Musa textilis*), African spinach, alfalfa, *Anthurium,* apricot, areca nut (betel nut, *Areca catechu*), artichoke, avocado, azaleas, bamboo, banana, barley, beans (*Phaseolus* spp.), beet (garden beet, sugar beet), betel vine (*Piper betel*), bird-of-paradise (*Strelitzia reginae*), black pepper (*Piper nigrum*), boxwood, Brazil peppertree (*Schinus terebinthifolius*), broad bean (faba bean, *Vicia faba*), butterfly ginger (ginger-lily, *Hedychium coronarium*), cabbage, cacao (cocoa, *Theobroma cacao*), calabash (*Crescentia cujete*), *Calathea,* canna, cantaloupe, cardamom (*Elettaria cardamomum*), carrot, castor bean (*Ricinus communis*), *Chamaedorea,* cherries, citrus (Plate 2P), clovers (*Trifolium*), cluster bean (*Cyamopsis tetragonoloba*), coconut palm (*Cocos nucifera*), coffee, corn (maize), cowpea (*Vigna unguiculata*), cucumber, *Dieffenbachia, Dizygotheca,* eggplant, English ivy (*Hedera helix*), *Epipremnum,* fig, gardenia, ginger (*Zingiber officinale*), grasses (forage grasses, turfgrasses, and wild species), guava (*Psidium*), *Hibiscus,* holly, *Jacaranda,* jasmine, kale, kava (yaqona, *Piper methysticum*), lettuce, loquat, lupines, magnolia, mango, *Maranta, Monstera,* muskmelon, *Nephthytis,* nutmeg (*Monodora*), okra, oleander, onions, palms, pea, peach, peanut, pear, pecan, pepper (garden or bell pepper and ornamentals, *Capsicum annuum*), periwinkle, persimmon, philodendrons, pigeon pea (red gram, *Cajanus cajan*), pineapple, pines, plantain (*Musa* × *paradisiaca*), plums, potato, pothos, pumpkin, radish, rice, rice bean (*Vigna umbellata*), sorghums, soybean, squashes, strawberry, *Strelitzia,* sugarcane, sweet potato, sword bean (*Canavalia gladiata*), *Syngonium,* tamarind (*Tamarindus indica*), taro (*Colocasia esculenta*), tea-plant (*Camellia sinensis*), tobacco, tomato, turmeric (*Curcuma domestica*), turnip, vanilla, watermelon, and many others, including weeds.

#### PARASITISM AND HABITAT

*Radopholus* species are migratory, polyphagous endoparasites in roots, tubers, and corms, usually in the cortex. Females lay eggs that hatch and form new populations along the migratory path. These nematodes are found in all stages in fibrous roots and in soil around roots. Adult females remain within roots until the tissues are badly damaged. They cause dark reddish brown to black lesions and extensive necrosis throughout the cortex. Cavities form and enlarge but do not cross the endodermis. The root system is restricted and may be reduced to a few short stubs. Other common symptoms are wilting and dieback of shoots.

Necrotic areas are readily invaded by bacteria and fungi, especially *Cylindrocarpon, Cylindrocladium,* and pathogenic species of *Fusarium* and *Sclerotium,* which cause increased necrosis and hasten root death.

These nematodes are a serious problem in banana plantations, because they destroy anchor roots, causing unthrifty growth, and plants can topple as a result, especially with a heavy load of fruit.

*R. citrophilus* causes spreading decline of citrus. Affected trees wilt rapidly during periods of low soil moisture (Plate 2P). The nematode burrows through the root cortex and invades the cambium and phloem in the stele. Wound gum forms in the cortex, and small lesions form later, filled with nematodes. The lesion surface cracks. Large numbers of nematodes (200 or more) can be found in a single lesion. Symptoms in citrus roots are most evident 25 to 30 cm be-

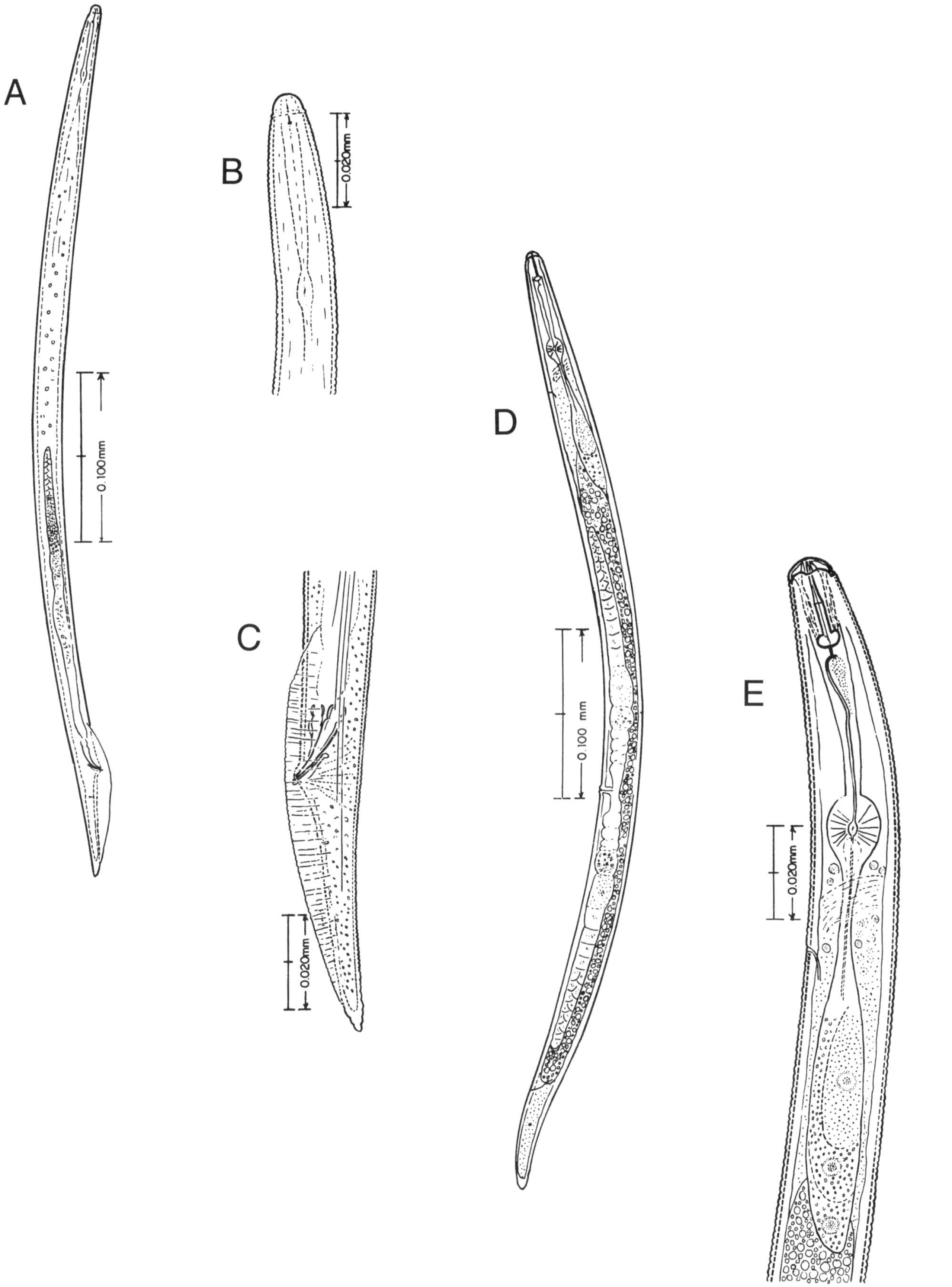


**Figure 3.44.** *Radopholus.* **A,** Male, full length. **B,** Male, anterior part. **C,** Male, posterior part. **D,** Female, full length. **E,** Female, anterior part. (Reprinted, by permission, from Taylor, 1971)

low the surface of the soil. Severely infected trees lack vigor, have sparse foliage and small leaves, produce fruit of reduced size and weight, and defoliate prematurely; they are chlorotic, stunted, and subject to severe dieback, and they may slowly die. Root tips are stubby and somewhat swollen. Fruit set is poor, and yields are low.

Similar effects occur in avocado and black pepper.

#### BODY

The body is vermiform and cylindrical, except at the extremities; adults are less than 1 mm long (a = 22–61) (Figure 3.44). Sexual dimorphism is marked in the anterior region. In females, the head region is low, typically rounded, and continuous with the body contour or slightly offset, and the cephalic framework is well developed. In males, the lip region is higher, often knoblike, and offset by a constriction, and cephalic sclerotization is reduced. The stylet and esophagus are reduced and indistinct in males. Annulations are well marked, and in most species the lateral fields contain four incisures. Deirids are absent. These nematodes lie more or less straight or slightly curved ventrally when killed by the application of gentle heat.

#### STYLET

In females the stylet is strong and 14–23 µm long, with prominent, rounded basal knobs in most species. The anterior portion of the stylet is longer than the posterior part. The spear knobs are anteriorly flattened or point forward. In males, the stylet is very slender and poorly developed, without basal knobs, as adult males do not enter roots and feed.

#### ESOPHAGUS

In females, the metacorpus (median bulb) is spherical to ovoid and heavily muscular and has a conspicuous valve. The length of the isthmus is generally about 1 body width. The basal esophageal gland lobe typically overlaps the intestine, mostly dorsally. The junction of the esophagus lumen and the intestine is very obscure. The excretory pore is at the level of the esophago-intestinal valve. In males, the esophagus is degenerate, without a prominent median bulb.

#### REPRODUCTIVE STRUCTURES

Females typically have two opposed, outstretched ovaries, which are sometimes very long, joining the vulva just behind the midpoint of the body. Some females have a single gonad and postvulval sac, more posteriorly located. The oocytes in the ovary are in single file except for a short region of reproduction. Eggs are about twice as long as the body diameter.

Males have a single, anteriorly outstretched testis, occupying one-fourth to one-third of the body length. The spicules are 18–22 µm long, slender, and bow-shaped (arcuate), with pointed distal ends. The gubernaculum is thin, troughlike, and less than half as long as the spicules; it protrudes from the cloaca.

#### VULVA

The vulva is located at or slightly posterior to the middle of the body (V = 47–58).

#### TAIL

In females, the tail is elongate (up to 5 to 7 anal body widths) and tapers to a rounded or almost pointed terminus. In males, the tail is elongate, conoid, and ventrally arcuate, and a bursa envelops about two-thirds or more of it. The phasmids are usually in the anterior part of the tail.

#### GENERA RESEMBLING *RADOPHOLUS*

*Hirschmanniella.* The esophageal glands overlap the intestine ventrally, rather than dorsally, in this genus.

*Pratylenchoides.* The esophagus slightly overlaps the intestine, and there is no marked sexual dimorphism in this genus.

*Pratylenchus.* The esophageal glands overlap the intestine ventrally, rather than dorsally, in this genus. *Pratylenchus* females have a single ovary, whereas those of *Radopholus* have two ovaries.

#### COMMENTS

There are about 30 species of *Radopholus,* most of which occur in Australasia. The genus is widely distributed in subtropical and tropical areas, especially in sandy soils. The major species include *R. citrophilus, R. similis,* and *R. oryzae.*

*R. citrophilus,* formerly the citrus race of *R. similis,* the cause of spreading decline, is difficult to control in citrus, as it follows roots in the soil to depths of almost 4 m. The life cycle generally requires 18 to 20 days at 24–27°C (75–80°F), but reproduction and root invasion can occur between 12 and 32°C (54 and 90°F). The disease can spread at an average rate of 15 m per year in deep sandy soils. It is found only in the deep, well-drained sandy soils of the Central Ridge region of Florida. This nematode can survive no longer than six months in soil in the absence of living roots, but its wide host range (ornamentals, many cultivated crops, and weeds) enables it to persist in the absence of citrus trees. Sampling for *R. citrophilus* is difficult in citrus, because of the deep vertical distribution of the nematode. Fibrous roots should be collected between 0.3 and 1.0 m beneath the soil surface. Nematodes can be extracted by incubation in jars.

*R. similis* is called the burrowing nematode because of its active movement and cell destruction within roots. One or more races or biotypes cause yellow or slow wilt of black pepper and root rot, blackhead, and toppling disease of bananas. Sexual reproduction is usually required, but parthenogenesis can occur.

*R. citrophilus* and *R. similis* have been extensively disseminated by the use of infected citrus and banana planting stock.

*R. oryzae* damages rice.

## *Rhadinaphelenchus cocophilus:* Coconut Red Ring Nematode

### Class Secernentea, order Aphelenchida, family Aphelenchoididae

#### HOSTS

*Rhadinaphelenchus cocophilus,* the coconut red ring nematode, feeds on palms, including cabbage palm (*Roystonea oleracea*), cabbage palmetto (Florida cabbage palm, *Sabal palmetto*), Canary Island date palm (*Phoenix canariensis*), coconut palm (*Cocos nucifera*), cocorite (*Mauritia caribea*), cohune (*Attalea cohune*), date palm (*Phoenix dactylifera*), gru-gru (*Acrocomia aculeata*), ita palm (*Mauritia flexuosa*), *Mauritia mexicana, Maximiliana maripa, Oenocarpus distichus,* oil palm (*Elaeis guineensis*), royal palm (*Roystonea regia*), and *Sabal* sp.

#### PARASITISM AND HABITAT

*R. cocophilus* is a migratory endoparasite in the roots, trunk, and petioles of palms in the West Indies (Grenada, Tobago, St. Vincent, and Trinidad) and the American tropics, from Mexico to Brazil. This nematode completes its life cycle in nine or 10 days. It survives in diseased palm tissue for no more than three months and in soil for up to 48 h. Both females and males are found in soil.

Juveniles (dauer larvae) are passively transported by the American or coconut palm weevil (*Rhynchophorus palmarum*), the sugarcane weevil (*Metamasius* sp.), and possibly the palm weevil (*Dynamis borassi*). Palm development is arrested by resistant third-stage juveniles, which enter larvae of palm weevils through the mouth or spiracles. Juveniles persist throughout the insect's development and enter the soft parenchymatous tissues of palms during oviposition, usually in a leaf axil in the crown. In adult weevils, nematodes can be found in the gut, body cavity, and ovipositor region. They may also be transported in bits of diseased tissue clinging to weevils.

Large nematode populations build up in cortical tissues of roots and in intercellular spaces in stems and leaves. A characteristic brick red or red-orange to reddish brown ring of discolored necrotic tissue, 2 to 4 cm wide, appears in coconut palm stems, 2.5 to 5 cm beneath the stem surface (Plate 2Q), and hence the name of the disease, *red ring.* In longitudinal section, the internal discolored tissue commonly appears as two bands joined in the bole, forming a U shape. The normally white cortex of the roots turns orange to faint pink and has a dry, flaky texture.

In coconut palms, infection causes a yellowing of the lower, oldest leaves, starting at the tip and spreading toward the base of the leaf. The leaves then turn brown, die, and may break across the petiole or the lower part of the rachis or become partly dislodged at the base and hang downward. Later, the upper leaves turn yellow, and the entire tree dies within two to four months (Plate 2Q). Trees are generally affected for two years before and two years after they come into bearing. Nuts are shed prematurely about the time leaf symptoms appear. The crown often topples over four to six weeks after symptoms first appear. The trunk, however, remains standing for several months, until it decays. The most characteristic symptoms are internal. Trees die when the growing point and inflorescence are invaded by nematodes. In some areas infected coconut and oil palms produce stunted leaves, a symptom known as little leaf.

In infected oil palms and dwarf cultivars of coconut palm, internal stem tissue is discolored brownish, but in oil palms the leaves turn dry and brown without yellowing. Diseased oil palms of the same age group take three to four years to die. Most nematodes are found outside the internal discolored zone.

#### BODY

Both sexes are long (0.84–1.2 mm) and usually very slender (a = 60–180) (Figure 3.45). The males resemble the females, except for secondary sex characteristics, but are somewhat shorter. The lips are high, angular and flattened to rounded. The adult head is slightly offset from the body, with a weak cephalic framework, and slightly narrower than the body. The cuticle is finely annulated, with four lateral incisures. The hemizonid is posterior to the excretory pore, separated from it by a distance of about 1 body width. Juveniles have a high, dome-shaped head, which is not offset from the body.

#### STYLET

The stylet is short (10.7–16 μm) and slender, with small, weak, rounded basal knobs, which may be obscure, especially on immature specimens.

#### ESOPHAGUS

The procorpus is elongate, cylindrical, and slender. The median bulb is ovate, twice as long as it is wide, and has a conspicuous oval valve. The dorsal esophageal gland orifice is in the median bulb. The esophageal glands are long and slender and overlap the intestine dorsally for a distance of 5 to 6 body widths.

#### REPRODUCTIVE STRUCTURES

Females have a single, anterior ovary, which is outstretched and not reflexed. The uterus is thick-walled and compact. The oocytes are large, several times longer than wide, and in a single file, with a long postuterine sac about three-fourths the distance to the anus.

Males have a single, anteriorly outstretched testis. The spicules are paired, usually large, thorn-shaped (free or partially fused), and arcuate, usually with a prominent, sharply pointed rostrum with disklike expansions and an apex that tends to continue the line of the transverse bar rather than that of the dorsal limb (Figure 3.45E). There is no gubernaculum.

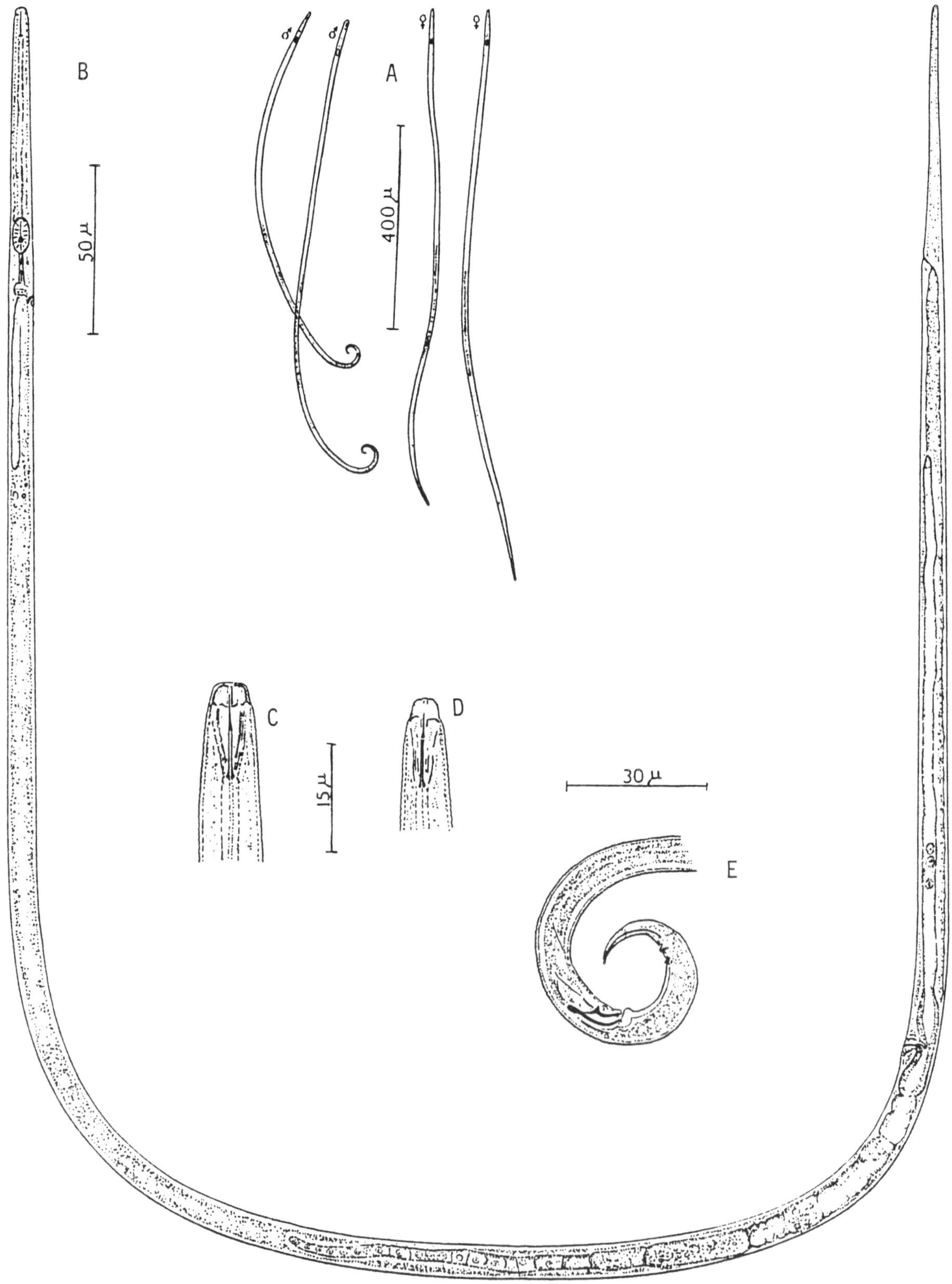


**Figure 3.45.** *Rhadinaphelenchus cocophilus.* **A,** Adult males (left) and females (right). **B,** Adult female. **C,** Female head. **D,** Male head. **E,** Male tail. (Reprinted, by permission, from Braithwaite and Siddiqi, 1975)

#### VULVA

The vulva, located posterior to the midsection (V = 65–69), is slitlike, sclerotized, and curved, with a wide, overlapping flap. The vagina slopes anteriorly.

#### TAIL

The female tail is elongate-subcylindrical, with a length of 10–17 anal body widths, narrowing to a usually broadly rounded, unstriated terminus. The male tail is pointed, acute, strongly curved ventrally, and sclerotized. It has a circular to oval bursa, which reaches the tail, and four pairs of ventro-submedian papillae. The terminus appears talonlike in lateral view.

#### GENERA RESEMBLING *RHADINAPHELENCHUS*

*Aphelenchoides*
*Bursaphelenchus*

#### COMMENTS

*R. cocophilus* causes serious problems in coconut palm. It was first found in Trinidad in 1918. Coconut palms four to seven years old are most susceptible. Infected trees should be burned.

To recover the nematode from diseased coconut tissue, chop the tissue into pieces about 1 cm thick, and place them in a large funnel of water with the stem closed at the end and the neck lightly plugged with cotton, which acts as a filter separating the tissue from 10 to 20 ml of clear water in the stem.

Alternatively, mince 15 g of palm tissue in a blender, and suspend it in 250 ml of water for 30 sec. Make up the resultant suspension to 1 L in a bottle, and let it stand for 30 min. Then pour the sedimented contents of the bottle over another container filled with water. After 30 min, discard the contents of the lower container. Finally, sieve the contents of the top bottle four times through a 60-mesh sieve.

To obtain a sample from a living tree, drive a stainless steel tube, sharpened at one end, into the tree, pointing downward at a 45° angle at a point in the stem selected for sampling. Place the extracted core (generally including a cylinder of necrotic red ring tissue) in a blender with 50 ml of water, and process it for 2 min. Then pour the contents of the blender into a dish, and let it stand for 20 min. Recover the nematodes by sieving. Ten grams of coconut stem tissue may contain 50,000 nematodes.

## *Rotylenchulus*: Reniform (Kidney-Shaped) Nematodes

### Class Secernentea, order Tylenchida, family Hoplolaimidae

#### HOSTS

*Rotylenchulus* species have several hundred hosts in diverse groups in warm climates worldwide, including abacá (*Musa textilis*), *Acacia,* acerola (West Indian cherry, *Malpighia glabra*), ageratum, alfalfa, arborvitae, artichoke, avocado, bamboo, banana, barley, beans (*Phaseolus* spp.), beet (garden beet, sugar beet), betel vine (*Piper betel*), black pepper (*Piper nigrum*), breadfruit (*Artocarpus altilis*), broad bean (faba bean, *Vicia faba*), broccoli, cabbage, cacao (cocoa, *Theobroma cacao*), cantaloupe, cardamom (*Elettaria cardamomum*), carrot, cashew (*Anacardium occidentale*), cassava (*Manihot esculenta*), castor bean (*Ricinus communis*), cauliflower, cherries, chickpea (*Cicer arietinum*), Chinese cabbage, cinnamon (*Cinnamomum verum*), citrus, clove (*Syzygium aromaticum*), clovers (*Trifolium*), coconut palm (*Cocos nucifera*), coffee, corn (maize), cotton, cowpea (*Vigna unguiculata*), cucumber, cumin (*Cuminum cyminum*), eggplant, fenugreek (*Trigonella foenum-graecum*), fig, finger millet (African millet, *Eleusine coracana*), garlic, gherkin, ginger (*Zingiber officinale*), grape, grasses (forage grasses, turfgrasses, and wild species), hyacinth bean (*Lablab purpureus*), indigo, jackfruit (*Artocarpus heterophyllus*), *Jacquemontia,* jute (*Corchorus capsularis* and *C. olitorius*), *Kalanchoe,* kale, kava (yaqona, *Piper methysticum*), lettuce, loquat, lupines, mango, mulberry, mung bean (*Vigna radiata*), muskmelon, nutmeg (*Monodora*), okra, olive, onions, papaya, passionfruit (*Passiflora edulis*), pea, peanut, pearl millet (*Pennisetum glaucum*), pigeon pea (red gram, *Cajanus cajan*), pineapple, pistachio, plantain (*Musa* × *paradisiaca*), royal poinciana (*Delonix regia*), potato, pumpkin, radish, rice, snake gourd (*Trichosanthes anguina*), sorghums, soybean, squashes, sugarcane, sunn hemp (*Crotalaria juncea*), sweet potato, Swiss chard, taro (*Colocasia esculenta*), teak (*Tectona grandis*), tea-plant (*Camellia sinensis*), tobacco, tomato, turmeric (*Curcuma domestica*), vanilla, watermelon, wheat, *Xanthosoma* spp., yams (*Dioscorea*), and numerous others.

Weed hosts include chicory (witloof, *Cichorium intybus*).

#### PARASITISM AND HABITAT

*Rotylenchulus* species are sedentary ectoparasites and semiendoparasites on roots of many plants, causing root and cortical necrosis. Severe infection can kill young plants. Older plants may be affected with leaf chlorosis or dwarfing or other decline, dieback, reduced production of flowers, decreased leaf size, leaf curling, and reduced yield. Infected sweet potato roots may develop large cracks in storage.

Mature females feed with only the neck embedded in roots. They are difficult to see, because the body is often covered by a gelatinous mass (containing about 50 eggs) and soil particulates, giving it a dark appearance. Females and their adhering egg masses are easily observed under a stereoscopic or dissecting microscope.

Juveniles, males, and immature females are found in soil. The fourth-stage juvenile female is the infective stage, penetrating roots and becoming sedentary. Males never enter plants. Reproduction without males is characteristic of some populations of *Rotylenchulus* species, but not all.

*R. macrodoratus* induces a single giant cell in the soybean endodermis or pericycle. This huge giant cell replaces

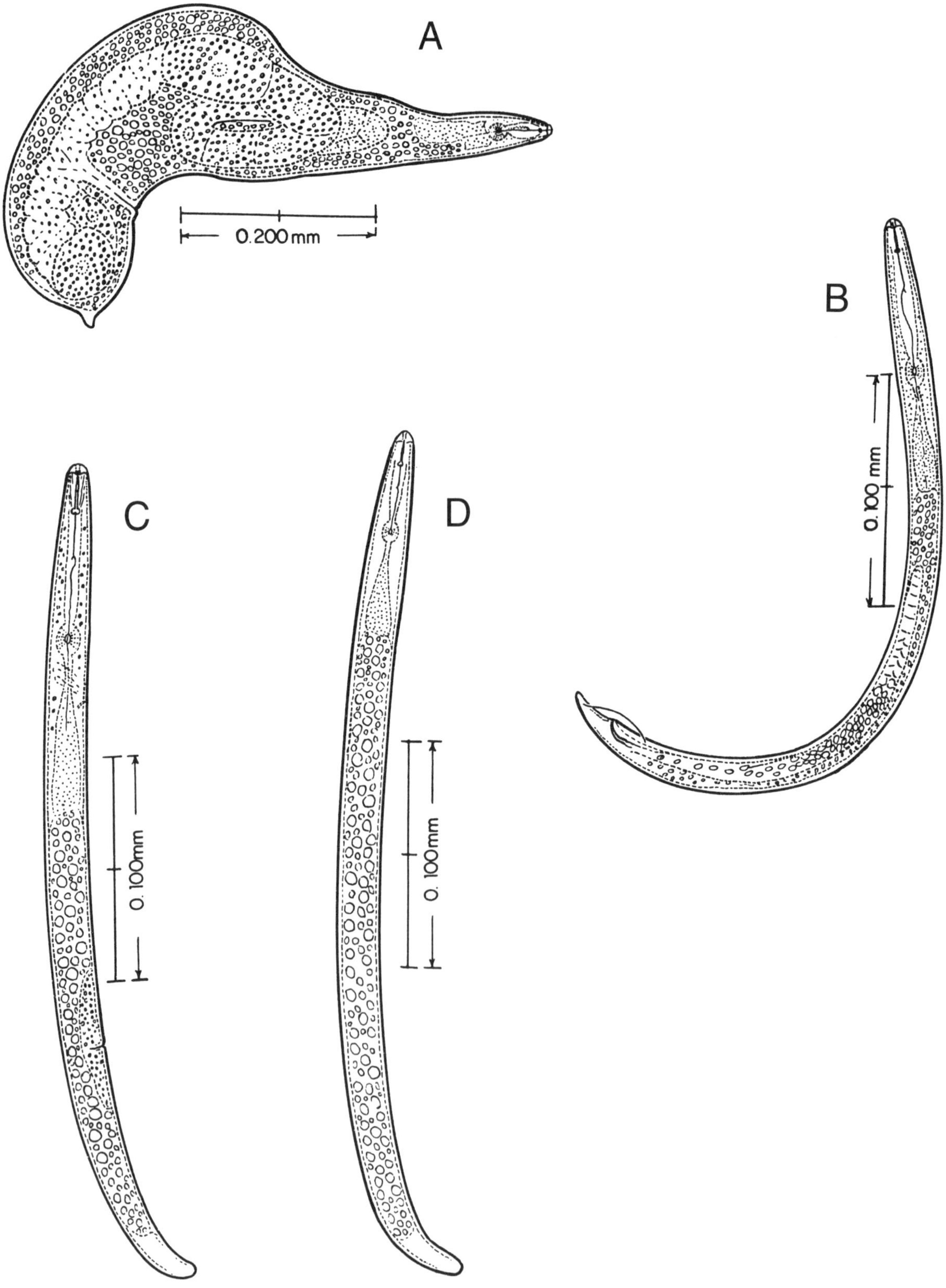


**Figure 3.46.** *Rotylenchulus.* **A,** Adult female. **B,** Adult male. **C,** Immature female. **D,** Juvenile. (Reprinted, by permission, from Taylor, 1971)

part of the stele. Similar single giant cells develop in other hosts invaded by sessile nematodes of different genera.

#### BODY

Immature females are free in soil, 0.3–0.5 mm long, and slender (Figure 3.46). The body often assumes an open spiral or C shape when killed by heat. Mature females are typically greatly swollen, reniform (kidney-shaped) to irregular in shape, 0.5–0.9 mm long, and ventrally arcuate (in *R. parvus,* the body is twisted, with the posterior part commonly crossing the neck region). The anterior portion of the body is usually embedded in root tissue, and the posterior, kidney-shaped part remains outside the root. Males are 0.33–0.5 mm long and vermiform (a = 24). Juveniles are 0.3–0.45 mm long and slender (a = 21). The head region of females is elevated, conoid or rounded, and not set off from the body; it has a moderately sclerotized labial framework exhibiting fine annules. Cephalic sclerotization is much weaker in males than in immature females. The head region of second-stage juveniles is rounded. The lateral field of males, immature females, and juveniles has four nonareolated incisures.

#### STYLET

The female stylet is slender and short (10–26 μm), with small, rounded basal knobs that slope posteriorly. Males have a weak stylet and stylet knobs. The juvenile stylet is short (13–15 μm) and lightly sclerotized and has rounded knobs.

#### ESOPHAGUS

The isthmus is long and narrow. In females and second-stage juveniles, the metacorpus (median bulb) is well developed, with a large, conspicuous valve and prominent triradiate lumen lining. The elongated gland lobe (with a length of 4 to 5 body widths) overlaps the intestine latero-ventrally, and occasionally dorsally, in a flattened lobe. The dorsal esophageal gland orifice is located posterior to the stylet knobs, separated from them by a distance of typically 0.6 to 1.9 times the stylet length. This position is diagnostic for *Rotylenchulus.* In males, the esophagus is degenerate, with a reduced median bulb and an indistinct valvular apparatus.

#### REPRODUCTIVE STRUCTURES

Females have two outstretched ovaries. In immature females, each ovary has a double flexure. In mature females, the ovaries are highly convoluted, occupying much of the body. The eggs are very large (those of *R. parvus* are 64 × 34 μm) and usually rounded; they are deposited in a gelatinous matrix.

Males have a single testis; elongate, slender, ventrally arcuate spicules; and a linear, nonprotruding gubernaculum.

#### VULVA

The vulva is a transverse slit posterior to the midsection, at a distance of about two-thirds the body length from the anterior end (V = 58–72). The vulval lips are prominent and protrude in swollen mature females.

#### TAIL

The adult female tail is conoid, with a length of usually 2 to 3 anal body widths, and has a rounded terminus containing a small plicate process. The male tail is conoid, pointed, and 1 to 1.5 times as long as the curved spicules; the immature female tail is similar. Males have a small, narrow bursa, which does not extend to the tail tip. Juveniles resemble immature females but have a stouter, more rounded tail. The phasmids are porelike and anterior to the middle of the tail.

#### GENERA RESEMBLING *ROTYLENCHULUS*

Males and immature females of the following genera resemble those of *Rotylenchulus:*

*Pratylenchus*
*Radopholus*
*Tylenchulus*

#### COMMENTS

*Rotylenchulus* contains 11 species. *R. reniformis,* the reniform nematode, an obligate parasite, is the most common and the most important, especially in subtropical and tropical soils. Some populations of *R. reniformis* include both males and females, while in others males are absent or very rare. Other major species are *R. borealis,* which has occasionally been reported in temperate soils in Europe, and *R. parvus,* which is found mostly in Africa.

*R. reniformis* can survive at least seven months in air-dried soil and up to 29 months in the absence of host plants by entering an anhydrobiotic state. The female, after entering the plant and feeding in a cell in the endodermis, can produce eggs within 11 days. The eggs are deposited in a gelatinous matrix, which often covers the entire body. Hatching is stimulated by root exudates of certain host plants. Second-stage juveniles emerge from the eggs. They do not feed. They undergo three molts in the soil and develop into young adults. A stimulus from growing host roots nearby is needed for the final molt. The life cycle of *R. reniformis* is completed in about 25 days if the newly mature female is able to immediately find host roots in which to feed. Males of *R. reniformis* are degenerate, with a poorly developed stylet and median bulb.

Infection by *R. reniformis* enhances the action of pathogenic, wilt-producing fungi, including *Rhizoctonia solani, Verticillium albo-atrum, V. dahliae, Phytophthora nicotianae* var. *parasitica, Macrophomina phaseoli, Fusarium oxysporum,* and other *Fusarium* species. Parasitism by this nematode also increases the severity of the viral diseases Brinjal mosaic and okra yellow vein mosaic.

## *Rotylenchus:* Spiral Nematodes (Boxwood Nematode and Related Species)

**Class Secernentea, order Tylenchida, family Hoplolaimidae**

### HOSTS

*Rotylenchus* species have numerous hosts, probably worldwide, including African violet, *Agathis,* andromeda (*Pieris*), apple, arborvitae, banana, Japanese barberry (*Berberis thunbergii*), *Basella,* beans (*Phaseolus* spp.), beeches, beet (garden beet, sugar beet), begonia, birches, black gram (urd, *Vigna mungo*), blueberry, boxwood, broccoli, Brussels sprouts, cabbage, cacao (cocoa, *Theobroma cacao*), camellia, cana brava (giant reed, *Arundo donax*), cardamom (*Elettaria cardamomum*), carnation, carrot, cassava (*Manihot esculenta*), castor bean, cauliflower, *Cedrus,* celeriac, celery, *Cephalotaxus,* chickpea (*Cicer arietinum*), cinnamon (*Cinnamomum verum*), citrus, conifers, corn (maize), cotton, cowpea (*Vigna unguiculata*), cranberry, Italian cypress (*Cupressus sempervirens*), deodar (*Cedrus deodara*), Douglas fir, elder (*Sambucus*), English ivy (*Hedera helix*), *Eucalyptus,* ferns, flax, gentian, geranium (*Pelargonium*), gooseberry, grape, grasses (forage grasses, turfgrasses, and wild species), guava (*Psidium*), heath (*Erica alba plena*), heather (*Calluna vulgaris*), holly, junipers, larches, leek, lettuce, lilac, lilies, mango, nutmeg (*Monodora*), oaks, papaya,

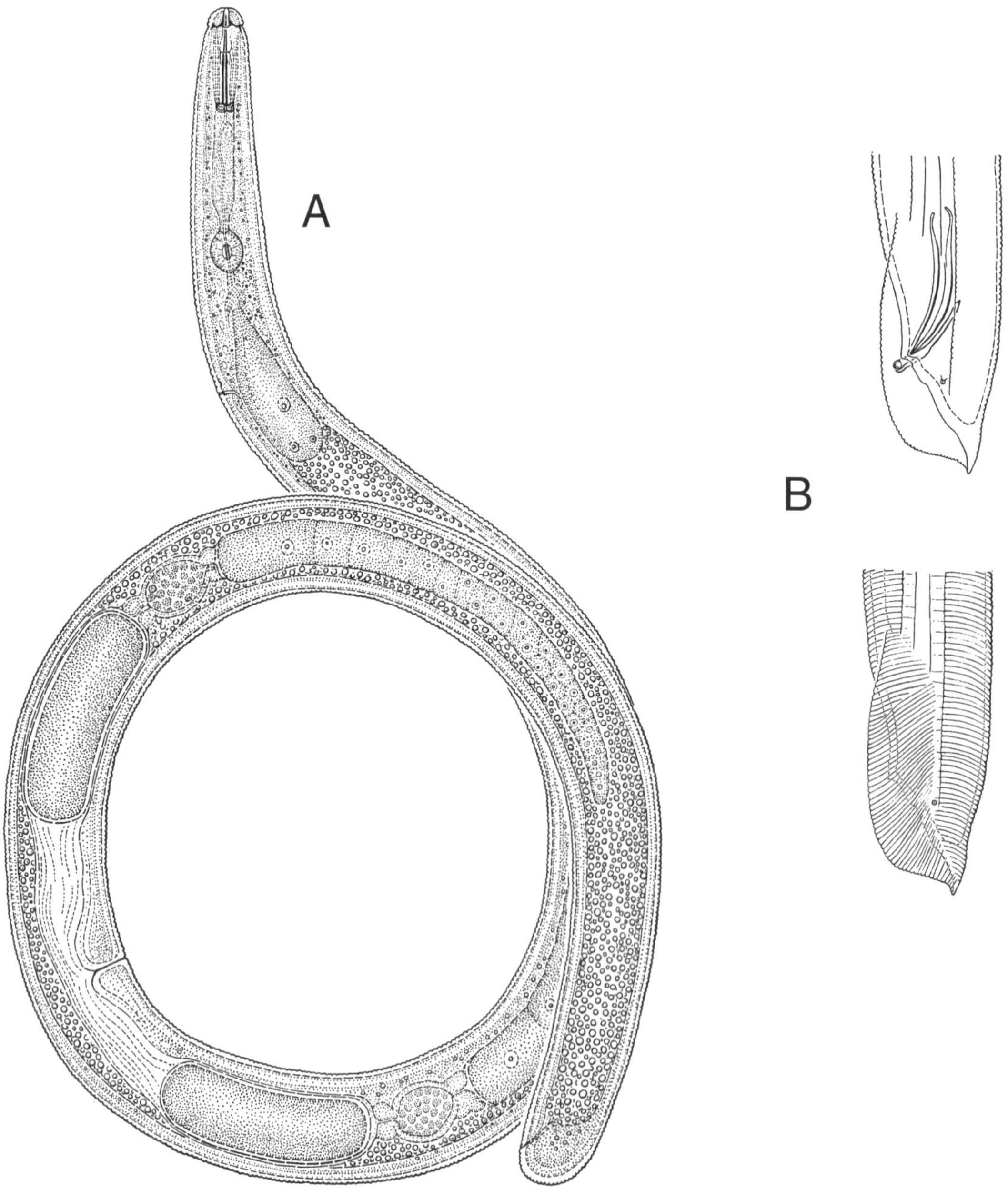


**Figure 3.47.** *Rotylenchus robustus.* **A,** Adult female. **B,** Male tails. (Reprinted, by permission, from Thorne, 1961)

pea, periwinkle, pineapple, pines, poinsettia, potato, privet, raspberry, roses, rye, sorghums, soybean, spinach, spruces, strawberry, Swiss chard, sycamore, tea, tobacco, tomato, turmeric (*Curcuma domestica*), turnip, wheat, woundwort (*Anthyllis vulneraria*), yams (*Dioscorea*), yew (*Taxus*), and others.

Pathogenicity has been demonstrated in relatively few of these plants.

#### PARASITISM AND HABITAT

All species of *Rotylenchus* are migratory ectoparasites on roots. Occasionally some individuals partially or completely enter roots along the sides and at the tips. Root lesions and rots are not unusual. Shoot growth and roots are reduced, and chlorosis and loss of yield may occur in some hosts. Nematode populations may reach high levels in coarse, sandy soils. These nematodes are regarded as polyphagous.

#### BODY

Females are 0.6–2.0 mm long and slender (a = 31) (Figure 3.47). Males are 0.8–1.4 mm long (a = 25–38). The lip region may have annulations or striations or may lack them. In females, the head is hemispherical and slightly offset or not offset. In males, the lips are more distinctly offset and elevated than in females. The cephalic framework is moderate to massive. One or more labial annules are subdivided by longitudinal striations, giving them a tiled appearance. The annules and longitudinal striations are well developed and slightly more than 1.5 µm apart. The lateral fields have four incisures. Phasmids are small and dotlike and located on or near the tail and near the anus. These nematodes usually assume a spiral form or a C shape when relaxed with gentle heat.

#### STYLET

The stylet is usually long (25–50 µm) and well developed, with broad, spreading, anteriorly indented basal knobs.

#### ESOPHAGUS

The procorpus is long and narrow. The opening of the dorsal esophageal gland is a short distance (usually less than one-fourth the stylet length) from the base of the stylet. The metacorpus (median bulb) is spherical, with a prominent valve. The isthmus is short. The esophageal glands overlap the anterior portion of the intestine dorsally and laterally for a distance of 1 to 2 body widths. The excretory pore is at the level of the esophago-intestinal junction.

#### REPRODUCTIVE STRUCTURES

Females have two outstretched ovaries, one extending anteriorly and the other posteriorly. Usually, each ovary has a spermatheca.

Males have a single outstretched testis; moderately developed, slightly bow-shaped (arcuate) spicules; and a cephalated gubernaculum bearing titillae.

#### VULVA

The vulva is near the midbody (V = 54–63).

#### TAIL

In females, the tail is bluntly rounded and short (less than 2 anal body widths); thc tail tip is hemispherical, rounded or conoid, and usually more convex dorsally, with annulations extending to the terminus. A small rounded projection may be present at the tip. In males, the tail is short and conical and enveloped by a prominent bursa.

#### GENERA RESEMBLING *ROTYLENCHUS*

*Aorolaimus*

*Helicotylenchus.* The esophageal gland orifice is farther from the stylet knobs in this genus than in *Rotylenchus.* The phasmids are small in both genera.

*Hoplolaimus.* The labial annules are tiled in this genus, as in *Rotylenchus.*

*Peltamigratus*

*Scutellonema.* The phasmids are larger in this genus than in *Rotylenchus,* and the lip region is set off from the body and sclerotized, unlike that of *Rotylenchus* and *Helicotylenchus.* The esophageal gland lobe overlaps the intestine dorsally, as in *Rotylenchus.*

#### COMMENTS

There are some 40 species of *Rotylenchus,* of which *R. buxophilus* and *R. robustus* are the most important. *R. robustus* attacks a wide variety of plants in North and South America, Europe, and India. *R. buxophilus* occurs on various plants, especially the roots of common or Turkish boxwood (*Buxus sempervirens*).

*Rotylenchus* species generally prefer light, sandy soils. In some species the sexes are separate, but in others males are unknown.

Other microorganisms, e.g., *Cylindrocarpon* and *Fusarium,* may enter wounds caused by these nematodes.

*Rotylenchus* may reduce the number of *Rhizobium* nodules on legumes.

## *Scutellonema:* Spiral Nematodes

### Class Secernentea, order Tylenchida, family Hoplolaimidae

#### HOSTS

*Scutellonema* species are almost cosmopolitan pathogens, especially in the tropics. Their numerous hosts include African violet, alfalfa, amaryllis, red anthurium (*Anthurium andraeanum*), apricot, arborvitae (*Thuja*), bamboo, banana, barley, beans (*Phaseolus* spp.), beet (garden beet, sugar beet), birches, bird-of-paradise (*Strelitzia reginae*), black pepper (*Piper nigrum*), blackberry, cacao (cocoa, *Theobroma cacao*), canna, carnation, cashew (*Anacardium occi-*

*dentale*), cassava (*Manihot esculenta*), *Cassia,* castor bean (*Ricinus communis*), cowpea (*Vigna unguiculata*), celeriac, cherries, chickpea (*Cicer arietinum*), chrysanthemum, citrus, clovers (*Trifolium*), cluster bean (guar, *Cyamopsis tetragonoloba*), coconut palm (*Cocos nucifera*), coffee, Congo jute (*Urena lobata*), corn (maize), cotton, cowpea (*Vigna unguiculata*), cranberry, *Crotalaria,* daylily, dogwood, eggplant, *Eucalyptus, Euphorbia,* fig, grape, grasses (pasture grasses, turfgrasses, and wild species), *Haemanthus, Hibiscus, Hippeastrum,* hop, hyacinth bean (*Lablab purpureus*), *Hymenocallis, Lantana,* lettuce, *Lippia, Lycoris,* mango, melons, mung bean (*Vigna radiata*), nectarine, oaks, oil palm (*Elaeis guineensis*), okra, oleander, passionfruit (*Passiflora edulis*), pawpaw, pea, peach, peanut, pear, pepper (garden or bell pepper, *Capsicum annuum*), pigeon pea (red gram, *Cajanus cajan*), pineapple, plums, potato, privet, *Pueraria,* red spiderlily (*Lycoris radiata*), rhododendrons, rice, rubber-tree (*Hevea brasiliensis*), *Sansevieria,* sesame (*Sesamum*), sorghums, soybean, strawberry, sugarcane, sunflower, sweet potato, sycamore, *Talinum,* teak (*Tectona grandis*), tea-plant (*Camellia sinensis*), tobacco, tomato, tuliptree (*Liriodendron tulipifera*), hairy vetch (*Vicia villosa*), yams (*Dioscorea*), and others, including weeds.

#### PARASITISM AND HABITAT

*Scutellonema* species are widely distributed migratory endo- and ectoparasites on roots and tubers, primarily in warmer regions. They destroy cortex cells during feeding, causing brown, necrotic lesions, which may be extensive. Growth of shoots and roots is reduced, and foliage becomes chlorotic. The damage makes host tissues more susceptible to infection by pathogenic fungi and bacteria. The number of *Rhizobium* nodules on infected legumes may be reduced.

In yams, *S. bradys* causes dry rot disease in the outer 1 to 2 cm of the tuber. Lesions form under the outer skin without external symptoms. The lesions are cream-colored to yellow at first, later turning light brown and then dark brown to black. The damage is similar to that caused by *Pratylenchus coffeae.* Extreme cracks occur in the outer skin, and parts of the outer skin flake away, exposing patches of dark brown dry rot. Tubers in storage may have a general decay.

#### BODY

Males and females have the same general appearance (Figure 3.48). They are 0.3–1.5 mm long (a = 22–36) and cylindrical except at the extremities. The lips are high and hemispherical or conoid (more bluntly conoid in females than in males), with a moderately cuticularized cephalic framework, and are slightly set off from the body by a constriction. The head is annulated and may be marked with longitudinal striae. The amphid apertures and four papillae are located just under the margin of the labial disk. The cuticular annulations are well marked in both sexes. The lateral fields are distinct and contain four or fewer incisures. The cuticle is areolated primarily in the region of the phasmids and anteriorly. The phasmids are a diagnostic character of the genus: they are enlarged, shieldlike scutellae, 3 to 4 µm across, usually located opposite each other at about the level of the anus (behind the vulva in *S. christiei*). These nematodes assume a continuous spiral or C shape when relaxed by heating and in soil extracts.

#### STYLET

The stylet is well developed (24–31 µm long) and heavily sclerotized, with prominent, rounded or oval basal knobs.

#### ESOPHAGUS

The procorpus is narrow. The metacorpus (median bulb) hemispherical and well developed, with a conspicuous valve. The isthmus is short. The esophageal gland lobe only slightly overlaps the anterior end of the intestine dorsally and dorsolaterally. The dorsal esophageal gland duct empties into the lumen of the esophagus near the stylet knobs.

#### REPRODUCTIVE STRUCTURES

Females have two outstretched ovaries, one extending anteriorly, the other posteriorly. Spermathecae are usually absent.

Males, where present, have a single outstretched testis. The spicules are well developed and slightly bow-shaped (arcuate). They rest on a large, troughlike gubernaculum bearing titillae.

#### VULVA

The vulva is a transverse slit slightly posterior to the midbody (V = 54–62).

#### TAIL

The female tail is short (0.5 anal body width) and bluntly rounded. The male tail is short and enveloped by a large bursa.

#### GENERA RESEMBLING *SCUTELLONEMA*

*Aorolaimus.* In this genus, the phasmids are enlarged and well anterior to the anus but posterior to the vulva and not opposite each other.

*Helicotylenchus*
*Hoplolaimus*
*Peltamigratus*
*Rotylenchus*

#### COMMENTS

There are 18 or more species of *Scutellonema,* which are widely distributed in neotropical regions and Africa. The major species include *S. aberrans, S. brachyurum, S. bradys, S. cavenessi,* and *S. clathricaudatum.* They are mainly ectoparasites and are found in soil and roots. Males are rare or absent in some species, e.g., *S. brachyurum,* and abundant in others, resembling the females in general appearance.

*Scutellonema* is associated with secondary fungal and bacterial pathogens that enter wounds, including those made by nematodes. Secondary decay organisms include *Botryodiplodia, Fusarium,* and *Erwinia.*

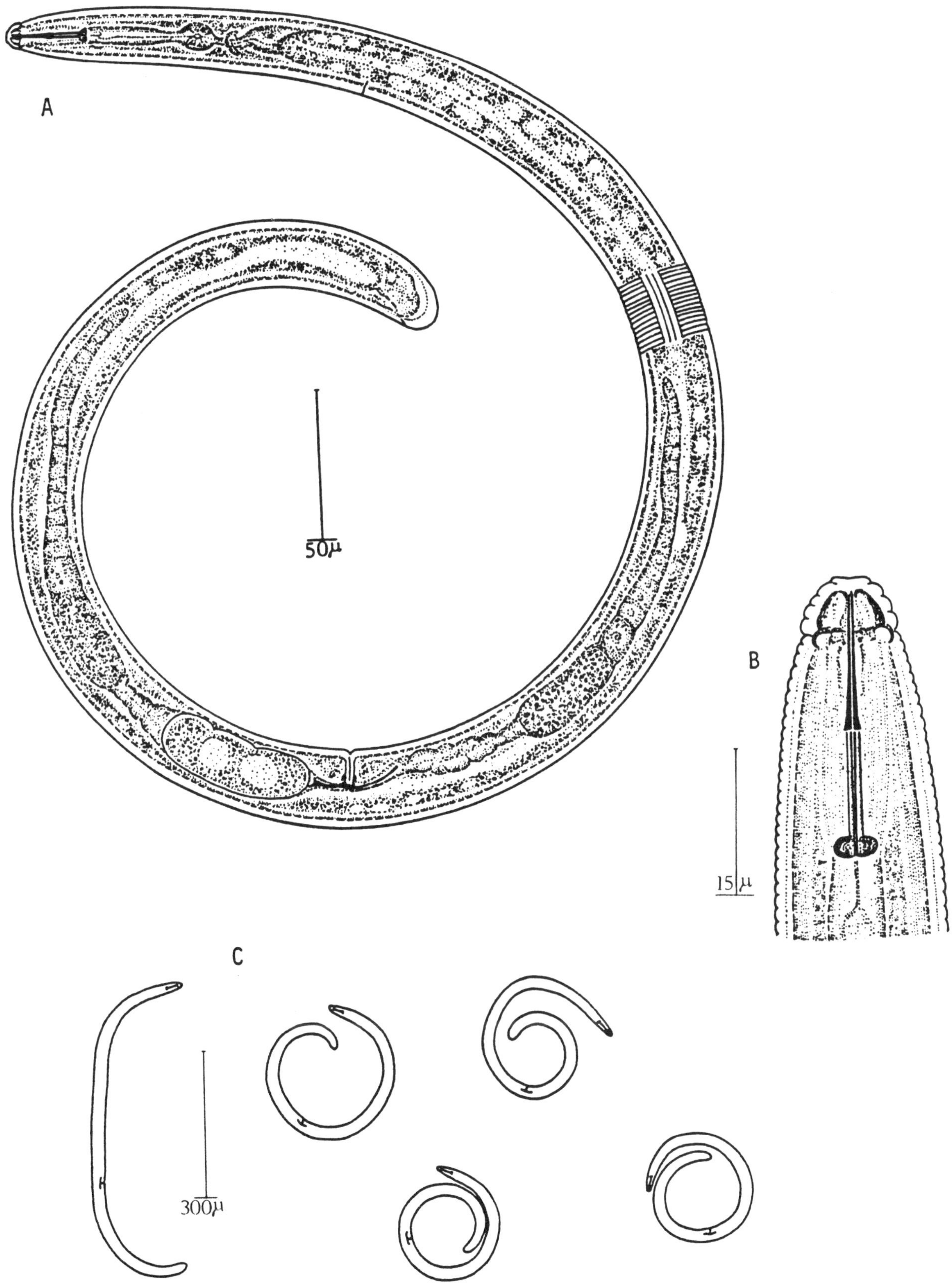


**Figure 3.48.** *Scutellonema brachyurum.* **A,** Adult female. **B,** Adult female, anterior part. **C,** Adult females in relaxed position. (Reprinted, by permission, from Siddiqi, 1974)

To diagnose *S. bradys* in yams, scrape away the outer 1 to 2 mm of the tuber surface, finely chop the tuber peelings, and tease them apart.

The life cycle of *S. cavenessi* in Senegal is adapted to alternating rainy and dry seasons. Fourth-stage juveniles and adults live in the soil, where they coil and become quiescent. They survive the dry season until the rains begin, when the juveniles molt to become adults. The nematodes mate, females deposit eggs, and second-stage juveniles hatch, invade roots, and continue to develop. Fourth-stage juveniles return to the soil.

## *Sphaeronema*

### Class Secernentea, order Tylenchida, family Paratylenchidae

#### HOSTS

Hosts of *Sphaeronema* species include California laurel (*Umbellularia californica*), citrus, manzanita (*Arctostaphylos* spp.), *Populus angustifolia,* and American sweetgum (*Liquidambar styraciflua*).

*S. sasseri* may cause a slow decline of Fraser fir (*Abies fraseri*) and red spruce (*Picea rubens*).

#### PARASITISM AND HABITAT

These nematodes are mostly sedentary endoparasites in roots. Females tend to occur in colonies in small lateral and feeder roots, where they cause a general necrosis of the cortex and bark. Galls are usually not formed.

#### BODY

Mature females are subspherical (0.1–0.21 mm) with a protruding neck region (Figure 3.49). The cephalic framework is delicate and obscure in both sexes but well sclerotized with hexaradiate symmetry in juveniles. Second-stage juveniles are 0.39–0.47 mm long (a = 25–32) and cylindrical, except at the extremities. The annules are obscure. The cuticle is 9 µm thick, with a coarse reticulate pattern after the last molt. Males are active. They are 0.4–0.47 mm long and slender (a = 33–45), tapering toward the extremities, and they have a smooth lip region, set off by a slight constriction.

#### STYLET

In females, the stylet is well developed but small (14–20 µm long), with small, sloping knobs. Second-stage juveniles have a strongly developed stylet, usually more than 20 µm long, with sloping basal knobs. Males have a degenerate stylet or no stylet and presumably do not feed.

#### ESOPHAGUS

In adult females, the esophagus is very strongly developed, with a large, spheroid median bulb and a conspicuous valve. The isthmus is very slender. The basal gland (terminal bulb) is ovoid and distinct. The excretory pore is posterior to the median bulb in mature females; it is opposite the nerve ring in juveniles. The esophagus is poorly developed (degenerate) in males.

#### REPRODUCTIVE STRUCTURES

Females have a single ovary, coiled within the body cavity. The uterus is unusually large and spheroid, with very thick muscular walls. Spermathecae are present. Eggs are extruded and accumulate in masses embedded in a gelatinous matrix, which may also contain juveniles and males.

Males have one testis, directed anteriorly; long, slender, bow-shaped (arcuate) spicules, with a conspicuous sheath; and a small gubernaculum.

#### VULVA

The vulva is very prominent, subterminal, and protruding.

#### TAIL

Adult females have no tail. The male tail tapers to a pointed to conoid terminus; there is no bursa. The juvenile tail is bluntly rounded and variable in form.

#### GENERA RESEMBLING *SPHAERONEMA*

*Globodera, Heterodera, Punctodera,* and *Sarisodera.* The cyst nematodes resemble *Sphaeronema* but form much larger cysts.

#### COMMENTS

*Sphaeronema* contains eight species. The type species is *S. californicum.* Females of this species have been found wholly embedded in roots of *Umbellularia californica. S. minutissimum* can be found in large numbers on citrus roots, with only the anterior parts of females embedded in root tissue. These nematodes are found in soil and roots and are sampled by conventional methods (see Chapter 2).

## *Trichodorus*

See *Paratrichodorus* and *Trichodorus:* Stubby-Root Nematodes.

## *Tylenchorhynchus:* Stunt or Stylet Nematodes

### Class Secernentea, order Tylenchida, family Tylenchorhynchidae

#### HOSTS

*Tylenchorhynchus* species feed on numerous hosts worldwide, including *Acacia, Acalypha,* ageratum, alfalfa, amaryllis, andromeda (*Pieris*), *Annona cherimola* (cherimalla, cherimoya), apple, ashes, azaleas, bachelor's-button (coneflower, *Centaurea cyanus*), banana, barley, *Basella,* beans (*Phaseolus* spp.), beeches, beet (garden beet, sugar beet),

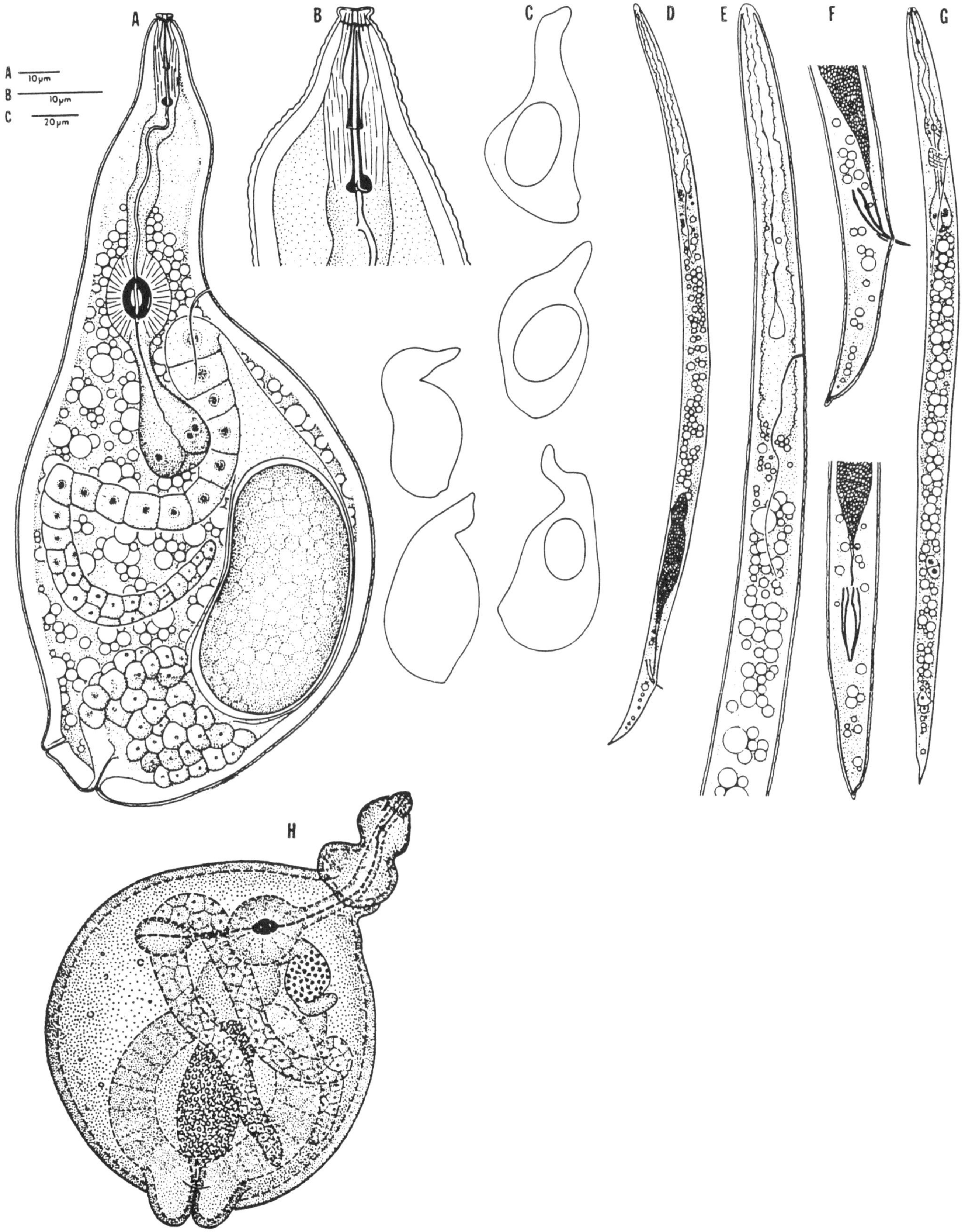


**Figure 3.49.** *Sphaeronema.* **A–G,** *S. sasseri.* **A,** Adult female. **B,** Adult female, anterior part. **C,** Outline of adult females, three with a characteristic large egg. **D,** Adult male, full length. **E,** Adult male, anterior part. **F,** Adult male, posterior part, lateral and ventral views. **G,** Second-stage juvenile. **H,** Adult female *S. californicum.* (**A–G** reprinted, by permission, from Eisenback and Hartman, 1985; **H** reprinted, by permission, from Raski and Shear, 1952)

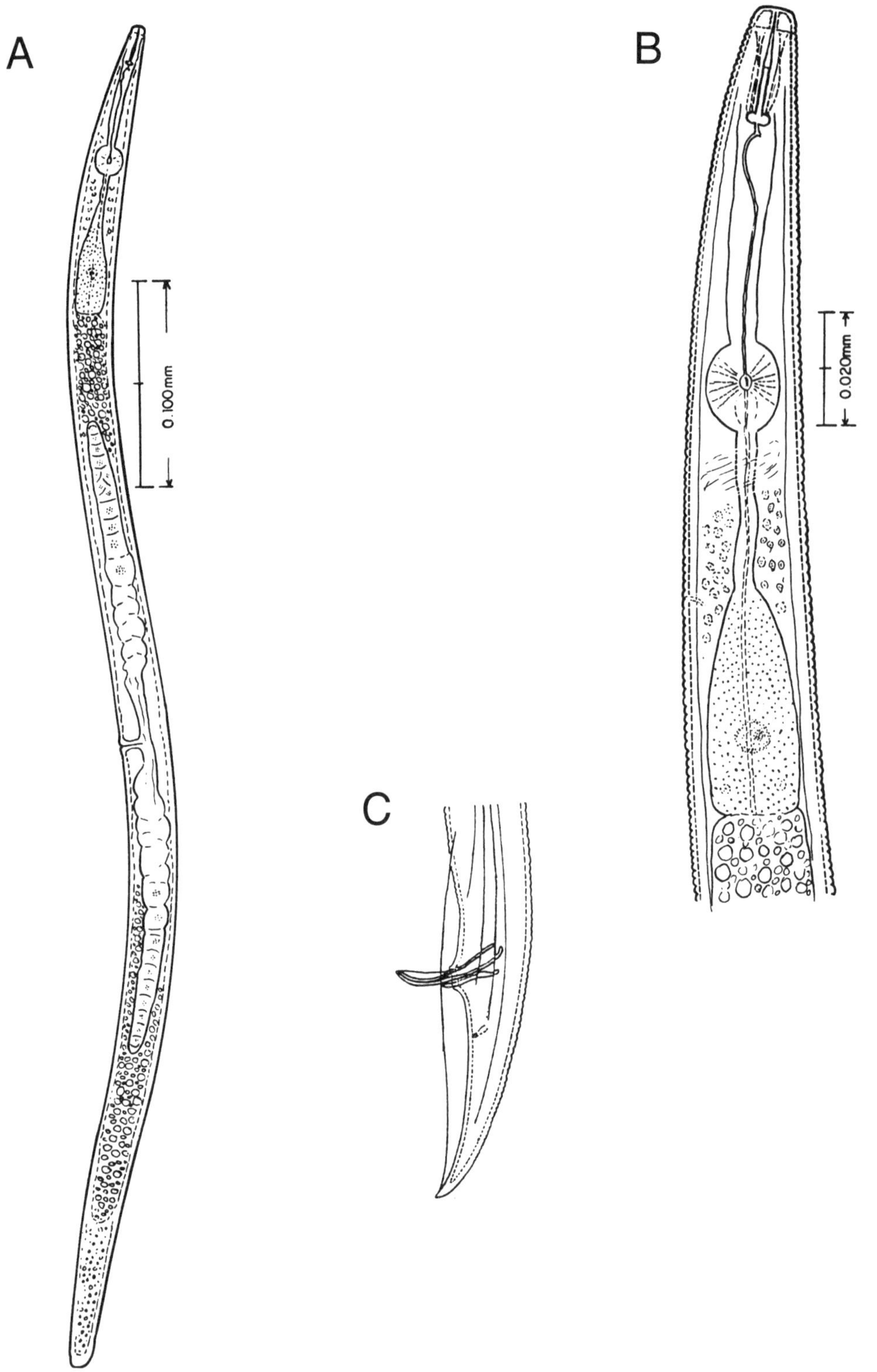


**Figure 3.50.** *Tylenchorhynchus.* **A,** Adult female. **B,** Female, anterior part. **C,** Male, posterior part. (**A** and **B** reprinted, by permission, from Taylor, 1971; **C** reprinted from Thorne, 1961)

birches, black gram (urd, *Vigna mungo*), black locust (*Robinia*), black pepper (*Piper nigrum*), blackberry, blueberry, boxwood, broad bean (faba bean, *Vicia faba*), broccoli, cabbage, cacao (cocoa, *Theobroma cacao*), *Calluna,* canna, cantaloupe, carnation, carrot, cauliflower, *Chamaecyparis,* cherries, chickpea (*Cicer arietinum*), citrus, clove (*Syzygium aromaticum*), clovers (*Trifolium*), cluster bean (*Cyamopsis tetragonoloba*), coconut palm (*Cocos nucifera*), coffee, coriander (parsley), corn (maize), cosmos, cotton, cowpea (*Vigna unguiculata*), cranberry, *Cryptomeria,* cucumber, cumin (*Cuminum cyminum*), dahlia, date palm (*Phoenix dactylifera*), dill, dogwood, Douglas fir, eggplant, elms, *Erica,* fenugreek (*Trigonella foenum-graecum*), ferns, fiddleleaf ficus (*Ficus lyrata*), fig, geranium (*Pelargonium*), gherkin, ginger (*Zingiber officinale*), grape, grass pea (*Lathyrus sativus*), grasses (pasture grasses, turfgrasses, and wild species), guava (*Psidium*), *Hamamelis,* heather (*Calluna vulgaris*), hemlock, holly, European hop (*Humulus lupulus*), horse gram (*Macrotyloma uniflorum*), hyacinth bean (*Lablab purpureus*), kenaf (Indian hemp, *Hibiscus cannabinus*), kohlrabi, jasmine, *Juncus,* junipers, lespedezas, lettuce, magnolia, mango, maples, mints, small mudar (*Calotropis procera*), mulberry, mung bean (*Vigna radiata*), muskmelon, *Myoporum,* narcissus, oaks, oat, okra, onions, palms, papaya, parsley, pawpaw, pea, peach, peanut, pear, pearl millet (*Pennisetum glaucum*), pecan, pepper (garden or bell pepper, *Capsicum annuum*), philodendrons, pigeon pea (red gram, *Cajanus cajan*), pines, pomegranate, poplars, potato, privet, radish, rape, raspberry, rhododendrons, rice, roses, rubber plant (*Ficus elastica*), *Rudbeckia,* rutabaga, rye, sedum, *Senecio,* sorghums, soybean, spinach, spruces, squash, strawberry, Sudangrass, sugarcane, sunflower, sweet potato, sweetgum (*Liquidambar*), sycamore, tobacco, tomato, tree-of-heaven (*Ailanthus altissima*), tulip, tuliptree (*Liriodendron tulipifera*), turmeric (*Curcuma domestica*), turnip, unicorn-flower (*Proboscidea louisianica*), hairy vetch (*Vicia villosa*), wheat, willows, yams (*Dioscorea*), yew (*Taxus*), and others.

These nematodes also have numerous weed hosts, including ball nightshade (*Solanum carolinense*), crabgrass, dock (*Rumex*), milkweed, plantain (*Plantago*), and shadscale (*Atriplex confertifolia*).

#### PARASITISM AND HABITAT

*Tylenchorhynchus* species are migratory ecto-, semiecto- or even endoparasites on young root tips. They are widely distributed, especially in North America, Europe, and northern Asia, with most root and soil samples containing one or more species. They are found in all stages in soil. Most species are bisexual and polyphagous.

These nematodes usually feed on epidermal cells in the vicinity of root hairs and the cell elongation region. An individual nematode feeds for less than 10 min in the same area. Usually only a large population of nematodes can weaken and destroy enough roots to cause stunting and chlorosis of shoots and a reduction in yield. Defoliation and wilting may occur. Roots are short and sparse, generally discolored, and shriveled, and they do not elongate normally. Root hair growth is stopped, and feeder roots may be brown, withered, reduced in size, and sometimes swollen near the tip.

#### BODY

Both sexes are small (0.5–1.0 mm long; some species are occasionally over 1 mm), slender (a = 24–47), and tapering slightly at both ends (Figure 3.50). The head region is typically rounded and continuous with the body contour or slightly set off, with three to seven thin annules and a generally weak to moderately sclerotized cephalic framework. The body cuticle is weakly annulated. The lateral fields have two to five incisures and are sometimes areolated (the cuticle is divided into blocks). Longitudinal ridges are sometimes present on the body. Deirids are often absent. The phasmids, located at about the middle of the tail, are conspicuous. The sexes are separate in most species, but males are unknown in some species. The males resemble the females, except for secondary sex characters. These nematodes lie more or less straight or slightly curved ventrally when killed by heat.

#### STYLET

The stylet is thin to slender and quite short (15–30 µm), with large, rounded, backward-sloping basal knobs.

#### ESOPHAGUS

The esophagus consists of a long, narrow, moderately developed procorpus, an oval to fusiform median bulb, and a narrow isthmus. The esophageal gland lobe forms a conspicuous pear-shaped terminal bulb, which does not overlap the intestine. The esophago-intestinal valve is well defined. The dorsal esophageal gland orifice is close to the end of the terminal bulb.

#### REPRODUCTIVE STRUCTURES

Females are typically amphidelphic, with two opposed, outstretched ovaries. Usually the uteri are clear and elongated. The spermathecae are rounded.

Males have a single, outstretched testis; small, slightly curved spicules; and a gubernaculum. The spicules and gubernaculum are typically tylenchoid (like the genus *Tylenchus*).

#### VULVA

The vulva is a transverse slit near the middle of the body (V = 50–67).

#### TAIL

The female tail is typically subcylindrical to conoid and slightly tapering, with a length of usually 2 to 3 anal body widths. It has a bluntly rounded terminus, and the cuticle is sometimes thicker in the distal portion. The male tail is elongate, conically pointed, and slightly arcuate. The bursa is well developed, extending from a point anterior to the cloaca to the tip of the tail, and trilobed in some species. The phasmids are riblike and extend into the bursa.

#### GENERA RESEMBLING *TYLENCHORHYNCHUS*

*Amplimerlinius* and *Merlinius.* The lateral field has six incisures in these genera, and the female tail is acute or subacute.

*Psilenchus.* Nematodes in this genus have a slender tail with a length of much more than 2 anal body widths.

*Quinisulcius.* Females have two ovaries, the esophagus does not overlap the intestine, and the lateral field is marked with fine lines in this genus.

*Tylenchus.* Females have only one ovary, the vulva is located in the posterior fourth of body, and the tail is sharply pointed in both sexes in this genus.

**Other related genera.** *Geocenamus, Nagelus, Paratrophurus, Trichotylenchus,* and *Triversus* are closely related to *Tylenchorhynchus.* In these genera, and also in *Merlinius,* females have two ovaries, but these nematodes are distinguished by the asymmetrical fine striation of the lip region, lack of labial framework, shape of the stylet knobs, attachment of the spear muscles, and presence of a vulval flap.

#### COMMENTS

There are over 140 species of *Tylenchorhynchus.* The genus has a wide geographic distribution, and most soil samples contain at least one species. The major species include *T. annulatus, T. claytoni, T. cylindricus, T. dubius, T. maximus, T. nudus, T. robustoides,* and *T. vulgaris.* These nematodes are probably susceptible to desiccation in all stages. They typically aggregate in tangled, wiggling masses, which are said to be "swarming."

Various species are adapted to particular ecological niches. *T. cylindricus* is found in arid soils of the western United States. *T. annulus* (*T. martini*) is widespread in subtropical and tropical regions, parasitizing mostly rice, sugarcane, and grasses. *T. dubius* attacks a wide range of hosts and is found mostly in sandy and sandy loam soils.

Fungi associated with species of *Tylenchorhynchus* include *Aphanomyces euteiches, Fusarium oxysporum, Phoma pinodella,* and *Rhizoctonia solani* in peas and *F. roseum* in Kentucky bluegrass (*Poa pratensis*).

## *Tylenchulus:* Citrus Root Nematode and Related Species

### Class: Secernentea, order Tylenchida, family Paratylenchidae

#### HOSTS

Hosts of *Tylenchulus* species include Carolina ash (*Fraxinus caroliniana*), blackberry, citrus (practically all species), grape, grasses (forage grasses, turfgrasses, and wild species), huckleberry, lilac, loquat, mango, mangosteen (*Garcinia mangostana*), olive, peach, pear, persimmon, rice, and wheat.

#### PARASITISM AND HABITAT

*Tylenchulus* species are strong, sedentary endoparasites on roots of citrus and other woody plants worldwide. Slender young females, males, and juveniles are found in soil, where they can be easily overlooked or mistaken for juveniles of other species. Mature females penetrate partway into roots, with their saccate posterior parts protruding but retaining a pointed tail. Their bodies are covered by a layer of soil particles held in place by gelatinous mucus. They are commonly found in clusters.

*T. semipenetrans,* the citrus nematode, causes slow decline of citrus, in contrast to the more severe spreading decline, caused by *Radopholus similis* (see the section on *Radopholus*). Trees affected by slow decline are chlorotic, with stunted terminal growth, twig dieback, and sparse foliage, especially in the upper crown, and they produce a reduced yield of small, nonuniformly sized fruit. The branch rootlets become swollen and have an irregular appearance. The cortex of heavily invaded feeder roots decays and readily sloughs off, exposing the central cylinder. This eventually results in the death of the affected portion of the root. Secondary microorganisms (e.g., *Fusarium* species) commonly invade feeding wounds, causing dark, necrotic lesions in the cortex and a gradual decline in vigor of the tree. As many as 100 nematodes can be found penetrating a 1-cm segment of citrus root.

#### BODY

These nematodes are small in all stages (0.2–0.5 mm long) (Figure 3.51). Immature females are vermiform, migratory, and ecto- and endoparasitic on roots. The lip region is continuous with the body contour, and the head skeleton is weak. Mature females are obese, sedentary semiendoparasites, embedding the anterior third of the body in root tissue. The neck is slender, distorted, and irregularly swollen; the swellings fit between the cortical cells of the host root. The posterior portion of the female body is long, narrow, typically swollen (elongate-saccate), and usually bent ventrally in the vulvular region, with a projecting tail. The posterior portion remaining outside the host is irregularly shaped and becomes covered with a gelatinous matrix, which is produced by a well-developed excretory cell and exuded from a pore. The excretory pore has prominent lips and a duct directed anteriorly. It is typically situated in a protuberance just anterior to the vulva (the distance from the pore to the anterior end is 68–85% of the body length). The head is not offset in either males or females. The cuticle is thin and annulated in the anterior region but smooth and very thick on the swollen body. The rectum and anus are absent or difficult to see in all stages, especially in juveniles. Males are short and vermiform.

#### STYLET

Females have a fairly robust stylet (11–15 µm long) with large, rounded basal knobs. Juveniles have a stylet about 13 µm long, with small basal knobs. The stylet is degenerate in males. Males and male juveniles do not feed.

#### ESOPHAGUS

In females, the procorpus is elongate-cylindrical, with a sclerotized lumen widening to a rounded median bulb, which is strongly muscular and spheroid and has a promi-

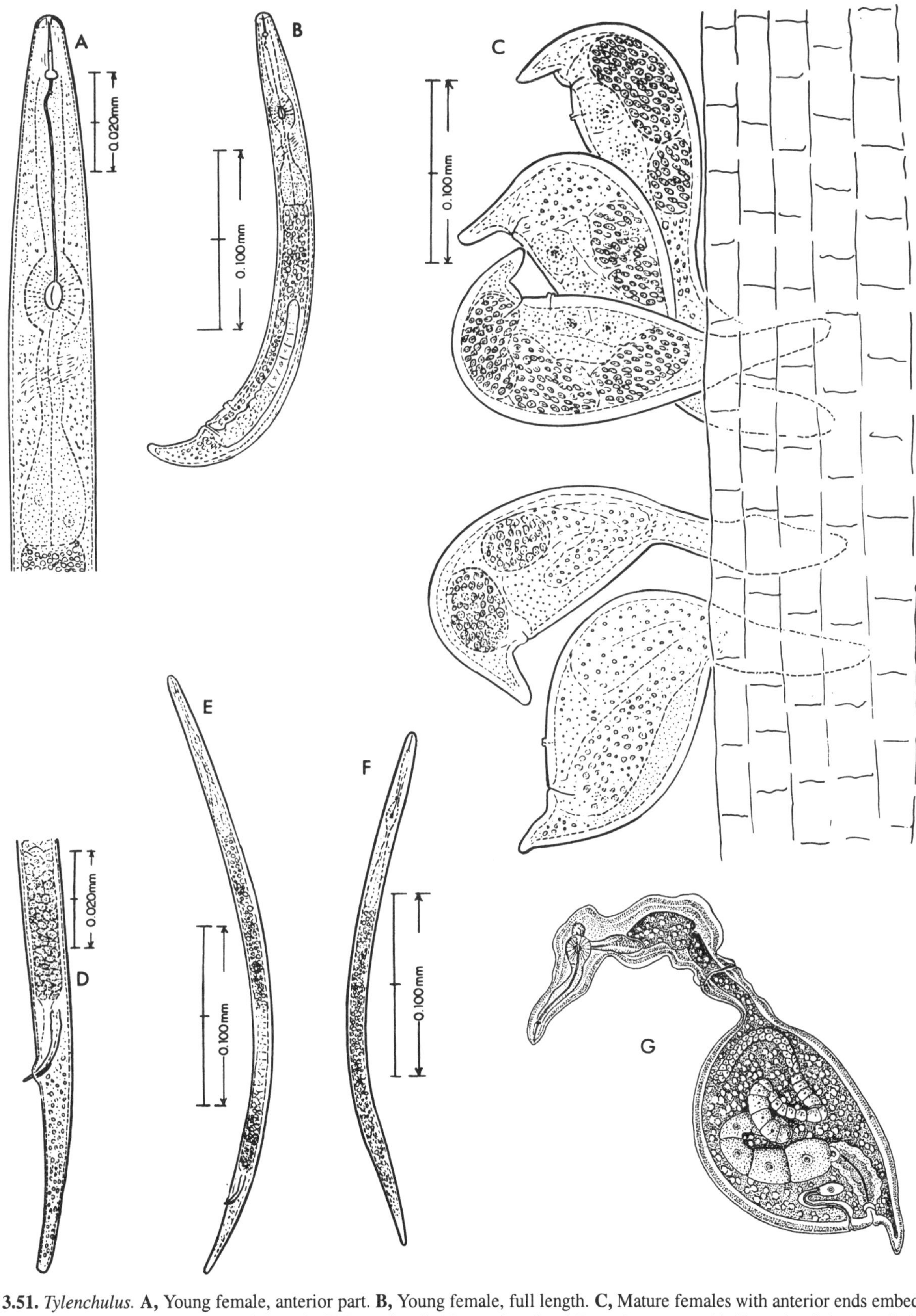


**Figure 3.51.** *Tylenchulus.* **A,** Young female, anterior part. **B,** Young female, full length. **C,** Mature females with anterior ends embedded in a root. **D,** Adult male, posterior part. **E,** Adult male. **F,** Juvenile. **G,** Mature female of *T. semipenetrans.* (**A–F** reprinted, by permission, from Taylor, 1971; **G** reprinted, by permission, from Ayoub, 1980)

nent valve. The isthmus is narrow, widening to a small, ovate terminal bulb. The posterior bulb does not overlap the intestine in juveniles, young males, and immature females. The distance from the stylet base to the opening of the dorsal esophageal gland is about 1 stylet length. The excretory pore is just anterior to the vulva. The esophagus is degenerate in males.

#### REPRODUCTIVE STRUCTURES

Females have a single, coiled, anteriorly outstretched ovary and a spermatheca. Eggs are retained in a gelatinous matrix produced by the excretory cell. Reproduction can occur with or without males.

Males have a single, anteriorly outstretched testis. The spicules are 14 to 15 µm long, narrow, ventrally curved, and guided by a gubernaculum.

#### VULVA

The vulva is a deep transverse slit with thick lips, located in the posterior end of the body (V ≈ 90).

#### TAIL

The female tail tapers to a broadly rounded tip. The male tail is elongate to conical and lacks a bursa.

#### GENERA RESEMBLING *TYLENCHULUS*

Cyst nematodes resemble mature females of *Tylenchulus.*

*Meloidogyne.* Juveniles of this genus resemble those of *Tylenchulus.*

*Nacobbus.* The excretory pore is located near the basal region of the esophagus in this genus.

*Rotylenchus*

*Trophotylenchulus.* This genus differs from *Tylenchulus* chiefly in the lip region of the female. In juveniles of *Trophotylenchulus* the lip region is modified to form a circumoral elevation, which does not occur in *Tylenchulus.* The excretory pore is nearer the vulva in *Tylenchulus* than in *Trophotylenchulus.*

#### COMMENTS

There are four species of *Tylenchulus: T. furcus, T. graminis, T. palustris,* and *T. semipenetrans,* which is by far the most widespread and important. Four races, strains, or biotypes of *T. semipenetrans* have been identified worldwide.

Mature females feeding on young roots are usually obscured by gelatinous egg masses to which soil particles adhere. For best observation, the egg masses should be removed. About 75 to 100 eggs are deposited in the mucoid egg mass.

The eggs of *T. semipenetrans* average 33 × 67 µm and hatch in 12 to 14 days at 24°C (75°F). The life cycle is completed in six to eight weeks, depending on the temperature. For *T. semipenetrans,* the optimum temperature for infection and reproduction is between 20 and 30°C (68 and 86°F), and infection and reproduction are generally favored by high soil moisture in fine-textured or organic soils. Below 15°C (59°F), the nematode becomes dormant.

Populations of *T. semipenetrans* can be found around all types of citrus plants virtually everywhere they are grown in the world. Some populations can attack other woody plants. They have been recovered from soil up to 4 m deep, but the greatest concentration of these nematodes is in the upper 30 cm of soil. Second-stage juveniles may arrest their development and survive in soil without host plants for several years. When host roots become available, they proceed to maturity.

Soil samples for citrus nematode evaluation should be collected where feeder roots are abundant, between the trunk and the outer drip line of the tree. Juveniles can be extracted by the Baermann funnel method or by incubating roots. To estimate the female population, count the nematodes in fibrous roots stained with acid fuchsin or after root trituration in a blender, at a magnification of 40×.

## *Xiphinema:* Dagger Nematodes

### Class Adenophorea, order Dorylaimida, family Longidoridae

#### HOSTS

*Xiphinema* species tend to be associated with woody plants. They have numerous hosts worldwide, including *Acacia, Adansonia,* alders, alfalfa, algarroba bean (*Ceratonia siliqua*), almond, *Annona,* apple, apricot, *Artocarpus,* ashes, asparagus, avocado, bamboo, banana, barley, beans (*Phaseolus* spp.), beeches, beet (garden beet, sugar beet), bird's-foot trefoil, black pepper (*Piper nigrum*), blackberry, blueberry, Boston ivy (*Parthenocissus tricuspidata*), boxwood, cabbage, cacao (cocoa, *Theobroma cacao*), camellia, cardamom (*Elettaria cardamomum*), carnation, carrot, cashew (*Anacardium occidentale*), cassava (*Manihot esculenta*), castor bean, celery, cherries, cinnamon (*Cinnamomum verum*), *Cistus,* citrus, clove (*Syzygium aromaticum*), clovers (*Trifolium*), coconut palm (*Cocos nucifera*), coffee, corn (maize), cotton, cowpea (*Vigna unguiculata*), cucumber, cypress (*Cupressus*), date palm (*Phoenix dactylifera*), dogwood, Douglas fir, eggplant, elder (*Sambucus*), elms, *Eucalyptus, Euonymus,* fig, firs, flax, garden balsam (*Impatiens*), geranium (*Pelargonium*), gherkin, ginger (*Zingiber officinale*), gladiolus, grape, grasses (pasture grasses, turfgrasses, and wild species), hackberry, hemlock, hickories, hop, hornbeam, hyacinth bean (*Lablab purpureus*), *Jacaranda,* jasmine, junipers, jute (*Corchorus capsularis*), larches, laurel (*Laurus*), laurel oak (*Quercus imbricaria*), lilies, litchi (lychee, *Litchi chinensis*), lupines, magnolia, mango, *Mapania,* maples, marjoram (*Origanum vulgare*), monk's-hood-vine (*Ampelopsis aconitifolia*), *Moraea,* mulberry, *Murraya,* muskmelon, narcissus, nectarine, nutmeg (*Monodora*), oaks, oat, oil palm (*Elaeis guineensis*), okra, oleander, olive, *Opuntia, Osmanthus,* palms, papaya, *Passiflora,* pea, peach, peanut, pear, pecan, pepper (garden or bell pepper, *Capsicum annuum*), periwinkle, petunia, philodendrons, pigeon pea (red gram, *Cajanus cajan*), pines, pistachio, planetree, plantain (*Musa* × *paradisiaca*), plums, poison ivy, poplars, potato, quince, raspberry, *Rheedia,* rhododendrons, rhubarb, rice, *Robinia,* roses, rubber-tree (*Hevea brasiliensis*), skunkbush (*Rhus trilobata*), sorghums, soybean, spruces, strawberry, Sudangrass, sugarcane, sumac, sweet potato, sweetgum (*Liquidambar*), sycamore, tea, tobacco, tomato, Transvaal daisy (*Gerbera jamesonii*), tulips, turmeric (*Curcuma domestica*), turnip, hairy vetch (*Vicia villosa*), *Vinca rosea,* walnut, watermelon, wheat, willows, yams (*Dioscorea*), yucca, and numerous others.

Weed hosts include ball nightshade (*Solanum carolinense*), milkweed, plantain (*Plantago*), and ragweed (*Ambrosia*).

Different species of *Xiphinema* have distinctive sets of hosts. Relatively few species are known to be pathogenic to plants.

#### PARASITISM AND HABITAT

*Xiphinema* species are cosmopolitan migratory ectoparasites on many plants. They are found in all stages in soil, especially soil with a medium texture. The most severe damage due to these nematodes usually occurs in sandy soils. They feed near the tips of small, actively growing feeder roots, causing reddish brown to black, slightly sunken lesions. Feeding causes death and discoloration of cortical cells and a dramatic reduction in root growth. Extensive root damage results in reduced shoot growth and yield.

The type of damage varies with the nematode species and the host. Common symptoms are severe plant stunting, chlorosis, and sparse, discolored, fine feeder roots. Other symptoms include varying degrees of root necrosis; the formation of clusters of short, stubby branch roots; and terminal swellings or galls on root tips. In roses, *X. diversicaudatum* causes root galls similar to those caused by root-knot nematodes (Figure 3.52).

#### BODY

The body is typically very long (1.3–5.5 mm or more) and slender (a = 34–114), without annulations (Figure 3.53). The lip region, with two circlets of six to 10 papillae, is low or hemispherical and continuous with the body contour or slightly offset by a constriction. Supplements (papillae) consist of an adanal pair and a ventromedian series. The cuticle appears smooth under a light microscope. The lateral chords are one-fourth the body width, with pores arranged in two lines. The amphid apertures are broad transverse slits leading posteriorly to a short, funnel-shaped pouch. Males and juveniles are similar to females. Males are common in some species (e.g., *X. diversicaudatum*) and rare in others (e.g., *X. americanum, X. index, X. italiae,* and *X. pachtaicum*), which are parthenogenic. The body assumes a wide C shape when killed by gentle heat.

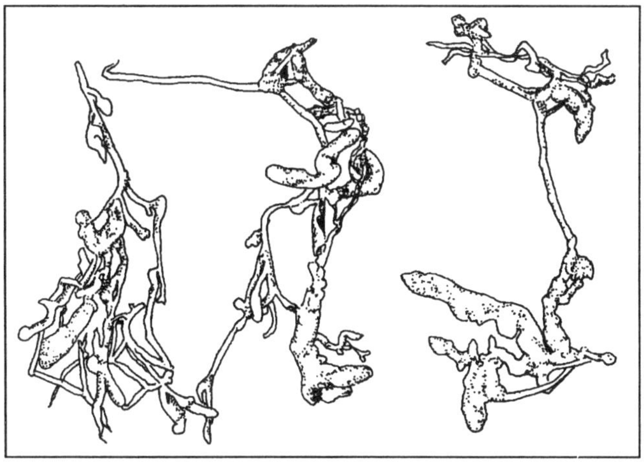

**Figure 3.52.** Roots of *Rosa multiflora* with galls produced by *Xiphinema diversicaudatum.*

#### STYLET

The stylet is very long (60–250 µm) and straight and, like that of *Longidorus* and other longidorids, consists of an odontostyle and an odontophore. The odontostyle (the anterior portion of the stylet) is a sclerotized, needlelike spear with a forked base, where it joins the nonsclerotized odontophore (stylet extension), which has three prominent, refractive basal flanges. There are no basal knobs. The stylet has a double guide ring, located near the junction of the odontostyle and the odontophore. The odontostyle penetrates root cells, and the odontophore contains nerve tissues adjacent to the food canal. The nerve tissues probably enable the nematode to discriminate between feeding sites within plant roots.

#### ESOPHAGUS

The procorpus is a long, narrow, flexible, cylindrical tube, which abruptly expands posteriorly to form a short, cylindrical bulb half as wide as the body at its base and two and one-half times as long. The esophagus is dorylaimoid, without a median bulb or valve. The dorsal esophageal gland nucleus is at the extreme anterior end of the bulb. The esophageal glands are within the terminal bulb and do not overlap the intestine. When the stylet is retracted, the anterior part of the esophagus is looped or coiled where it joins the expanded terminal bulb, to allow the thrusting of the stylet when it is protruded. Intestinal cells are packed with coarse, refractive granules. A prerectum is present.

#### REPRODUCTIVE STRUCTURES

Females have one or two ovaries, which are usually paired and reflexed, one reduced and extending anteriorly, the other posteriorly. The gonad structure is variable, and

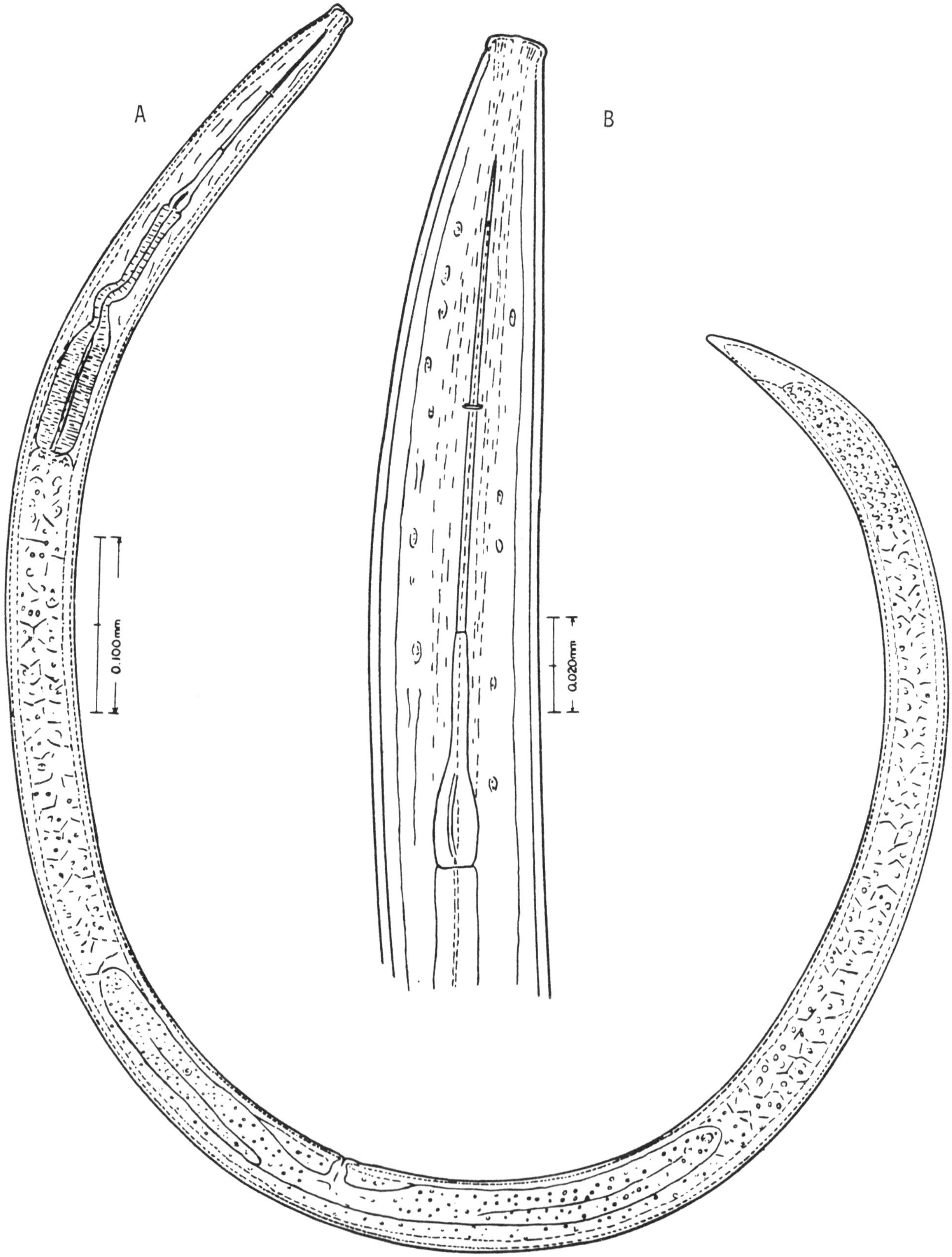


**Figure 3.53.** *Xiphinema.* **A,** Adult female. **B,** Female, anterior part. (Reprinted, by permission, from Taylor, 1971)

different species have their characteristic features. Eggs are 4 to 5 body widths in their long dimension.

Males have two opposed, outstretched testes. The spicules are strong and arcuate and have small lateral guiding pieces.

#### VULVA

The vulva is a broad, transverse slit, usually near the middle of the body (V = 38–60) or near the posterior end of the esophagus when only one ovary is present (V = 24), as in *X. chambersi* (which is pathogenic on strawberry).

#### TAIL

The tail varies from short and rounded to elongate-conoid and occasionally is filiform with a dagger-like point. Most tails of both males and females are conoid or broadly rounded or have a blunt ventral or terminal projection.

#### GENERA RESEMBLING *XIPHINEMA*

*Longidorus* and *Paralongidorus.* Nematodes in these genera lack flanged knobs and have a guiding ring near the anterior end of the stylet.

#### COMMENTS

There are 170 or more species of *Xiphinema.* The major species include *X. americanum* (most commonly found in the United States), *X. diversicaudatum, X. index, X. italiae, X. pachtaicum,* and *X. rivesi.*

Soil populations of these nematodes are often high. Under ideal conditions, a life cycle is completed in 22 to 27 days, but it may take a year or more, depending on the soil structure, temperature, moisture, host species, and growth of the host plant. For individuals in some *Xiphinema* species, the life cycle takes three years. Members of some species can survive in soil without a host plant for up to three years. Some females live for more than five years.

In temperate climates, egg laying occurs in spring and early summer under annual plants when the production of new roots is at its peak.

Some species (*X. americanum, X. bricolensis, X. californicum, X. diversicaudatum, X. index, X. intermedium, X. italiae, X. rivesi,* and *X. tarjanense*) transmit nepoviruses (polyhedral, isodiametric particles about 29 to 30 µm in diameter), including Arabis mosaic, cherry rasp leaf, grapevine fanleaf, peach rosette mosaic, strawberry latent ringspot, tobacco ringspot, and tomato ringspot (grape yellow vein) viruses (see Appendix, Table A.1). Nepovirus diseases are usually diagnosed in the field by observation of rosette symptoms in the foliage and by removal of bark from the graft union to check for stem pitting (constriction disease). The acquisition feeding period (the duration of feeding in which a nematode acquires a virus from an infected host) may be an hour or less, and a nematode can transmit a virus for weeks and possibly months. When the esophageal lining is discarded during molting, the nematode loses its ability to transmit the virus or viruses. A population of a *Xiphinema* species may transmit a particular strain of a virus, while another population of the same species transmits other strains of the virus.

Some species of *Xiphinema* reproduce sexually. Most species are parthenogenic, however, and males are rare or absent in these species.

## References


Agrios, G. N. 1997. Plant Pathology. 4th ed. Academic Press, San Diego.

Ayoub, S. M. 1980. Plant Nematology: An Agricultural Training Aid. Rev. ed. NemaAid Publications, Sacramento, Calif.

Barker, K. R., Carter, C. C., and Sasser, J. N., eds. 1985. An Advanced Treatise on *Meloidogyne.* Vol. 2, Methodology. Department of Plant Pathology, North Carolina State University, and U.S. Agency for International Development, Raleigh, N.C.

Barker, K. R., Pederson, G. A., and Windham, G. L., eds. 1998. Plant and Nematode Interactions. American Society of Agronomy, Madison, Wisc.

Braithwaite, C. W. D., and Siddiqi, M. R. 1975. *Rhadinaphelenchus cocophilus.* C.I.H. Descriptions of Plant-Parasitic Nematodes, set 5, no. 72. S. Wilmott, P. S. Gooch, M. R. Siddiqi, and M. Franklin, eds. Commonwealth Institute of Helminthology, St. Albans, Herts., U.K.

Decraemer, W. 1995. The Family Trichodoridae: Stubby Root and Virus Vector Nematodes. Kluwer Academic Publishers, Boston.

Dropkin, V. H. 1989. Introduction to Plant Nematology. 2nd ed. John Wiley & Sons, New York.

Ebsary, B. A. 1991. Catalog of the Order Tylenchida (Nematoda). Research Branch, Agriculture Canada, Ottawa.

Eisenback, J. D., and Hartman, K. M. 1985. *Sphaeronema sasseri* n. sp. (Tylenchulidae), a nematode parasitic on Fraser fir and red spruce. J. Nematol. 17:346–354.

Eisenback, J. D., Hirschmann, H., Sasser, J. N., and Triantaphyllou, A. C. 1981. A Guide to the Four Most Common Species of Root-Knot Nematodes (*Meloidogyne* species) with a Pictorial Key. International Meloidogyne Project. Department of Plant Pathology, North Carolina State University, Raleigh.

Evans, K., Trudgill, D. L., and Webster, J. M., eds. 1993. Plant Parasitic Nematodes in Temperate Agriculture. CAB International, Wallingford, Oxon., U.K.

Fortuner, R., ed. 1988. Nematode Identification and Expert System Technology. Plenum Press, New York.

Franklin, M. T. 1965. A root-knot nematode, *Meloidogyne naasi,* n. sp., on field crops in England and Wales. Nematologica 11:79–86.

Franklin, M. T. 1978. *Aphelenchoides* and related genera. Pages 172–187 in: Plant Nematology. 3rd ed. J. F. Southey, ed. Ministry of Agriculture, Fisheries and Food, GD1. Her Majesty's Stationery Office, London.

Frederick, J. J., and Tarjan, A. C. 1989. A Compilation of the Genus *Pratylenchus,* Filipjev, 1936 (Nemata: Pratylenchidae). Rev. Nematol. 12:243–256.

Gerber, W., and Taylor, A. L. 1988. A simple technique for mounting whole root-knot nematodes. Nematol. Abstr. 60:153–167.

Golden, A. M., Epps, J. M., Riggs, R. D., Duclos, L. A., Fox, J. A., and Bernard, R. L. 1970. Terminology and identity of infraspecific forms of the soybean cyst nematode (*Heterodera glycines*). Plant Dis. Rep. 54:544–546.

Handoo, Z. A., and Golden, A. M. 1989. A key and diagnostic compendium to the species of the genus *Pratylenchus,* Filipjev, 1936 (lesion nematodes). J. Nematol. 21:202–218.

Hesling, J. J. 1978. Cyst nematodes: Morphology and identification of *Heterodera, Globodera* and *Punctodera.* Pages 125–155 in: Plant Nematology. 3rd ed. J. F. Southey, ed. Ministry of Agriculture, Fisheries and Food, GD1. Her Majesty's Stationery Office, London.

Hooper, D. J. 1961. A redescription of *Longidorus elongatus* (de Man, 1876) Thorne & Swanger, 1936, (Nematoda, Dorylaimidae) and descriptions of five new species of *Longidorus* from Great Britain. Nematologica 6:237–257.

Hunt, D. J. 1993. Aphelenchida, Longidoridae and Trichodoridae: Their Systematics and Bionomics. CAB International, Wallingford, Oxon., U.K.

Inagaki, H. 1979. Race status of five Japanese populations of *Heterodera glycines.* Jpn. J. Nematol. 9:1–4.

Jairajpuri, M. S., and Ahmad, W. 1992. Dorylaimida: Free-Living, Predaceous, and Plant-Parasitic Nematodes. E. J. Brill, Leiden.

Jenkins, W. R., and Taylor, D. P. 1967. Plant Nematology. Reinhold, New York.

Jepson, S. B. 1987. Identification of Root-Knot Nematodes (*Meloidogyne* species). CAB International, Farnham Royal, England.

Khan, M. W., ed. 1993. Nematode Interactions. Chapman and Hall, New York.

Lamberti, F., and Taylor C. E., eds. 1986. Cyst Nematodes. Plenum Press, New York.

Loof, P. A. A. 1974. *Criconemoides morgensis.* C.I.H. Descriptions of Plant-Parasitic Nematodes, set 3, no. 42. S. Wilmott, P. S. Gooch, M. R. Siddiqi, and M. Franklin, eds. Commonwealth Institute of Helminthology, St. Albans, Herts., U.K.

Luc, M., Sikora, R. A., and Bridge, J., eds. 1990. Plant Parasitic Nematodes in Subtropical and Tropical Agriculture. CAB International, Wallingford, Oxon., U.K.

Maggenti, A. 1981. General Nematology. Springer-Verlag, New York.

Maggenti, A. 1983. Nematode higher classification as influenced by species and family concepts. Pages 25–40 in: Concepts in Nematode Systematics. Systematics Assoc., Spec. Vol. 22. A. R. Stone, H. M. Platt, and L. F. Khalil, eds. Academic Press, San Diego.

Maggenti, A. 1991. Nemata: High classification. Pages 147–187 in: Manual of Agricultural Nematology. W. R. Nickle, ed. Marcel Dekker, New York.

Mai, W. F., and Mullin, P. G. 1996. Plant-Parasitic Nematodes: A Pictorial Key to Genera. With photographs by H. H. Lyon and K. Loeffler. 5th ed. Comstock Publishing, Cornell University Press, Ithaca, N.Y.

Mamiya, Y., and Kiyohara, T. 1972. Description of *Bursaphelenchus lignicolus* from pinewood and histopathology of nematode infested trees. Nematologica 18:120–124.

Muller, R., Gooch, P. S., Siddiqi, M. R., and Hunt, D. J., eds. 1985. C.I.H. Descriptions of Plant-Parasitic Nematodes. Set 8. Commonwealth Institute of Helminthology, St. Albans, Herts., U.K.

Nickle, W. R., ed. 1984. Plant and Insect Nematodes. Marcel Dekker, New York.

Nickle, W. R., ed. 1991. Manual of Agricultural Nematology. Marcel Dekker, New York.

Raski, D. J., and Shear, S. A. 1952. *Sphaeronema californicum* nov. gen. nov. spec. an endoparasite of the roots of certain plants. Proc. Helminthol. Soc. Wash. 19:77–80.

Riggs, R. D., and Schmitt, D. P. 1988. Complete characterization of the race scheme for *Heterodera glycines.* J. Nematol. 20:392–395.

Riggs, R. D., and Wrather, J. A., eds. 1992. Biology and Management of the Soybean Cyst Nematode. American Phytopathological Society, St. Paul, Minn.

Santos, M. S. N. De A. 1967. *Meloidogyne ardenensis* n. sp. (Nematoda: Heteroderidae), a new British species of root-knot nematode. Nematologica 13:593–598.

Sasser, J. N., and Carter, C. C., eds. 1985. An Advanced Treatise on *Meloidogyne.* Vol. 1, Biology and Control. International Meloidogyne Project. Department of Plant Pathology, North Carolina State University, Raleigh.

Sher, S. A. 1968. Revision of the genus *Hirschmanniella* Luc and Goodey, 1963 (Nematoda: Tylenchoidea). Nematologica 14:243–275.

Siddiqi, M. R. 1974. *Scutellonema brachyurum.* C.I.H. Descriptions of Plant-Parasitic Nematodes, set 4, no. 54. S. Wilmott, P. S. Gooch, M. R. Siddiqi, and M. Franklin, eds. Commonwealth Institute of Helminthology, St. Albans, Herts., U.K.

Smart, G. C., and Khuong, N. B. 1988. Illustrated Key for the Identification of Common Nematodes in Florida. University of Florida, Gainesville.

Southey, J. F., ed. 1978. Plant Nematology. 3rd ed. Ministry of Agriculture, Fisheries and Food, GD1. Her Majesty's Stationery Office, London.

Taylor, C. E., and Brown, D. J. F. 1981. Nematode-virus interactions. Pages 281–301 in: Plant Parasitic Nematodes. Vol. 3. B. M. Zuckerman and R. A. Rohde, eds. Academic Press, San Diego.

Taylor, A. L. 1971. Introduction to Research on Plant Nematology: An FAO Guide to the Study and Control of Plant-Parasitic Nematodes. Rev. ed. Food and Agricultural Organization of the United Nations, Rome.

Taylor, A. L., and Sasser, J. N. 1978. Biology, Identification and Control of Root-Knot Nematodes (*Meloidogyne* species). Department of Plant Pathology, North Carolina State University, and U.S. Agency for International Development, Raleigh, N.C.

Thorne, G. 1949. On the classification of the Tylenchida, new order (Nematoda, Phasmidia). Proc. Helminthol. Soc. Wash. 16(2): 37–73.

Thorne, G. 1961. Principles of Nematology. McGraw-Hill, New York.

Thorne, G., and Schuster, M. L. 1956. *Nacobbus batatiformis,* n. sp. (Nematoda: Tylenchidae), producing galls on the roots of sugar beets and other plants. Proc. Helminthol. Soc. Wash. 23(2):128–134.

U.S. Department of Agriculture. USDA Nematode Collection, Nematode Database Search Page: http://www.nem.barc.usda.gov/database/search.cfm

Veech, J. A., and Dickson, D. W. 1987. Vistas on Nematology. Society of Nematologists, Hyattsville, Md.

Wilmott, S., Gooch, P. S., Siddiqi, M. R., and Franklin, M., eds. 1972–1977. C.I.H. Descriptions of Plant-Parasitic Nematodes. Sets 1–7. Commonwealth Institute of Helminthology, St. Albans, Herts., U.K.

## APPENDIX

# Nematode Vectors of Plant Viruses

> Nematode parasitism of a host plant is a complex and dynamic interaction involving the attraction of the nematode to the roots, followed by penetration of the root and the recognition of host tissues suitable for the establishment of a feeding site. (Taylor and Brown, 1997)

Of the four orders of nematodes containing species that feed on plant roots—the Aphelenchida, Dorylaimida, Triplonchida, and Tylenchida, according to Maggenti (1983) and Coomans (1996)—only species in the Dorylaimida and Triplonchida are able to transmit harmful viruses to plants (Taylor and Brown, 1997).[1] The two nematode groups containing vectors are the longidorids (species of *Longidorus, Paralongidorus,* and *Xiphinema*), in the Dorylaimida, and the trichodorids (species of *Paratrichodorus* and *Trichodorus*), in the Triplonchida. Other genera of longidorids lacking virus-transmitting ability include *Xiphidorus* and *Longidoroides.* Non-virus-transmitting genera of trichodorids are *Allotrichodorus* and *Monotrichodorus,* which have been described only in South America.

Both longidorids and trichodorids have well-developed labia, papillose cephalic sensilla, a stirrup-shaped amphidial pouch, stoma armed with a mural tooth or hollow spear, and an esophagus divided into a slender anterior part and an elongated or pear-shaped posterior portion containing secretory glands (Taylor and Brown, 1997).

Sixteen species of longidorid nematodes (eight species of *Longidorus,* one of *Paralongidorus,* and seven of *Xiphinema*) have been implicated in transmitting plant viruses (Table A.1). These genera are in the order Dorylaimida, suborder Dorylaimina, which is distinguished by a hollow, protrusible spear (composed of an anterior odontostyle and a posterior odontophore) and an esophagus with a narrow anterior part and an expanded muscular-glandular posterior portion.

Longidorid nematodes transmit nepoviruses, which are isometric particles about 28 to 30 nm in diameter. About one-third of the 37 members of the nepoviruses are known to be transmitted by nematodes. Most nepoviruses have a very wide host range among cultivated and wild plant species, are commonly seed- and pollen-borne, and are readily sap-transmissible from woody or herbaceous hosts to herbaceous test plants (bait plants).

Eleven trichodorids (seven species of *Paratrichodorus* and four of *Trichodorus*) are also vectors of plant viruses (Table A.1). *Paratrichodorus* and *Trichodorus* are classified in the order Triplonchida, suborder Diphtherophorina. The two genera are distinguished by a slightly expanded labial region bearing cephalic sensilla, amphids in an inverted stirrup shape, a spear in the form of a solid dorsal tooth, and an elongate, narrow esophagus that expands posteriorly into a short, pyriform bulb (Taylor and Brown, 1997).

Trichodorid nematodes transmit tobraviruses, which are straight, tubular rods about 23 nm in diameter and mostly 45–115 or 180–210 nm long, depending on the isolate. There are only three tobraviruses, all of which are transmitted by nematodes.

In 1958, *Xiphinema index* was the first nematode shown to be a natural vector of grapevine fanleaf nepovirus (Hewitt et al. 1958). During the 1960s, 1970s, and early 1980s more than 100 reports of associations between nematodes and plant viruses were published. However, many of these reports did not contain sufficient evidence to justify the conclusion that a particular virus is transmitted by a particular nematode species. The principal reasons for not establishing transmission by nematodes were the following:

1. The host plants were not systemically infected with the virus.
2. Contamination of plant roots with more than one pathogen was a common problem.
3. The methods used to describe the nematode–virus associations were not deemed adequate, or proper controls were not in place.
4. The virus or nematode species was not correctly identified.

Laboratory methods and criteria for properly assessing the transmission of nepoviruses by longidorid nematodes were established by Trudgill et al. (1983), and similar criteria for the transmission of tobraviruses by trichodorid

[1] Some authorities recognize only three orders—the Aphelenchida, Dorylaimida, and Tylenchida. Two genera (*Paratrichodorus* and *Trichodorus*) containing species capable of transmitting plant viruses are placed in the Dorylaimida in Chapter 3; they are placed in the Triplonchida by some authorities.

**Table A.1.** Longidorid and trichodorid nematode species proved to be vectors of nepoviruses and tobraviruses[a]

| Nematode vector | Virus | Hosts |
|---|---|---|
| Species transmitting nepoviruses | | |
| *Longidorus apulus* | Artichoke Italian latent (AILV), Italian strain | Artichoke (*Cynara cardunculus*) |
| *L. arthensis* | Cherry rosette (CRV) | Cherry |
| *L. attenuatus* | Tomato black ring (TBRV), German-English strain | A wide host range, including alfalfa, artichoke, asparagus, beans, cabbage, celery, cucumber, gladiolus, grape, leek, lettuce, onion, narcissus, peach, *Pelargonium,* phlox, potato, pumpkin, raspberry, strawberry, sugar beet, tomato, tulip, and turnip |
| *L. diadecturus* | Peach rosette mosaic (PRMV) | Peach, plum |
| *L. elongatus* | Raspberry ringspot (RRSV), Scottish strain | A wide host range, including anemone, beets, blackberry, cherry, gooseberry, grape, narcissus, phlox, potato, raspberry, red currant, and strawberry |
| | TBRV, Scottish strain | See *L. attenuatus* |
| *L. fasciatus* | AILV, Greek strain | Artichoke, chicory |
| *L. macrosoma* | RRSV, English strain | See *L. elongatus* |
| *L. martini* | Mulberry ringspot (MRSV) | Mulberry (only in Japan) |
| *Paralongidorus maximus* | RRSV, grapevine strain | Raspberry and red currant |
| *Xiphinema americanum* | Cherry rasp leaf (CRLV)<br>PRMV<br>Tobacco ringspot (TRSV)<br>Tomato ringspot (ToRSV) | A wide host range, including almond, American spearmint, anemone, apple, artichoke, ash, autumn crocus, beans, begonia, blackberry, blueberry, cherry, chickweed, crocus, cucumber, dandelion, dogwood, elderberry, forsythia, geranium, gladiolus, grape, hyacinth, hydrangea, impatiens, iris, lily, narcissus, nectarine, orchids, peach, *Pelargonium, Plantago lanceolata,* plum, raspberry, roses, soybean, strawberry, tobacco, tulip, and many wild plants |
| *X. diversicaudatum* | Arabis mosaic (ArMV), type strain and hop strain | A wide host range of woody and herbaceous plants, including barley, begonia, carnation, celery, cherry, crocus, cucumber, gladiolus, grape, hops, hyacinth, lily, narcissus, olive, peach, raspberry, rhubarb, roses, strawberry, tulip, and white clover |
| | Strawberry latent ringspot (SLRSV) | Black currant, celery, cherry, olive, peach, plum, raspberry, rose, strawberry |
| *X. index* | Grapevine fanleaf (GFLV) | Grape |
| *X. intermedium*[b] | TRSV, ToRSV | See *X. americanum* |
| *X. italiae* | GFLV | Citrus, conifers, grape, mulberry, olive, raspberry |
| *X. rivesi*[b] | CRLV, PRMV, TRSV, ToRSV | See *X. americanum* |
| *X. tarjanense*[b] | TRSV, ToRSV | See *X. americanum* |
| Species transmitting tobraviruses[c] | | |
| *Paratrichodorus allius* | Tobacco rattle (TRV) | More than 400 woody and herbaceous plants, weeds, and other wild plants, including anemone, artichoke, aster, cereals, corn (maize), crocus, cyclamen, freesia, gladiolus, grasses, hyacinth, lettuce, lily, narcissus, orchids, pepper (*Capsicum*), petunia, phlox, potato, spinach, tobacco, and tulip |
| *P. anemones* | Pea early-browning (PEBV) | Alfalfa, pea |
| | TRV | See *P. allius* |
| *P. minor* (syn. *P. christiei*) | TRV | See *P. allius* |
| *P. nanus* | TRV | See *P. allius* |
| *P. pachydermus* | PEBV, TRV | See *P. allius* and *P. anemones* |
| *P. tansaniensis* (syn. *P. allius*) | TRV | See *P. allius* |
| *P. teres* | PEBV, TRV | See *P. allius* and *P. anemones* |
| *P. tunisiensis* | TRV | See *P. allius* |
| *Trichodorus cylindricus* | TRV | See *P. allius* |
| *T. primitivus* | PEBV, TRV | See *P. allius* and *P. anemones* |
| *T. similis* | TRV | See *P. allius* |
| *T. viruliferus* | PEBV, TRV | See *P. allius* and *P. anemones* |

[a] Adapted from Taylor and Brown (1997).
[b] Identification of species belonging to the *X. americanum* group confirmed after the review of this group by Lamberti and Bleve-Zacheo (1979) (unequivocal identification of the vector species was not previously available).
[c] A third tobravirus, pepper ringspot virus (PepRSV), has been reported to systemically infect crops (including pepper, tomato, and artichoke) and wild hosts (e.g., *Bidens pilosa*) in Brazil. There is little information on what species of *Paratrichodorus* and *Trichodorus* are involved as vectors.

nematodes (species of *Paratrichodorus* and *Trichodorus*) were established by Brown et al. (1989, 1995). These criteria are the following:

1. Systemic infection of the bait host plant must be demonstrated.
2. Laboratory experiments should be done only with handpicked nematodes after positive identification of the species.
3. Appropriate conditions should be included to show, without question, that a specific nematode is the vector.
4. The virus isolated should be fully characterized.

Nematode-virus associations that satisfy these stringent criteria are presented in Table A.1. Associations between nematodes and viruses in which virus transmission has not been demonstrated according to these criteria are presented in Table A.2.

**Table A.2.** Longidorid–nepovirus and trichodorid–tobravirus associations not fulfilling criteria for assessing transmission of viruses by nematodes[a]

| Nematode | Virus | Reason for not fulfilling criteria[b] | Hosts |
|---|---|---|---|
| Associations with nepoviruses | | | |
| *Longidorus attenuatus* (syn. *L. apulus*) | Artichoke Italian latent (AILV) | 3 | Artichoke |
| *L. breviannulatus* | Brome mosaic bromovirus | | Bromegrass, bean, cucumber, tobacco |
| *L. caespiticola* | Arabis mosaic (ArMV) | 1 | A very wide host range of woody and herbaceous plants, including barley, begonia, carnation, crocus, celery, cherry, cucurbits, gladiolus, grape, hops, hyacinth, lily, narcissus, olive, peach, raspberry, rhubarb, roses, strawberry, sugar beet, tulip, and white clover |
| | Raspberry ringspot (RRSV) | 1 | A wide host range, including anemone, beets, blackberry, cherry, gooseberry, grape, narcissus, phlox, potato, raspberry, red currant, and strawberry |
| *L. diadecturus* | Tomato black ring (TBRV) | 1 | A wide host range, including alfalfa, artichoke, asparagus, beans, cabbage, celery, cucumber, gladiolus, grape, leek, lettuce, onion, peach, *Pelargonium,* phlox, potato, pumpkin, raspberry, strawberry, sugar beet, tomato, tulip, and turnip |
| *L. elongatus* | Carnation ringspot dianthovirus | 1, 2 | A wide host range, including carnation and sweet William |
| *L. euonymus* | Euonymus mosaic necrovirus | 1, 2 | European spindle-tree (*Euonymus europaeus*) |
| *L. leptocephalus* | RRSV | 1 | See *L. caespiticola* |
| *L. macrosoma* | Brome mosaic bromovirus | 1, 2 | See *L. breviannulatus* |
| | Carnation ringspot | | See *L. elongatus* |
| | Pear stony pit ilarvirus | 1–3 | Pear |
| | Prunus necrotic ringspot ilarvirus | 1, 2 | Almond, apricot, cherry, peach, plum, roses |
| *L. profundorum* | RRSV | 1, 2 | See *L. caespiticola* |
| *Longidorus* sp. | Potato virus U (PVU) | 2 | Potato |
| *Paralongidorus maximus* | ArMV | 1 | See *L. caespiticola* |
| | Strawberry latent ringspot (SLRSV) | 1 | A wide host range, including black currant, celery, cherry, gladiolus, lily, olive, narcissus, phlox, peach, plum, raspberry, roses, and strawberry |
| *Xiphinema americanum sensu lato* | Euonymus ringspot necrovirus | 2, 3 | European spindle-tree (*Euonymus europaeus*) |
| | Potato black ringspot (PBRSV) | 2, 3 | Potato |
| | Raspberry leaf crinkle | 2, 3 | A wide host range, including almond, apple, artichoke, blackberry, blueberry, cherry, chickweed, cucumber, dandelion, gladiolus, grape, hydrangea, narcissus, nectarine, peach, orchids, *Plantago lanceolata,* plum, raspberry, strawberry, and tobacco |
| *X. bakeri* | ArMV | 1 | See *L. caespiticola* |
| *X. basiri* (syn. *X. ifacolum*) | Cowpea mosaic comovirus | 2, 3 | Cowpea, mung bean, soybean |
| *X. brevicolle* | Tomato ringspot (ToRSV) | 2, 3 | See *X. americanum* |
| *X. diversicaudatum* | Cherry leafroll (CLRV) | 1, 2 | Blackberry, cherry, elm, rhubarb |
| | RRSV | 1, 2 | See *L. caespiticola* |
| | Brome mosaic bromovirus | 1, 2 | See *L. breviannulatus* |
| | Carnation ringspot dianthovirus | 1, 2 | See *L. elongatus* |
| *X. europaeum* (syn. *X. coxi*) | ArMV | 1, 2 | See *L. caespiticola* |
| | Brome mosaic bromovirus | 1, 2 | See *L. breviannulatus* |
| | CLRV | 1, 2 | See *L. diversicaudatum* |
| | SLRSV | 2, 3 | See *P. maximus* |
| | Tobacco ringspot (TRSV) | 1, 2 | A wide host range, including American spearmint, anemone, apple, ash, autumn crocus, begonia, blackberry, blueberry, cherry, crocus, cucumber, dogwood, elderberry, forsythia, grape, geranium, hyacinth, hydrangea, impatiens, iris, lily, narcissus, roses, soybean, tobacco, and tulip |
| *X. index* | ArMV | 2 | See *L. caespiticola* |
| | Grapevine chrome mosaic (GCMV) | 2 | See *L. diadecturus* |
| *X. vuittenezi* | CLRV | 1, 2 | See *X. diversicaudatum* |
| | Grapevine fanleaf (GFLV) | 2, 3 | Grape |
| Associations with tobraviruses | | | |
| *Paratrichodorus minor* | Papaya mosaic potexvirus | 1, 2 | Papaya |
| *P. porosus* | Tobacco rattle (TRV), California strain | 1–3 | A very wide host range, including anemone, artichoke, aster, cereals, crocus, cyclamen, freesia, gladiolus, grasses, hyacinth, iris, lettuce, lily, narcissus, orchids, pepper, petunia, phlox, potato, spinach, tobacco, and tulip |
| *P. teres* | TRV, Oregon strain | 1–3 | See *P. porosus* |
| *Trichodorus hooperi* | TRV, Scotland strain | 2 | See *P. porosus* |
| *T. sparsus* | TRV, Netherlands strain | 2, 3 | See *P. porosus* |

[a] Adapted from Taylor and Brown (1997).

[b] Reasons for not fulfilling criteria for assessing transmission of viruses by nematodes: 1) the plants tested were not systemically infected with virus, and contamination of plant roots by the virus was possible; 2) the methods used were inadequately described, or controls were inadequate or inappropriate; 3) the virus or the nematode species was not correctly identified.

## Virus Transmission Test Procedures

To establish that a nematode is a vector of a virus, the following conditions must be met (Taylor and Brown, 1997):

1. The nematode must ingest the virus from a source plant.
2. It must retain the virus, which it later ejects into a bait plant.
3. It must infect the bait plant with the virus.
4. Contamination of the bait plant with the virus from some other source and transmission by an alternative vector have been shown not to have occurred.

Further refinements in methodology are constantly occurring as new techniques become available, especially in molecular biology. This allows investigations into why only certain species of nematodes are vectors of specific viruses and why certain populations of a vector species differ in their ability to transmit different serological isolates of a virus.

A highly specific relationship takes place between a nematode vector and the virus or viruses it transmits. This specificity is a genetically determined recognition process between vector and virus. It is a consequence of the nature, site, and mechanism of virus particle retention within the nematode vector. This is determined by the nematode's ability to retain virus particles at specific sites within the feeding apparatus for a period of days, weeks, or months. *Paratrichodorus, Trichodorus,* and *Xiphinema* species can retain their associated viruses for periods of up to eight to 12 months when stored in moist, plant-free soil. Viruses are less persistent in *Longidorus* species, usually being retained for less than two months.

In laboratory-greenhouse studies, the rate of transmission of viruses by their nematode vectors is affected by temperature, soil moisture, the bait plant, the population and developmental stages of the nematode, the virus isolate, and even the size of the pot.

Nematode vectors that feed on plant roots can transmit viruses in all developmental stages. In general, increasing the acquisition and transmission access periods from hours to several weeks increases the frequency of transmission. For example, *X. americanum* was shown to transfer a virus in an hour or less, but it took four days of access to bait plants to achieve 100% transmission. An increase in the rate of transmission with increasing access time probably reflects the increasing proportion of nematodes that have fed on the source plant and not necessarily the intrinsic efficiency of the vector (Taylor and Brown, 1997). Only one virus-infective nematode per pot is often sufficient for transmission.

## Sites of Virus Retention in Nematode Vectors

As shown by electron microscopy, viruses are attached mainly to the cuticular lining of the odontostyle and the anterior region of the alimentary tract of nematode vectors. They are shed with the cuticle when the nematode molts. Virus particles are not found in eggs. The following retention sites have been identified (Taylor and Brown, 1997):

1. the inner surface of the odontostyle (in *Longidorus apulus, L. arthensis, L. elongatus, L. fasciatus,* and *L. macrosoma*)
2. the cuticle lining the lumen of the odontophore and the esophagus (in *Xiphinema americanum sensu lato, X. diversicaudatum,* and *X. index*)
3. the esophageal lumen, with virus particles adsorbed to the cuticular lining along the entire length of the lumen (tobacco rattle virus in *Paratrichodorus pachydermus* and *Trichodorus similis*)
4. the triangular section of the lumen adjacent to the anterior end of the guide ring (in *L. macrosoma* and *X. diversicaudatum*)

Differences in the efficiency of transmission by nematode vectors are related to the intimate interaction between the nematode and the virus strain, such as the flow of secretions from the esophageal glands during feeding and failure to release sufficient virus particles to effect transmission. In some cases, transmission may occur only when the virus isolate and the nematode vector population originate from the same locality.

Determinants of vector transmissibility of certain nepo- and tobraviruses (e.g., raspberry ringspot, tomato black ring, and tobacco rattle) are carried in the RNA-2 segment of the virus genome. RNA-2 determines the protein coat of the virus. The specificity of transmission also depends on the specific adsorption of virus particles at retention sites in the nematode. Dissociation of virus particles probably occurs slowly when secretions of the esophageal glands are ejected forward through the lumen of the esophagus and feeding apparatus during the early stages of feeding.

The primary requirement of virus transmission is compatibility between virus and vector at the site of retention in the nematode, which involves the surface of the virus protein and one or more genetic determinants (Taylor and Brown, 1997).

The mechanisms of virus retention within nematode vectors is presently unknown. Surface charges are believed to be involved, since virus particles are retained during ingestion and released by pH changes induced by salivation. Another possibility for some nematodes is that the virus coat protein may have lectin-like properties, with ingested carbohydrates being involved both in retention and dissemination of virus particles. During feeding, most ingested virus particles pass directly into the alimentary tract, and only a small proportion is retained.

## References

Brown, D. J. F., Ploeg, A. T., and Robinson, D. J. 1989. A review of reported associations between *Trichodorus* and *Paratrichodorus* species (Nematoda: Trichodoridae) and tobraviruses with a description of laboratory methods for examining virus transmission by trichodorids. Rev. Nematol. 12:235–241.

Brown, D. J. F., Robertson, W. M., and Trudgill, D. L. 1995. Transmission of viruses by plant nematodes. Annu. Rev. Phytopathol. 33:223–249.

Coomans, A. 1996. Phylogeny of the Longidoridae. Russ. J. Nematol. 4:51–59.

Hewitt, W. B., Raski, D. J., and Goheen, A. C. 1958. Nematode vector of soil-borne fanleaf virus of grapevines. Phytopathology 48:586–595.

Lamberti, F., and Bleve-Zacheo, T. 1979. Studies on *Xiphinema americanum sensu lato* with descriptions of fifteen new species (Nematoda, Longidoridae). Nematol. Mediterr. 7:51–106

Lamberti, F., and Roca, F. 1987. Present status of nematodes as vectors of plant viruses. Pages 321–328 in: Vistas on Nematology. J. A. Veech and D. W. Dickson, eds. Society of Nematologists, Hyattsville, Md.

Lamberti, F., Taylor, C. E., and Seinhorst, J. W., eds. 1975. Nematode Vectors of Plant Viruses. Vol. 3. Plenum Press, New York.

Maggenti, A. 1983. Nematode higher classification as influenced by species and family concepts. Pages 25–40 in: Concepts in Nematode Systematics. Systematics Assoc., Spec. Vol. 22. A. R. Stone, H. M. Platt, and L. F. Khalil, eds. Academic Press, San Diego.

Taylor, C. E., and Brown, D. J. F. 1981. Nematode-virus interactions. Pages 281–301 in: Plant Parasitic Nematodes. Vol. 3. B. M. Zuckerman and R. A. Rohde, eds. Academic Press, San Diego.

Taylor, C. E., and Brown, D. J. F. 1997. Nematode Vectors of Plant Viruses. CAB International, Wallingford, Oxon., U.K.

Trudgill, D. L., Brown, D. J. F., and McNamara, D. G. 1983. Methods and criteria for assessing the transmission of plant viruses by longidorid nematodes. Rev. Nematol. 6:133–141.

# Glossary

For an all-inclusive listing of terms used by plant pathologists and in plant diagnostic clinics, see M. C. Shurtleff and C. W. Averre III, *A Glossary of Plant Pathological Terms* (American Phytopathological Society, St. Paul, Minn., 1997). See also the glossary in the *International Code of Zoological Nomenclature,* 3rd ed. (International Trust for Zoological Nomenclature, London, 1985).

**a** In de Man's measurement code, the length of the body of a nematode divided by (usually) its greatest width, often expressed as a range (e.g., a = 50–60). A large number indicates a slender body.

**abbr.** Abbreviation.

**accession number** The number given to a culture or specimen in a collection of cultures of microorganisms.

**-aceae** The suffix attached to the stem of a name (usually a genus name) to form the name of a taxonomic family of plants.

**acquisition access period** The period during which a vector has access to a source of inoculum.

**acquisition feeding period** The period during which a vector (e.g., a nematode) feeds on a virus-infected plant to acquire the virus for subsequent transmission (become viruliferous).

**acquisition threshold period** The minimum feeding period needed for a vector to acquire an infective dose of a pathogen (usually a virus).

**acute** Pertaining to disease in which symptoms develop suddenly (as opposed to chronic disease) and are generally more severe than chronic symptoms; having a sudden onset and a sharp rise within a short time. See **chronic**.

**adanal** Located near the anus (used with reference to a bursa that does not envelop the entire tail).

**adj.** Adjective.

**adpressed** See **appressed**.

**agar, agar-agar** A nonnitrogenous, gelatinous mixture of dried polysaccharides, some anionic, extracted from certain red algae (seaweeds, class Rhodophyceae) and used in semisolid media on which microorganisms are cultured. Agar forms a gel with water (below about 40°C) and is usually used at a concentration of 1.5–2.0% to solidify media. Some nematodes in pure culture can complete their life cycles by feeding on fungi growing on agar media.

**aggressiveness** The relative ability of plant pathogens to colonize and damage plants. See **virulence**.

**ala** (pl. **alae**) An expansion or winglike projection formed by a longitudinal thickening of the cuticle in nematodes. There are three types:

- **cervical alae,** in the anterior region of nematodes parasitic on animals
- **caudal alae,** forming the bursa, in the posterior region of males in many genera
- **longitudinal alae,** extending the length of the body sublaterally; most nematodes have four of these alae

**-ales** The suffix attached to the stem of a genus name to form the name of a taxonomic order of plants.

**alimentary tract** The tubular, food-carrying passage extending from the mouth to the anus. In nematodes, it includes the oral aperture, stoma and stylet, esophagus, intestine, prerectum, rectum, and anus.

**ambifenestrate** Having two openings in the vulval cone, as in *Heterodera* species with a vulval bridge, which separates the openings.

**amphid** A chemosensory organ in nematodes. Amphids, also called lateral organs, occur laterally in pairs in the anterior region. They are composed of the following structures:

- **amphidial duct,** a channel that connects the amphidial pore to the sensilla pouch
- **amphidial gland,** a microvillus organ enveloping the sensilla pouch
- **amphidial nerve,** a nerve innervating the amphid
- **amphidial pore,** the opening of the amphid, through an aperture in the cuticle
- **amphidial pouch,** a chamber, often cup-shaped in outline, usually just behind the amphidial pore; also called the amphidial pocket or amphidial chamber
- **sensilla pouch,** a chamber containing nerve processes, located posterior to the amphidial pouch

In the class Secernentea, the amphidial pores are small openings on the lips. In the class Adenophorea, the pores are postlabial and are circular, spiral, or cyathiform (pocketlike).

**amphidelphic** Having two ovaries, one directed anteriorly and the other posteriorly.

**amphimixis** Reproduction by the union of an egg and a sperm; true bisexual reproduction. adj. **amphimictic**. In amphimictic nematode populations, males and females are usually present in about equal numbers.

**ampulla** (pl. **ampullae**) A membranous, flask-shaped sac or swelling; a widening in an anatomical canal, forming a reservoir.

**anabiosis** A state of suspended animation of organisms induced by desiccation and terminated by hydration.

**anaerobe** A microorganism able to grow in the absence of free (molecular) oxygen, usually in an environment with a high level of carbon dioxide. See **facultative anaerobe**; **obligate anaerobe**.

**anaerobic** Living or occurring in the absence of oxygen.

***Anguina*** The genus of seed gall and leaf gall nematodes.

**anhydrobiosis** See **anabiosis**.

**aniso-** (prefix) Unequal.

**annual** Lasting only one year or one growing season.

**annulation** Any of the transverse depressions (striae) formed at regular intervals in cuticle in nematodes; a ring- or collar-like fold.

**annule** The thickened interval between two transverse striae in the cuticle in nematodes; a ring- or collar-like fold.

**antagonism** The action of a microorganism in producing toxic metabolic products that kill, injure, or inhibit the growth of another microorganism in close proximity; generally, counteraction between organisms or groups of organisms.

**anteriad** Directed forward, as opposed to posteriad.

**anterodorsal** In the front and at the tip or upper side of the body.

**anteromesal** In the front and along the middle of the body.

**anteroventral** In the front and underneath or on the lower side of the body.

**anti-** (prefix) Against.

**antrorse** Directed upward or forward, as opposed to retrorse.

**anus** The posterior opening of the alimentary tract. In nematodes, it forms a slitlike opening on the ventral surface of the body.

**aperture** An opening, such as a pore, slit, or crack.

**aphelenchoid** Resembling nematodes in the genus *Aphelenchus* or *Aphelenchoides* or the superfamily Aphelenchoidea. See **esophagus**.

***Aphelenchoides*** The genus of bud and leaf (foliar) nematodes, including *A. besseyi,* causing summer crimp (summer dwarf) in strawberry and white tip in rice; *A. fragariae,* causing spring crimp (spring dwarf) in strawberry; and *A. ritzemabosi*, the chrysanthemum nematode.

**apical** At or near the end, tip (apex), or outermost part.

**apogamy** A type of apomixis in which the embryo is produced from an unfertilized female gamete or an associated cell. See **parthenogenesis**.

**apolysis** Separation of the cuticle from the hypodermis or epidermis during molting of nematodes, often preceded by a period of immobility (lethargus).

**apomixis** Nonsexual reproduction. See **apogamy**; **parthenogenesis**.

**apophysis** A swelling or expansion, like that in organ Z in *Xiphinema* species.

**appressed** Closely flattened out; closely applied to or pressed against a surface.

**arch** The curved portion of the perineal pattern dorsal to the lateral field or lines in females of *Meloidogyne.*

**arcuate** Strongly curved, like a bow; archlike.

**areola** (pl. **areolae**) A portion of the cuticle (typically, the lateral fields) delimited by transverse and longitudinal markings in some nematodes.

**areolate, areolated** Marked out or divided into little areas, usually by cracks or crevices (used with reference to the cuticle of some nematodes).

**areolation** Transverse annulation extending into the lateral fields of some nematodes.

**aseptic** Free of microorganisms.

**asymmetrical** Lacking symmetry about a definite axis.

**automixis** Reproduction by parthenogenesis or self-fertilization. adj. **automictic**. In automictic nematode populations, males are usually rare or absent.

**avirulent** Nonpathogenic; lacking virulence; unable to cause disease.

**awl nematode** Any member of the genus *Dolichodorus.*

**axenic culture** Artificial growth of a single species in the absence of other organisms; a germ-free, pure, uncontaminated culture. See **gnotobiotic culture**.

**axial** Arranged around a point or axis.

**b** In de Man's measurement code, the length of the body of a nematode divided by the distance from the anterior end to the junction of the esophagus and the intestine, often expressed as a range. A low number indicates a long head region.

**Baermann funnel technique** A method of extracting nematodes from soil, screening residue, plant tissue, or other matter; sample material is placed on porous paper or cloth supported in a water-filled funnel, and nematodes by their own action move out of the sample and through the paper or cloth, settling in the funnel stem, where they are drawn off by tubing. Many modifications of this technique have been devised.

**basal** Related to, located at, or forming near the base of a structure; near the point of attachment of an appendage.

**basal bulb** A muscular or glandular swelling of the posterior end of the esophagus in nematodes, abutting the intestine, and in some species containing a valvular structure; also called the terminal, posterior, or end bulb or cardiac bulb.

**basal flanges** The knoblike expansions at the base of the odontostylet.

**basal gland lobe** The glandular basal region of the esophagus, overlapping the intestine, in some nematodes.

**basal knobs** See **stylet knobs and flanges**.

**basal region of esophagus** The posterior portion of the esophagus in nematodes, taking various forms in different species, such as a distinct bulb, a lobe overlapping the intestine, or a cylindrical structure.

**basal ring** The circular base of the cephalic framework in nematodes.

***Belonolaimus*** The genus of sting nematodes.

**bi-** (prefix) Two; twice; twofold.

**bifid** Divided into two equal lobes or parts by a median cleft.

**bilateral symmetry** The condition of having a body with two halves that are exact counterparts (mirror images) of each other. Organisms with bilateral symmetry are characterized by ventral and dorsal sides marked by openings.

**binucleate** Containing two nuclei.

**bio-** (prefix) Pertaining to life.

**bioassay** A qualitative and quantitative test involving the response of a living cell or organism to biologically active substances under standardized conditions, often used to measure the infectivity of a pathogen or the toxicity of a substance.

**biological form, biological race** See **physiologic race**.

**biotic** Of or pertaining to living organisms; animate.

**biotroph** A parasitic organism that obtains its energy from cells of a living host, commonly as an obligate parasite. adj. **biotrophic**. See **necrotroph**.

**biotype** A group of organisms belonging to the same species and differing from other members of the species in one or a few biochemical, physiological, or behavioral properties; a group of individuals with a like genetic makeup; a subdivision of a pathological race; a geographical isolate or clone of a vector species. See **physiologic race**.

**biserial, biseriate** In two series or rows.

**bloat** A disease of onions, caused by *Ditylenchus dipsaci*.

**body pores** A series of minute, slightly submedial or lateral depressions or openings traversing each side of the body of nematodes in the superfamily Dorylaimoidea.

**bright-field microscope** A microscope that uses transmitted light.

**bristle** See **seta**.

**buccal aperture** The anterior opening of the digestive system in nematodes; also called the oral aperture.

**buccal capsule** See **stoma**.

**bulb nematode** *Ditylenchus dipsaci*, also called the stem and bulb nematode.

**bulla** (pl. **bullae**) A dark-colored, irregularly shaped structure near the vulval fenestra or underbridge of some nematodes in the family Heteroderidae.

**burrowing nematode** *Radopholus similis*.

**bursa** (pl. **bursae**) A structure consisting of a pair of caudal alae (lateral fin- or winglike extensions of the cuticle at or near the posterior end) in some male nematodes, used to clasp the female during copulation. The bursa is adanal in some species and surrounds the tail tip in other species. It is separated by ribs in some nematodes.

**c** In de Man's measurement code, the length of the body of a nematode divided by the length of the tail, from the anus or cloaca to the terminus, usually expressed as a range. A low number indicates a long tail.

**c′** In de Man's measurement code, the length of the tail of a nematode, from the anus or cloaca to the terminus, divided by width of the body at the anus or cloaca. A low number indicates a short tail.

**caecum** A cavity open at one end, such as the blind end of the lumen in nematodes, especially an outpocketing of the intestine at the anterior end.

**cap cell** An apical cell at the distal end of the female and male gonads of nematodes; the primordial germ cell.

**capitulum** A sclerotized guiding piece for the gubernaculum, on the ventral cloacal surface of some nematodes in the family Hoplolaimidae; a medial ventral sclerotization of the spicular pouch.

**cardia** (p. **cardiae** or **cardias**) A valvular apparatus, frequently muscular, connecting the esophagus with the intestine, and functioning to prevent regurgitation; also called the cardiac valve or esophago-intestinal valve.

**cardiac bulb** See **basal bulb**.

**cardiac glands** The three glandular bodies located at the base of the esophagus in nematodes.

**cardiac valve** See **cardia**.

**cardinal temperatures** The minimum, optimum, and maximum temperatures for growth of an organism.

**caudal** Pertaining to the tail or the posterior region.

**caudal alae** Alae forming the bursa, situated in the posterior portion of some male nematodes.

**caudal glands** The three to five hypodermal glands located in the tail of some nematodes in the class Adenophorea, emptying subterminally or terminally through a pore (the spinneret), and secreting a fluid that hardens to form an attachment thread in water.

**caudal supplements** Papillate, glandular structures on the ventral surface of the caudal region of male nematodes in the class Adenophorea, possibly serving the same purpose as a bursa.

**caudalid** A narrow, clear, bandlike structure below the cuticle (similar to a hemizonid or hemizonion), located slightly anterior to the anus in some nematodes in the order Tylenchida, possibly corresponding to the paired anolumbar commissures linking the preanal ganglion to the lumbar ganglia.

**causal agent** The organism or agent (bacterium, fungus, phytoplasma, nematode, virus, viroid, toxin, etc.) that incites a given disease or injury; a biotic or abiotic agent capable of causing disease.

**cell** The structural and functional unit of all plant and animal life. The essential feature of a cell is living protoplasm limited by a membrane and containing organelles with diverse functions. In plants, the cell is surrounded by the cell wall.

**cell culture** A culture in a liquid or soft gel medium in vitro.

**cell sap** The liquid in the vacuoles of plant cells.

**cell wall** The protective, resistant, but permeable polysaccharide structure secreted on the exterior surface of the cell membrane in plants and some other organisms.

**cephalic** Pertaining to or located in the head region.

**cephalic framework** The sclerotized, subcuticular framework that provides rigidity in the head region of nematodes. In stylet-bearing nematodes, the stylet protractor muscles are attached to the cephalic framework, which axially forms a stylet guide.

**cephalic papillae, cephalic setae** Sensory organs of nematodes, typically arranged in an inner circlet of six, grouped around the oral opening, or a circlet of 10 on the outer contour of the lips. There are many variations on this arrangement. See **seta**.

**cephalid** Either of the two (posterior and anterior) highly refractive structures extending completely around the body in the cephalic region of some nematodes, possibly functioning as part of the nervous system; also called the hypodermal commissure.

**cervical** Pertaining to or located in the neck region.

**cervical papillae** See **deirid**.

**cheilorhabdion** A type of rhabdion forming the cuticularized walls of the cheilostom. See **rhabdion**.

**cheilostom** The most anterior part of the stoma in nematodes.

**chemotaxis, chemotropism** Movement or growth of an organism in response to a chemical stimulus, often a reaction to food material.

**chlorosis** Paling, yellowing, or whitening of normally green tissue, characterized by the partial to complete destruction of chlorophyll, due to infection by a virus, lack or unavailability of a nutrient (e.g., iron, manganese, zinc, nitrogen, boron, or magnesium), lack of oxygen in waterlogged soil, alkali injury, or other factors. adj. **chlorotic**. See **yellowing**.

**chord** or **cord** A longitudinal internal thickening of the hypodermis in nematodes. Most nematodes have four main chords (one dorsal, two lateral, and one ventral) extending the length of the body.

**Christie–Perry extraction technique** A method of isolating nematodes from soil, combining the wet-sieving and Baermann funnel techniques.

**chronic** Slow-developing, persistent, or recurring; pertaining to disease in which symptoms appear over a long period (as opposed to acute disease) or an infection that lingers, often without symptoms; pertaining to a condition of long duration. See **acute**.

**chrysanthemum nematode** *Aphelenchoides ritzemabosi*.

**circum-** (prefix) All around; round about.

**circumesophageal commissure** See **nerve ring**.

**circumfenestrate** Having a single opening in the vulval cone, as in *Heterodera* species lacking a vulval bridge.

**circumoral elevation** The thickened membrane surrounding the oral opening of nematodes in which the lips are rudimentary or absent.

**cirrhus** (pl. **cirrhi**) or **cirrus** (pl. **cirri**) A cephalic appendage in certain nematodes.

**citrus nematode** *Tylenchulus semipenetrans*.

**cloaca** A duct or chamber formed by the junction of the digestive and reproductive systems, emptying through the anus, in male nematodes; the hindgut.

**clover cyst nematode** *Heterodera trifolii*.

**co-** (prefix) Together; with.

**coarse root** A pathological condition in which most of the lateral roots of a plant are killed by nematode attack, resulting in an open system of mostly main roots.

**coconut red ring nematode** *Rhadinaphelenchus cocophilus*.

**coelom** The main body cavity of a higher animal containing organs (occasionally misused to refer to the pseudocoelom in nematodes). See **pseudocoelom**.

**coelomyarian** Pertaining to a type of muscle cell in nematodes, in which the protoplasmic portion extends into the body cavity and the fibrillar part extends along the hypodermis.

**colonization** The period following infection during which a pathogen becomes established in its host. See **infection**.

**colonize** To establish an infection in a host or in one of its parts.

**comminute** To grind or crush into small particles. See **macerate**; **mince**; **triturate**.

**commissure** A bundle of nerve fibers connecting ganglia in lateroventral and rectal structures of nematodes.

**cone** See **vulval cone**.

**constriction** A sharply pronounced groove at the junction of the lip region and the cervical region in nematodes.

**contaminant** A substance or microorganism that renders an otherwise pure system unclean or unsuitable for its intended purpose. Bacteria, fungi, and mites are common culture contaminants.

**cord** See **chord**.

**corn cyst nematode** *Heterodera zeae*.

**corpus** The anterior, usually cylindrical part (the procorpus) of the esophagus in nematodes, the basal portion of which is swollen, forming the median bulb (metacorpus), in many species.

**cotype** A strain (specimen) of a species described by the author of the original description of the taxon, when the author did not designate a holotype or paratype; a syntype. See **type**.

**crenate** Having a scalloped or round-toothed margin.

**crenation** A rounded projection, one of a series of such projections forming the edge of an object or structure (used with reference to nematode tails that are annulated to the terminus, as opposed to smooth).

**criconematoid** Resembling nematodes in the genus *Criconema* and related plant-parasitic forms. See **esophagus**.

***Criconemella*** A genus of ring nematodes.

**crustaformeria** A glandular region in the distal part of the uterus in nematodes, which may play a role in the formation of the egg envelope; also called the quadricolumella.

**cultivar** (abbr. **cv.**) A cultivated variety of a plant; the commercial or popular name of closely related plants of common origin, which differ from other members of the same species in certain minor details (e.g., form, color, flower, or fruit) and retain their distinguishing features when reproduced, sexually or asexually. See **variety, botanical**.

**culture** [1] (v.) To artificially grow and propagate microorganisms or plant tissue on a prepared nutrient medium. [2] (n.) The growth of an organism or a group of organisms artificially maintained on such food material. See **pure culture**.

**culture medium** A solid or liquid preparation of nutrients, usually containing various inorganic salts, sugars, amino acids, and antibiotics, on which microorganisms or plant cells are grown. See **medium**.

**culture plate** A shallow, circular, covered dish of thin glass or plastic in which microorganisms are grown in pure culture in the laboratory.

**cuticle** The noncellular exterior covering of nematodes. adj. **cuticular**.

**cv.** See **cultivar**.

**cyathiform** Pocket-like or cup-shaped, but with a flaring margin (used with reference to the amphids of certain nematodes). See **amphid**.

**cylindrical** Round in cross section and of equal width throughout the length of the body (used with reference to a type of esophagus in nematodes). See **esophagus**.

**cyst** The dark, hardened body or oxidized cuticle of a dead adult female of *Globodera, Heterodera, Punctodera,* or *Sarisodera* species, which protects the eggs that may be contained within it.

**cyst nematode** Any member of the genera *Globodera, Heterodera, Punctodera,* and *Sarisodera.*

**cystoid nematode** Any member of the genus *Meloidodera.*

**cyto-** (prefix) Pertaining to cells.

**cytoplasm** The living substance (protoplasm) of a cell other than the nucleus and cell wall.

**dagger nematode** Any member of the genus *Xiphinema.*

**damage threshold** The lowest pest population density at which the pest causes damage.

**dark-field illumination** A method of microscope illumination in which the illuminating beam is a hollow cone of light formed by an opaque disk (stop) placed at the center of the substage condenser lens to prevent direct light from entering the objective. The specimen is viewed with light striking it at a low angle, scattered or diffracted by it, and appears brightly illuminated against a dark background.

**dauer larva** The quiescent, obligatory stage of juveniles of some nematode species, specialized for dispersal and long-term survival. The dauer larva is often enclosed in the cast cuticle of the preceding stage, sometimes in a relationship with a vector host, and has an altered metabolism that survives moisture stress and freezing and prevents further development and aging.

**de Man's measurement code** A system of measurements expressing the sizes and positions of nematode body parts (body length, relative length of the esophagus, relative length of the tail, position of the vulva, etc.); also called de Man's formulas or de Man's indices. See **a**; **b**; **c**; **c′**; **L**; **o**; **O**; **T**; **V**.

**decant** To pour a liquid from a vessel gently, so as not to roil and remove substances remaining in the vessel.

**decay** Disintegration or decomposition of plant tissue or other substrates by bacteria, fungi, or possibly other microorganisms and forces.

**decline** Reduction in vigor of a perennial planting as a result of chronic disease; gradual reduction in the health and vigor of a plant or planting that is slowly dying.

**decomposition** Degradation into simpler compounds; rotting of colonized plant tissue, usually by microorganisms.

**DEGO** See **dorsal esophageal gland orifice**.

**deirid** A porelike organ in many nematodes in the family Tylenchidae. Deirids, also called cervical papillae, occur in pairs in the lateral fields, in the vicinity of the nerve ring, and are believed by some to be sensory organs.

**dentate** Bearing teeth; in the form of teeth.

**denticles** Minute teeth, projections, or prickles in the stoma of some nematodes.

**depressor ani** The H–shaped muscle that dilates the rectum and elevates the posterior lip of the anus in nematodes, thus permitting defecation.

**desiccate** To dry up; to remove water from a substance. n. **desiccation**.

**DGO** See **dorsal esophageal gland orifice**.

**diagnosis** (pl. **diagnoses**) [1] Identification of the nature and cause of a disease. [2] A shortened version (in Latin) of a taxonomic description of a species or other taxon.

**diapause** A survival stage entered by certain nematodes when conditions are unfavorable for their continued development.

**dichotomous key** A natural key to taxa, with two alternatives at each step, leading eventually to a single taxon. This type of key supposedly reflects a degree of phylogenetic relationship between taxa.

**didelphic** Having two complete ovaries.

**dieback** Progressive death of leaves, shoots, branches, or roots, generally starting at the tip. v. **die back**.

**differential hosts** A set of hosts (species, cultivars, inbreds, or isogenic lines) whose distinctive symptoms in response to a given disease agent are used for identification of the agent, and especially for determining the physiologic race of the agent. See **indicator plant**.

**digonic** Pertaining to hermaphroditic reproduction (occurring in some nematodes) in which sperm and eggs are produced by different gonads in a single individual.

**dimorphic** Having two distinct shapes or forms, such as the male and the female forms of some nematodes. n. **dimorphism**.

**dioecism** The condition in which males and females are separate, sexually distinct individuals. adj. **dioecious**.

**diorchic** Having two complete testes.

**diplo-** (prefix) Two; twice; double.

**diplogasteroid** See **esophagus**.

**discoid** Disk-shaped.

**disease, of plants** Any abnormal or malfunctioning process that interferes with normal growth and development (e.g., structure and function), economic value, or aesthetic quality and leads to the development of symptoms. A continuous, often progressive condition in which any part of an affected plant is abnormal (e.g., in structure, function, or economic value) or the normal activity of the cells or organs of the plant is interrupted. Biotic diseases are caused by pathogens; abiotic diseases are caused by inanimate agents. See **injury**.

**disease complex** A disease caused by the action of two or more disease agents (such as concurrent parasitism by two or more pathogens), often manifested in symptoms not caused by any of the agents acting alone, and producing a total effect greater than the sum of the effects of each agent acting alone (an example of synergism).

**disease cycle** The sequence of events in disease development, including the stages of development of the pathogen and the effect of the disease on the host; the chain of events occurring between the time of infection and the final expression of disease. See **life cycle**.

**disease incidence** A measure of the number of individuals affected by a given disease in a host population (distinguished from disease severity).

**disease intensity** The amount or severity of disease in an individual.

**disease prevalence** The proportion of plants or crops affected by a disease in a defined geographical area.

**disease severity** A measure of damage done by a given disease in a host population (distinguished from disease incidence).

**disease signature** All of the factors and events associated with a disease (e.g., in plant diseases, the time of occurrence, distribution in the field, weather, soil type, spectrum of plants involved, and internal and external symptoms).

**distal** Away from center of the body; toward the end farthest from the center of body; far from the point of attachment or origin. See **proximal**.

**distribution** [1] Geographical distribution. See **range**. [2] The spread of a pathogen to an area outside its previous geographical range.

***Ditylenchus*** A genus of tylenchoid nematodes, including *D. angustus,* the rice stem nematode; *D. destructor,* the potato rot nematode; and *D. dipsaci,* the stem and bulb nematode

**diverticulum** The pouch or lobe containing the esophageal glands of nematodes.

**dormancy** A state of reduced physiological activity; a resting state; quiescence. adj. **dormant**.

**dorsal** On, near, or toward the back or upper surface of the body of an organism or one of its parts. See **ventral**.

**dorsal esophageal gland** A salivary gland located in the dorsal region of the esophagus, the only dorsally located of the three to nine salivary glands in nematodes (the others are in the subventral position in the basal portion of the esophagus).

**dorsal esophageal gland orifice** (abbr. **DEGO** or **DGO**) The opening of the dorsal esophageal gland, which empties into the lumen of the esophagus in nematodes; also called the dorsal esophageal gland outlet.

**dorsal gland** See **dorsal esophageal gland**.

**dorsolateral** At the top and to the side of the body.

**dorsomesal** At the top and along the midline of the body.

**dorsoventral** From the top to the bottom or from the upper to the lower surface of the body.

**dorsoventral axis** The imaginary median line extending from the dorsal to the ventral side.

**dorylaim** A common name for any nematode in the order Dorylaimida.

**dorylaimoid** See **esophagus**.

**dwarf nematode** *Aphelenchoides fragariae,* also called the spring crimp nematode.

**dwarfing** Underdevelopment of a plant or plant organs, caused by a disease agent, faulty nutrition, or unfavorable environmental conditions.

**e-, ex-** (prefix) From; out of; without; not having.

**ecdysis** (pl. **ecdyses**) The shedding of the nematode cuticle, which may be partially or completely reabsorbed following molting.

**eclosion** The act of hatching from an egg.

**economic threshold** The pest population density at which control measures should be implemented to prevent an increasing pest population from reaching the economic injury level; the population density of a pest or pathogen or the damage level at or above which the value of crop losses, in the absence of management efforts, would exceed the cost of management practices (especially the use of pesticides).

**ecto-** (prefix) Outside; external.

**ectoparasite** A parasite that feeds on a host from outside, without entering the host. See **endoparasite**; **parasite**.

**eelworm** See **nematode**.

**egg mass, egg matrix** The gravid female of *Meloidogyne* species with the external matrix containing extruded eggs; in the narrow sense, the extruded eggs contained within the matrix. There is no membrane envelope surrounding the eggs in the matrix.

**ejaculatory duct** The muscular terminal portion of the vas deferens.

**ejaculatory glands** Glands associated with the ejaculatory duct, secreting an adhesive cement functional in copulation.

**electron microscope** An instrument in which a focused beam of electrons produces a greatly enlarged image (at magnifications greater than 100,000×) of objects too small to be seen with a compound light microscope. See **scanning electron microscope**.

**elutriator** An apparatus used for extracting nematodes from soil by washing. See **Christie–Perry extraction technique**; **Seinhorst extraction technique**.

**encyst** To form a cyst or encasement. n. **encystment**. See **cyst**.

**end bulb** See **basal bulb**.

**endo-** (prefix) Inside; within.

**endomycorrhiza** (pl. **endomycorrhizae**) An association between a fungus and a plant root in which fungal hyphae invade cortical cells of the root; also called a vesicular-arbuscular mycorrhiza. Nematodes can influence the amount of mycorrhizae on roots. See **mycorrhiza**.

**endoparasite** A parasite that enters a host and feeds within it. See **ectoparasite**; **parasite**.

**ento-** (prefix) Inside.

**epi-** (prefix) Upon.

**epicuticle** The outermost layer of the cuticle in nematodes.

**epithelium** A membrane-like layer of cells lining the internal or external surface of an organ, tissue, or cavity in the body of an animal.

**epithet** The second element of the Latin binomial of a species; the portion of the scientific name of an organism following the genus name, in some cases including a third or fourth term (designating a subspecies). The genus name and the epithet are latinized and italicized. See **genus**; **Latin binomial**; **species**.

**equatorial** Around the middle.

**erumpent** Breaking through the surface of a host or substratum in the course of development; bursting forth.

**-escent** (suffix) Beginning; slightly.

**esophageal bulb** Any of the several enlargements of the esophageal wall, which are muscular or glandular or both.

**esophageal gland** Any of the three to nine elongated salivary glands (one dorsal and two to eight subventral), consisting of simple or branched tubules, located in the esophagus in nematodes.

**esophageal gland lobe** The basal, glandular portion of the esophagus, usually containing three glands, which overlaps the anterior end of the intestine in many plant-parasitic nematodes.

**esophageal lumen** The canal of the esophagus in nematodes.

**esophago-intestinal valve** The valve between the esophagus and the intestine in nematodes; also called the cardia or cardiac valve.

**esophagus** (pl. **esophagi**) The portion of the alimentary tract between the stoma (buccal capsule) and the anterior portion of the intestine in nematodes. It is differentiated into as many as four distinct parts: the procorpus, metacorpus, isthmus, and basal region. Several types are distinguished:

- **aphelenchoid** (as in *Aphelenchoides*), consisting of four parts (procorpus, metacorpus, isthmus, and basal region); the metacorpus is rhomboid and exceeds two-thirds the width of the body; the basal region is tylenchoid; the dorsal gland outlet is located just anterior to the valve of the metacorpus, rather than near the stylet base
- **bulboid** (as in *Achromadora*), consisting of two parts: a long, slender anterior portion and a base with a distinct, muscular bulb; a **plectoid** esophagus (as in *Plectus*) is similar but has a basal bulb with a true valve
- **criconematoid** (as in *Criconemella*), consisting of three parts (corpus, isthmus, and basal region); the procorpus and metacorpus are fused into a large corpus, with a large valve in the metacorpus; the basal region is a reduced bulb; the dorsal gland outlet is located near the base of the stylet
- **cylindrical** (as in *Mononchus*), consisting of a single structure, which is approximately the same diameter throughout its length
- **diplogasteroid** (as in *Diplogaster*), consisting of four parts (procorpus, metacorpus, isthmus, and basal region); no valves are present; the basal region is a glandular bulb
- **dorylaimoid** (as in *Dorylaimus*), consisting of two parts: a long, slender, anterior portion and a long, thicker, bulbar portion, which is muscular and glandular
- **neotylenchoid** (as in *Neotylenchus*), consisting of three parts (procorpus, isthmus, and basal region); the metacorpus is not present or is only slightly developed and lacks a valve; the basal region is tylenchoid; the dorsal gland outlet is located near the base of the stylet
- **rhabditoid** (as in *Rhabditis*), consisting of three parts; the procorpus and metacorpus are fused into a large corpus; the basal region is a muscular bulb with a "butterfly" valve
- **tylenchoid** (as in *Tylenchorhynchus* and *Pratylenchus*), consisting of four parts (procorpus, metacorpus, isthmus, and basal region); the metacorpus is round and less than two-thirds the width of the body; the basal region is a distinct bulb (in *Tylenchorhynchus*) or con-

sists of highly glandular lobes overlapping the intestine (in *Pratylenchus*); the dorsal gland outlet is located near the base of the stylet

**eu-** (prefix) True.

**ex-** (prefix) From; out of; without; not having.

**excretory canal** A tubelike structure, lined with cuticle, in the lateral chord or pseudocoelom in nematodes, which collects excretory products and conveys them to the excretory duct and pore.

**excretory pore** The exterior opening of the excretory system in nematodes, by which waste products are emitted, usually located on the ventral side of the body, generally near the basal region of the esophagus; also called the cervical gland orifice.

**exo-** (prefix) Outside.

**exotic** (n.) An organism introduced into an area where it is not indigenous.

**extrude** To push out; to emit to the outside. n. **extrusion**. adj. **extruded**.

**facultative** Able to live under more than one set of conditions; not obligate.

**facultative aerobe** An organism that normally lives without atmospheric oxygen but is capable of utilizing it when it is available.

**facultative anaerobe** An organism that normally lives in the presence of atmospheric oxygen but is capable of living anaerobically. See **anaerobe**; **obligate anaerobe**.

**facultative parasite** An organism that is normally saprophytic but is capable of living as a parasite under certain conditions and can be cultured on a laboratory medium. See **obligate parasite**; **parasite**.

**facultative saprobe, facultative saprophyte** An organism that is normally parasitic but can under proper conditions live saprophytically during part of its life cycle. See **obligate saprobe**; **saprobe**; **saprophyte**.

**fallow** Not cultivated or not planted for one or more seasons (used with reference to cropland); maintained without plant growth.

**false root-knot nematode** Any member of the genus *Nacobbus*.

**family** A taxonomic category containing related genera, tribes, or subfamilies grouped according to resemblance and similarity; a rank above genus and below order, suborder, and superfamily. The names of nematode families end with the suffix -idae.

**fenestra** (pl. **fenestrae**) A window-like or transparent spot; a thin-walled region of the vulval cone in some nematodes in the family Heteroderidae.

**fenestrate** Having a fenestra.

**fiber layer** The inner layer of the cuticle in nematodes.

**fibrous** Containing, consisting of, or resembling fibers (tough, string-like tissue).

**field capacity, field moisture capacity** The maximum amount of water that a soil can hold against the force of gravity.

**filiform** Thread- or hairlike; long and slender.

**fissure** A cleft or groove separating adjacent tissues, anatomical parts, or organs.

**fix** To harden and preserve specimens (e.g., nematodes and plant tissue), especially for permanent mounting and microscopic study.

**fixation** The preservation of biological structures for microscopic examination by killing organisms in a suitable chemical or physical condition so as to avoid changes in structure.

**fixative** A substance (generally, a mixture of chemicals) that hardens and preserves specimens.

**flange** A rib, rim, or expansion developed for strength, guidance, or attachment to a body part or organ; a basal expansion of the odontostylet in nematodes.

**flexure** A turn or fold.

**floccose** Cottony soft, like cotton flannel; downy; woolly; having small, soft, cottony or woolly tufts.

**flocculose** Having small tufts of soft woolly hair; delicately cottony.

**fluorescence microscope** A compound light microscope in which the image is formed by fluorescence emitted from the specimen when it is illuminated with ultraviolet light.

**flush** Sudden growth or development.

**foliar nematode** Any of various nematodes that attack leaves, particularly species of *Aphelenchoides*.

**form** (Latin *forma*) A subdivision of a species below the rank of variety; a biotype or race. The term is infrequently used in recent literature.

**-form** (suffix) Shaped.

**fossores** Modified structures of the head and dorsal and ventral lips of nematodes, fitted for "digging," as in *Diplogaster coronata*.

**framework** See **cephalic framework**.

**free-living nematode** A nematode not known to be parasitic on higher plants or animals.

**furcate** Forked; branching like a fork.

**fusiform** Spindle-shaped; slender and tapered at each end.

**fusoid** Somewhat fusiform; tapering toward each end.

**gall** An abnormal swelling or localized outgrowth (tumor) on leaves, stems, roots, or other plant parts, often more or less spherical or fusiform, composed of unorganized cells produced by the plant as a result of attack by a fungus, bacterium, nematode, insect, mite, or other agent.

**gamete** A differential sex cell or haploid nucleus (containing a single set of chromosomes, denoted $1n$) capable of fusion with a gamete of the opposite mating type in sexual reproduction to form a zygote.

**ganglion** (pl. **ganglia**) In nematodes, a group of nerve cell bodies (dorsal, lateral, amphidial, retrovascular, etc.) with conspicuous nuclei, forming a nerve center.

**gel** A jellylike colloidal mass; a coagulated colloid.
**gelatinous** Resembling gelatin or jelly; viscid.
**genera** Plural of **genus**.
**generation time** The length of time necessary for an organism to complete its life cycle.
**genital papillae, genital setae** Tactile or sensory organs located on the tail of male nematodes, in a preanal, postanal, or caudal position.
**genital primordium** The initial cells of the reproductive system.
**genital setae** See **genital papillae**.
**genital tract** The reproductive system. In nematodes, the principal components are the ovary, oviduct, spermatheca, uterus, vagina, and vulva in females and the testis, vas efferens, vas deferens, cloaca, and anus in males.
**genus** (pl. **genera**) A taxonomic category containing closely related (structurally or phylogenetically) species. The genus name is the first element in a Latin binomial or trinomial of a species or subspecies. Genus names are latinized, capitalized, and italicized. adj. **generic**. See **epithet**; **Latin binomial**.
**germinal zone** A region of the ovary or testis in which rapid division of germ cells occurs.
**giant cell** A large, multinucleate mass of protoplasm formed by abnormal cell enlargement in which nuclear division occurs but not cell division, in plants infected by certain nematodes (e.g., root-knot nematodes). See **nurse cell**; **syncytium**.
***Globodera*** A genus of round cyst nematodes, including *G. pallida,* the white or pale potato cyst nematode; *G. rostochiensis,* the golden potato cyst nematode; and *G. solanacearum* and *G. tabacum,* the tobacco cyst nematodes.
**gland** An organ or a group of cells that secrete substances.
**globate, globoid, globose, globular, globulose** Globe-shaped; spherical or nearly so.
**glottoid apparatus** A sclerotized, valvular structure at the base of the stoma in nematodes; the metastom of free-living nematodes, composed of highly modified metarhabdions (one dorsal and two subventral).
**gnotobiotic culture** A culture in which all the living components are known. See **axenic culture**.
**golden nematode** *Globodera rostochiensis,* also called the golden potato cyst nematode.
**gonad** A sex organ that produces gametes, such as the ovary (producing ova) and the testis (producing sperms).
**gonoduct** The oviduct, uterus, and spermatheca of female nematodes.
**gonopore** The exterior opening of the reproductive system in nematodes, namely, the vulva in females and the anus or cloacal opening in males.
**gravid** Containing an egg or eggs; capable of depositing eggs.
**growth zone** A region of the ovary or testis in which the oogonia or spermatozoa gradually increase in size.
**gubernaculum** The grooved, sclerotized accessory piece located posterior to the spicules of some male nematodes, functioning to guide the spicules toward the male gonopore.
**guide ring, guiding ring** The cuticularized, sleevelike structure that surrounds and guides the stylet of nematodes in certain genera in the superfamily Dorylaimoidea. Its position varies in different genera, from near the apex to the posterior portion of stylet.

**haemolymph** The watery, lymph-like nutritive fluid in the pseudocoelom of nematodes.
**haustrulum** The spheroidal cavity of the valvular apparatus of the esophagus in nematodes, which functions as a pump.
**head** The portion of the body anterior to the base of the stoma or stylet in nematodes.
**head cap** A distinctly offset head region.
**heavy soil** A fine-textured clayey soil. See **light soil**.
***Helicotylenchus*** A genus of spiral nematodes.
**helminthology** The branch of zoology dealing with the study of worms, especially parasitic worms (nematodes, in the broad sense).
**hemi-** (prefix) Half; in part.
***Hemicycliophora*** The genus of sheath nematodes.
**hemizonid** A small, lenslike structure or nerve commissure situated between the cuticle and the hypodermal layer in nematodes, on the ventral side of the body, just anterior to the excretory pore, and generally believed to be associated with the nervous system. See **commissure**.
**hemizonion** A structure associated with the hemizonid, usually smaller than the hemizonid and posterior to it.
**hermaphrodite** An individual bearing functional male and female reproductive organs.
**hetero-** (prefix) Other; different. See **homo-**.
***Heterodera*** The genus of lemon-shaped cyst nematodes, including *H. glycines,* the soybean cyst nematode; and *H. schachtii,* the sugar beet nematode.
**heterogeneous** Differing in structure, tissues, qualities, or nature. See **homogeneous**.
**heterogonic cycle** The life cycle of nematodes in which eggs of parasitic females develop into free-living nematodes, the offspring of which become parasitic; also called an indirect cycle. See **homogonic cycle**.
***Hirschmanniella*** The genus of rice root nematodes.
**histochemical** Pertaining to chemical reactions in tissues.
**hol-, holo-** (prefix) Whole, entire, all.
**hologonic** Producing germ cells over the entire length of the gonad. See **telogonic**.
**holomyarian** Pertaining to the arrangement of the somatic musculature of nematodes with only two chords or no chords, in which the musculature is divided into two fields or is continuous.
**holotype** A specimen (strain or isolate) designated or indicated as the type of a taxonomic group by the author of

the original description of the taxon at the time of publication.

**homo-** (prefix) One and the same; common; like. See **hetero-**.

**homogeneous** Similar in certain characteristics, such as chemical or physical properties. See **heterogeneous**.

**homogonic cycle** The life cycle of nematodes in which eggs of parasitic females hatch as free-living juveniles that develop directly into parasitic forms; also called a direct cycle. See **heterogonic cycle**.

**homonym** A taxonomic designation rejected because the same name has already been used in a different sense (e.g., with reference to a different type).

***Hoplolaimus*** The genus of lance nematodes.

**host** A living organism (e.g., a plant) harboring or invaded by a living parasite, from which the parasite obtains part or all of its nourishment, and in or on which the parasite can reproduce.

**host range** The range of hosts attacked by a given pathogen or parasite.

**host-specific** Attacking or having a pathogenic association with only certain species of hosts.

**hyaline** Transparent or nearly so; translucent; clear or colorless.

**hybrid** The offspring of two individuals of different genetic character (differing in one or more heritable characteristics); the offspring of a cross between two species resulting in viable progeny. v. **hybridize**.

**hydrogen ion concentration** See **pH**.

**hyper-, hypero-** (prefix) Above; beyond; over.

**hyperparasite** An organism parasitic on another parasite.

**hyperplasia** An abnormal increase in the number of cells in a tissue or organ; cell proliferation. See **hypertrophy**.

**hypersensitivity** Rapid death of cells at the point of attack by a pathogen (e.g., a nematode) resulting in a limitation on the spread of infection.

**hypertrophy** Abnormal enlargement of cells in a tissue or organ. See **hyperplasia**.

**hypo-** (prefix) Under; lower.

**hypodermal commissure** See **cephalid**.

**hypodermal glands** The glands formed by the hypodermis in nematodes (e.g., caudal, lateral hypodermal, amphidial, and phasmidial glands).

**hypodermis** The thin cellular layer of the body wall beneath the cuticle in nematodes; the epidermis, in some nematodes; also called the subcuticle or subcuticular layer. Hypodermis cells secrete the cuticle, and thickenings of the hypodermis form the dorsal, lateral, and ventral chords extending the length of the body.

**immune** Not affected by or responsive to disease; exempt from infection because of inherent properties (e.g., in plants, a rough outer wall, hairiness, characteristics of natural openings, a waxy coating, or a thick cuticle).

**immunity** The natural or acquired state of being immune, or exempt from infection. See **resistance**.

**in toto** In entirety.

**in vitro** (Latin, "in glass") Pertaining to biological reactions occurring in an artificial environment; in culture; outside a living host. See **in vivo**.

**in vivo** (Latin, "in a living thing") Pertaining to biological reactions occurring in a host; within a living organism, as in laboratory testing of disease agents. See **in vitro**.

**incidence** See **disease incidence**.

**incisure** A longitudinal cuticular cleft dividing the lateral fields in nematodes; also called an involution or lateral line. Incisures are usually separated by narrow refractive bands of cuticle.

**incubate** To hold in a steady environment to await development; to allow a microorganism to grow undisturbed under controlled conditions.

**incubation period, incubation stage** The period between infection of a host and the appearance of disease symptoms.

**indexing** A procedure for testing for the presence of a particular pathogen in susceptible plants.

**indicator plant** A plant that reacts distinctively to a certain pathogen or environmental factor, developing characteristic symptoms, and thus used in the detection and identification of the causal agent.

**infection** The process in which a pathogen enters (invades or penetrates) and establishes a parasitic (often pathogenic) relationship with a host. Once infection has been effected, colonization may begin. See **colonization**; **infestation**.

**infection court** A site in or on a host (e.g., in plants, a root, shoot, leaf, flower, or fruit) at which invasion by a pathogen can occur.

**infectious** Capable of causing infection; capable of spreading a pathogen from host to host.

**infective stage** The stage in the life cycle of a nematode in which it is capable of invading host tissue and establishing infection.

**infestation** Contamination, as with microorganisms, of soil of other substrates without infection or parasitism; the presence of a large number of pests (especially insects, mites, nematodes, weeds, or rodents) in soil, in a field, or on plant surfaces, containers, or tools; dispersal in soil or other substrates. Infestation is not to be confused with infection by microorganisms, which occurs only in living hosts. See **infection**.

**inflect** To curve or bend inward or toward the axis. adj. **inflexed**.

**injury** Momentary (transitory) damage caused by, e.g., the action of a chemical, physical, or electrical agent or an adverse environmental factor. See **disease, of plants**.

**innervation process** A neuromuscular process that extends to the motor nerve in muscle cells in nematodes.

**inoculate** [1] To artificially introduce a pathogen into an infection court in order to induce disease. [2] To introduce a microorganism into a culture medium (loosely, and incorrectly, used with reference to the transfer of living cells of a microorganism to any place in which they will grow and develop). n. **inoculation**.

**inoculation feeding period** The period during which a vector feeds on a test plant in a pathogen transmission experiment; the test feeding period.

**inoculation threshold period** The minimum feeding time needed for an inoculative vector to transmit a disease agent to a susceptible test plant.

**inoculum** (pl. **inocula**) A pathogen or its parts (e.g., fungal spores or mycelium, nematodes, or virus particles, etc.) used to inoculate a host and induce disease; the portion of a pathogen that is brought into contact with a host.

**inter-** Between; among; during.

**intercellular** Between cells.

**intersex** An individual more or less intermediate in phenotype between male and female, displaying secondary male or female characters, and functioning as one or the other or neither sex, but never as both.

**intestine** The simple, mostly straight, lumen-lined tube, composed of a single layer of epithelial cells, between the esophagus and the rectum in nematodes, in which food is digested and nutrients are absorbed; the gut. It is connected to the proctodeum, rectum, or cloaca by the intestino-rectal valve, which is surrounded by a sphincter muscle.

**intestino-rectal valve** The valve, controlled by a sphincter muscle, connecting the intestine to the proctodeum, rectum, or cloaca in nematodes.

**intra-** Inside; within.

**intracellular** Inside or through a cell or cells.

**introrse** Inward, toward the direction of the central axis of the body.

**intruded** Pushed or projecting forward.

**invaginated** Enclosed in a sheath.

**invasion** The growth or spread of a pathogen into a host and establishment of the pathogen within the host. v. **invade**.

**involution** See **incisure**.

**iso-** (prefix) Equal; the same.

**isolate** [1] (v.) To separate a culturable organism from an infected host and grow it in the absence of other organisms. [2] (n.) A specimen of an organism from a defined source. Collections of a pathogen made at different times.

**isthmus** The narrow portion of the esophagus between the metacorpus and the basal region or posterior bulb in nematodes.

**juvenile** [1] Animal form in which there is little dimorphism between immature and adult stages. [2] In nematodes, any of the (usually four) developmental stages between the embryo and the adult.

**knot** A localized abnormal swelling; a gall.

**Koch's postulates** Four rules devised by bacteriologist Robert Koch to establish proof of the pathogenicity of a microorganism: 1) consistent association of the suspected causal agent with a disease syndrome; 2) isolation of the agent and growth of it in pure culture; 3) reproduction of the disease syndrome in previously healthy hosts inoculated with the isolated agent; and 4) reisolation of the agent (identical to the inoculum) from the inoculated host.

**L** [1] Liter (the abbreviation is usually capitalized to avoid confusion with the numeral 1). [2] In de Man's measurement code, the mean length of the body of a nematode, in millimeters, usually expressed as a range.

**labia** Plural of **labium**.

**labial** Having lips; pertaining to the lips.

**labial disk** A raised, more or less circular portion of the cuticle of the most anterior annule of some nematodes, delimited posteriorly by the first transverse striation.

**labium** (pl. **labia**) Lip. adj. **labial**. See **lip**.

**lance nematode** Any member of the genus *Hoplolaimus*.

**larva** (pl. **larvae**) A postembryonic stage of animal development, markedly different in form from the adult stage. Larvae (e.g., caterpillars and tadpoles) undergo metamorphosis to enter the adult state. The term is inappropriately used to designate the juvenile stages of nematodes, which do not undergo metamorphosis.

**latency** See **latent infection**.

**latent** Present but not manifested or visible.

**latent infection** Infection without visual symptoms. See **masked infection**.

**latent period, latent phase** The period after acquisition of a pathogen (e.g., a virus) by a vector but before the vector becomes infectious (capable of transmitting the pathogen). A latent period usually follows acquisition in vectors capable of persistent transmission.

**lateral** At the side.

**lateral fields** The longitudinal thickenings of the cuticle covering the lateral hypodermal chords of nematodes. In some nematodes, the lateral fields are divided by longitudinal striae (incisures) or transverse markings (areolations).

**lateral guiding pieces** A pair of very small, linear, cuticularized pieces forming a structure lateral to the distal portion of the spicules and joining with the muscular sheath surrounding the anus, in dorylaims and some related nematodes.

**lateral line** See **incisure**.

**lateral pores** The openings of the lateral hypodermal glands in nematodes, forming minute pores in the cuticle.

**laterodorsal** Situated 45° from the dorsomedian line and perpendicular to the anteroposterior axis of the body; submedian.

**lateroventral** Situated 45° from the ventromedian line and perpendicular to the anteroposterior axis of the body; submedian; to the side (away from the midline of the body) and below.

**Latin binomial** The scientific name of a species, composed of two parts, the first designating the genus and the second (the epithet) designating the species. See **epithet**; **genus**; **species**.

**leaf gall nematode** Any member of the genus *Anguina* producing galls on leaves.

**leaf nematode** Any member of the genus *Aphelenchoides* attacking leaves.

**lectotype** A syntype selected and designated through publication to serve as the type of a taxonomic group after publication of the original description of the taxon.

**leptoderan** Characterizing caudal alae that do not meet posterior to the terminus (tip) of the tail.

**lesion** A well-marked, localized area of diseased or disorganized tissue; a wound.

**lesion nematode** Any member of the genus *Pratylenchus*. Species in this genus are more properly called root-lesion nematodes and in Europe are called meadow nematodes.

**life cycle, life history** The cyclical progression of stages in the growth and development of an organism, from a given stage of development in one generation to the same stage in the next generation. See **disease cycle**.

**light soil** A popular term for a coarse-textured sandy soil, which is thereby easy to till but prone to nematode problems. See **heavy soil**.

**lipo-** (prefix) Pertaining to fat.

**lip** Any of the cuticular structures surrounding the mouth opening in nematodes. Most nematodes have six lips (two subdorsal, two lateral, and two subventral), which may be fused in pairs.

**lobe** A generally rounded form; a rounded projection of a bodily organ.

***Longidorus*** The genus of needle nematodes.

**longitudinal alae, longitudinal ridges** Raised areas of the cuticle of nematodes, extending the length of the body on all sides, outside the lateral fields.

**longitudinal striae** Superficial lines in the cuticle of nematodes, extending the length of the body and separating the longitudinal ridges. See **stria**.

**lumen** (pl. **lumina**) [1] The central cavity of a cell or structure. [2] The canal or duct of the esophagus in nematodes.

**lysis** Breakdown or dissolution of cells.

**macerate** To separate the cells of a tissue, so that they drift apart or collapse; to soften by soaking or soft rot; to waste away. See **comminute**; **mince**; **triturate**.

**macro-** (prefix) Long; large; great.

**macroscopic** Visible to the unaided eye.

**manual transmission** The spread or introduction of inoculum into an infection court by hand manipulation. See **mechanical transmission**.

**manubrium** The enlarged proximal portion of the spicule in male nematodes.

**marginal muscle fibers** Muscle fibers of the esophagus in nematodes, located at the apices of the triradiate lumen and having a suspensory and skeletal function.

**masked infection** An infection in which symptoms are absent under certain conditions (e.g., in plant diseases, light, temperature, nutrition, or the age of the plant) but may appear under other conditions. See **latent infection**.

**matrix** [1] A substance in which some other object develops. [2] The gelatinous substance in which eggs of *Meloidogyne* species and some other nematodes are deposited, forming an egg mass. [3] The middle layer of the cuticle in nematodes.

**meadow nematode** Any member of the genus *Pratylenchus*. Species in this genus also called root-lesion nematodes.

**mechanical transmission** The spread or introduction of inoculum (e.g., a nematode, bacterium, or virus) into an infection court (especially a wound) by artificial means, accompanied by physical disruption of host tissue, or by natural means in the field, when leaves rub together or roots come in contact. See **manual transmission**.

**median** In the middle; lying along the midline of the body.

**median bulb** See **metacorpus**.

**medium** (pl. **media**) A substrate (either solid or liquid, organic or inorganic, defined or undefined) providing a base for growth and nutrition of organisms (e.g., plants or microbes).

***Meloidodera*** The genus of cystoid nematodes.

***Meloidogyne*** The genus of root-knot nematodes.

**mes-, meso-** (prefix) Middle.

**mesenteron** The intestine in nematodes.

**mesh size** The number of openings per linear inch (2.54 cm) in standard screens used to separate solid particles by size. The width of the openings is approximately 860 µm in a 20-mesh screen; 230 µm in a 60-mesh screen; 104 µm in a 150-mesh screen; 74 µm in a 200-mesh screen; 43 µm in a 325-mesh screen; and 25 µm in a 500-mesh screen.

**meson** The midline of the body or an imaginary plane dividing the body into right and left halves.

**mesorhabdion** See **rhabdion**.

**mesostom** See **protostom**.

**met-, meta-** (prefix) Changed in form or position; between, among; with; after.

**metabolism** The totality of the chemical processes (physiological activities) occurring in a living organism; the process by which cells or organisms utilize nutrients to build living matter and structural components or to break down cellular material to obtain energy and simple substances to perform special functions.

**metabolite** Any chemical component of a reaction series or process; a product of metabolism.

**metacorpus** The enlarged, usually ovate, posterior portion of the corpus (in the middle portion of the esophagus) in nematodes, frequently containing a valve; also called the median bulb or middle bulb.

**metarhabdion** See **rhabdion**.

**metastom** See **protostom**.

**metatype** A specimen compared with the holotype of a

taxonomic group and determined by the author of the original description of the taxon to be conspecific with the holotype.

**micro-** (prefix) One-millionth; small, minute. See **micrometer**.

**microbe** A microscopic organism composed of one to many cells and capable of reproduction by growth and division. See **microorganism**.

**micrometer** (abbr. **µm**) A unit of length equal to one-thousandth of a millimeter (1 µm = 0.001 mm); formerly called a micron.

**microorganism** A microscopic organism, or microbe. Microorganisms include all unicellular prokaryotes and eukaryotes and some multicellular eukaryotes, such as bacteria, fungi, nematodes, microscopic algae, and protozoa. Viruses are not microorganisms.

**micropyle** A minute opening in the membrane of a nematode egg, through which a spermatozoon enters. adj. **micropylar**.

**microscopic** Too small to be seen or described (characterized) without the aid of a microscope.

**middle bulb** See **metacorpus**.

**migratory ectoparasite** A nematode that feeds on a host from outside while remaining capable of free movement. See **ectoparasite**.

**migratory endoparasite** A nematode that enters a host and feeds within it, moving freely through host tissue, and never becoming sedentary. See **endoparasite**; **sedentary endoparasite**.

**millimeter** (abbr. **mm**) A unit of measurement equal to one-thousandth of a meter (1 mm = 0.001 m = 0.1 cm = 1,000 µm = 0.03937 in., or about 1/24 in.).

**mince** To cut into small pieces. See **comminute**; **macerate**; **triturate**.

**mm** See **millimeter**.

**molt** or **moult** The process of shedding or casting off a cuticle or body encasement (exoskeleton) in the transition from one developmental stage to the next.

**mono-** (prefix) One.

**monodelphic** Having one ovary or female genital tube.

**monoecious** Having both male and female sex organs in a single individual; hermaphroditic.

**monogenic** Producing offspring of only one sex.

**mononchoid** Resembling nematodes in the genus *Mononchus.*

**monorchic** Having one testis or male genital tube.

**monoxenic culture** A culture containing an organism growing in the presence of another organism.

**moribund** Dying.

**morphology** The study or science of the form, structure, and development of organisms.

**morphometrics** Body measurements.

**motile** Exhibiting or capable of independent movement from place to place.

**moult** See **molt**.

**mouth** See **stoma**.

**mucro** A short, stiff or sharp point or tip projecting abruptly from an organ, such as the projections on the tip of the tail of certain nematodes.

**mucronate** Bearing one or more mucros.

**mucus** Slime; the gelatinous matrix of a nematode.

**mult-, multi-** (prefix) Many; much; a great number.

**multinucleate** Having more than one nucleus per cell; polynucleate.

**mural tooth** A cutting organ attached to or derived from the pharyngeal wall but formed farther back in the esophagus of some non-plant-parasitic nematodes.

**mutant** An individual possessing one or more new, heritable characteristics as a result of mutation. See **strain**.

**mutation** An abrupt, heritable change in an individual as a result of an accidental alteration of a gene or chromosome or an increase in chromosome number; the process by which such changes occur.

**mycorrhiza** (pl. **mycorrhizae** or **mycorrhizas**) A symbiotic association between the mycelium of a fungus (typically nonpathogenic or weakly pathogenic) and the roots of a higher plant. Mycorrhizae may aid in the uptake of certain nutrients (primarily phosphorus) by the host plant. There are two main types, ectotrophic and endotrophic. adj. **mycorrhizal**. See **endomycorrhiza**.

**myofilaments** The thick and thin muscle elements of the contractile region of muscle cells in nematodes.

**myriocytous** Pertaining to nematodes having a very large number of cuboidal cells (over 8,224) in the intestinal tract.

**n.** Noun.

***Nacobbus*** The genus of false root-knot nematodes.

**neck** The portion of the body occupied by the esophagus in nematodes.

**necrosis** (pl. **necroses**) Localized or general death and disintegration of plant cells, tissues, or parts of an organism, usually resulting in browning or blackening of tissue due to the oxidation of phenolics. In plants, necrosis is commonly a symptom of desiccation, injury, or diseases caused by fungi, nematodes, viruses, or bacteria. adj. **necrotic**.

**necrotroph** A parasitic microorganism that obtains its energy from dead cells, typically host cells that it has killed. adj. **necrotrophic**. See **biotroph**.

**needle nematode** Any member of the genus *Longidorus.*

**nematode** Eelworm; an unsegmented, wormlike animal (phylum Nematoda or Nemata), a tubular roundworm with a cuticle and a hydrostatic skeleton, parasitic in or on plants or animals or free-living as a saprophyte in moist soil, decaying matter, or water, and abundant in many soils; also called a nema. Most nematodes are microscopic. Almost all plant-parasitic nematodes pierce plant cells with a stylet and suck juices. Nematodes transmit microorganisms and viruses to plants and cause wounds by which bacteria and fungi enter plants.

**nematode wool** A mass of desiccated juveniles of *Ditylenchus dipsaci,* formed in some plant tissues.

**nematology** The science or study of nematodes.

**nematophagous** Feeding on nematodes. Some fungi are nematophagous.

**neotylench** A common name for any nematode in the family Neotylenchidae.

**neotype** A specimen (strain or culture) selected as the type of a taxonomic group subsequent to the original description of the taxon, when the primary types are known to have been lost or destroyed.

**nepovirus** (abbr. **NEPO**) Any of the nematode-transmitted polyhedral viruses.

**nerve ring** The center of the nervous system in nematodes, encircling the esophagus (near the esophageal-intestinal junction, in most nematodes), and composed largely of nerve fibers and associated ganglia; also called the circumesophageal commissure.

**NETU** Any of the nematode-transmitted tubular (rod-shaped) viruses.

**nonhost resistance** Disease resistance in an organism (e.g., a plant) that either lacks something required by the pathogen or contains one or more substances incompatible with the pathogen.

**nonpathogenic** Incapable of causing disease.

**nurse cell** A modified plant parenchymal cell formed at the feeding site of certain plant-parasitic nematodes (e.g., *Tylenchulus* spp.) and required for continued development of the nematode. Nurse cells are of normal size but have thickened walls, dense cytoplasm, an enlarged nucleus, and no central vacuole. See **giant cell**; **syncytium**.

**o** In de Man's measurement code, the distance between the dorsal esophageal gland opening and the stylet knobs divided by the length of the stylet and multiplied by 100.

**O** In de Man's measurement code, the percentage of the body of a female nematode occupied by each ovary (the length of the ovary divided by the length of the body and multiplied by 100); the value is usually given as a superscript to the left or right (according to the gonadal arrangement) of the V value.

**obligate** Necessary; essential; obliged; able to live only under certain restricted conditions.

**obligate anaerobe** An organism able to live only in the absence of atmospheric oxygen. See **anaerobe**; **facultative anaerobe**.

**obligate parasite** A parasite that can grow and multiply in nature only on or in living tissue and cannot be cultured on an artificial medium. See **facultative parasite**; **parasite**.

**obligate saprobe, obligate saprophyte** An organism that can develop only on dead organic matter. See **facultative saprobe**; **saprobe**; **saprophyte**.

**ocellus** (pl. **ocelli**) An eyelike spot (pigment spot) in certain nematodes.

**ocular micrometer disk** A ruled glass plate in the eyepiece of a microscope, used for measuring microscopic objects.

**odontium** (pl. **odontia**) A labial tooth situated on the esophageal wall of the stoma in certain nematodes,which is formed in the esophagus and is moved forward during molting.

**odontophore** See **stylet extension**.

**odontostyle, odontostylet** See **stylet**.

**oligochaete** A segmented worm in the class Oligochaeta.

**oligocytous** Pertaining to nematodes having a small number of rectangular cells (up to 128) in the intestinal epithelium.

**onchiostyle, onchiostylet** See **stylet**.

**onchium** (pl. **onchia**) A pharyngeal tooth formed in the stoma in certain nematodes, as opposed to an odontium, which develops in the esophagus and is then moved forward during molting.

**oocyte** A cell giving rise to an egg; a female germ cell.

**operculum** (pl. **opercula** or **operculums**) The less sclerotized area at the pole of a hard-shelled egg of some animal parasites (including nematodes), allowing the escape of the juvenile.

**opisthodelphic** Having a uterus or uteri directed posteriorly; having a single ovary situated posterior to the vulva.

**oral aperture** The anterior entrance of the stoma in nematodes; the mouth; also called the buccal aperture, opening, or orifice.

**organ Z** A muscular structure located between the oviduct and the uterus in certain species of *Xiphinema.*

**organelle** A structure or body, usually delimited by a membrane, within a cell (e.g., a mitochondrion or Golgi apparatus).

**organism** A living plant, animal, or microbe.

**ornate** Adorned.

**ova** Plural of **ovum**.

**oval** Broadly elliptical to nearly circular; egg-shaped, with the larger end at the base.

**ovary** The female reproductive organ, which produces ova (eggs). See **testis**.

**ovejector** or **ovijector** The elongated, generally heavily muscularized structure that aids in the expulsion of eggs from the vagina in female nematodes.

**overseason** To live from one planting season to the next.

**oversummer** To survive the summer.

**overwinter** To survive the winter.

**oviduct** The short, usually tubular, thick-walled part of the female reproductive system between the ovary and spermatheca or uterus in nematodes.

**ovijector** See **ovejector**.

**oviparous** Producing eggs that hatch after expulsion from the body.

**ovoviviparous** Producing thin-shelled eggs that hatch within the uterus.

**ovum** (pl. **ova**) The reproductive cell of a female organism; an egg.

**papilla** (pl. **papillae**) A tactile sensory organ in nematodes. Papillae occur in various regions of the body, e.g., at labial, cephalic, and cervical sites.

**parasite** An organism (e.g., a bacterium, fungus, phytoplasma, nematode, insect, or mite), virus, or viroid living with, in, or on another living organism (a host), obtaining all or a portion of its food from the host, and in some cases providing a benefit in return, but more often causing disease in the host; a biotroph or necrotroph. See **facultative parasite**; **obligate parasite**; **pathogen**.

***Paratylenchus*** The genus of pin nematodes.

**paratype** A specimen designated or indicated as being other than the holotype by the author of the original description of the taxon.

**parthenogenesis** A type of asexual reproduction in which an egg (female gamete) develops into a new individual without fertilization by a sperm (male gamete); apomictic development of haploid cells. adj. **parthenogenetic, parthenogenic**. See **apomixis**.

**Pasteur pipette** A short length of glass tubing with one end drawn out into a sealed capillary and the other end plugged with cotton and sterilized. The tip of the capillary is broken for use.

**pasteurization** A process by which selected pathogenic microorganisms (but not all beneficial microorganisms) in soil or other media are killed by heating to a prescribed temperature for a specified time, without materially changing the natural characteristics of the substance treated.

**pathogen, pathogene** A microorganism or agent (e.g., a bacterium, fungus, nematode, phytoplasma, virus, or viroid) capable of causing disease in a particular host (suscept) or range of hosts. Most pathogens are parasites, but there are a few exceptions.

**pathogenesis** The sequence of events in disease development from the time of infection to the final reaction in the host; the production and development of disease.

**pathogenic** Causing or capable of producing disease.

**pathogenicity** The ability of a pathogen to cause (incite) disease; the state or condition of being pathogenic. See **virulence**.

**pathology** See **plant pathology**.

**pathotype** [1] A variant or subdivision of a pathogenic species. [2] The typical reaction induced by a given pathogen in a given host.

**ped** A unit of soil structure, such as a soil aggregate, crumb, prism, block, or granule. Peds are commonly present in uncleaned lots of seed and can contain plant pathogens.

**pellet** The material concentrated at the bottom of a centrifuge tube after centrifugation.

**peloderan** Characterizing caudal alae that meet posterior to the terminus (tip) of the tail.

**penetration** The initial invasion of (entrance into) a host by a pathogen. See **infection**.

**perennial** [1] (n.) A plant that grows indefinitely from year to year and usually produces seed each year. [2] (adj.) Lasting for years.

**perforate** (adj.) Pierced through.

**perineal pattern** The fine, wavy, fingerprint-like pattern formed by cuticular striae surrounding the vulva and anus, especially in mature females of *Meloidogyne* species, in which the pattern is more or less an identifying characteristic of the species.

**persistent transmission** A type of transmission of pathogens (especially viruses) by vectors that remain inoculative for a long time, often for the life span of the vector. Typically, the vector acquires the pathogen from an infected host only after a long acquisition period (measured in minutes, hours, or days, rather than seconds), and acquisition is usually followed by a latent period, during which the vector is unable to transmit the pathogen. Nematode vectors can often transmit viruses immediately after acquisition and lose the capacity to transmit when they molt. Nonmolting dagger and stubby-root nematodes retain the capacity to transmit longer than needle nematodes.

**pest** Any organism (e.g., a bacterium, fungus, insect, mite, nematode, or weed) injurious or detrimental to a beneficial plant, plant product, animal, or human.

**petri dish, petri plate** See **culture plate**.

**pH** A measure of acidity and alkalinity, on a logarithmic scale of 0 to 14; pH 7 indicates a neutral solution, containing the hydrogen ion ($H^+$) and the hydroxide ion ($OH^-$) in equal concentrations; pH $< 7$ indicates an acidic solution (containing a greater concentration of $H^+$); and pH $> 7$ indicates an alkaline solution (containing a greater concentration of $OH^-$). On the logarithmic scale, a difference of one unit represents a 10-fold change in the hydrogen ion concentration ($pH = -\log(H^+ \text{ concentration})$).

**-phage** (suffix) One that eats.

**pharyngeal bulb** A muscular swelling of the esophagus wall around the buccal capsule in nematodes.

**pharynx** The muscular esophagus in nematodes (according to some authors), containing radial muscles, pharyngeal glands, and valves that prevent the regurgitation of food; and connected to the intestine by the esphagointestinal wall. There are three major morphological types. adj. **pharyngeal**. See **esophagus**.

**phase-contrast microscope** A compound microscope with an annular diaphragm in the front focal plane of the substage condenser and a phase plate at the rear focal plane of the objective, to make visible the differences in phase or optical path in transparent or reflecting media.

**phasmid** A usually minute, porelike structure on the tail in nematodes in the class Secernentea, believed to serve a sensory function; also called a precaudal gland. Phasmids generally occur in pairs on opposite sides of the lateral fields.

**phylloplane** [1] The surface of leaves. See **rhizoplane**; **spermoplane**. [2] The nonparasitic flora of the leaf surface.

**physiologic race** A group of organisms constituting a subdivision of a pathogenic species, resembling other races or other members of the species in morphology, but differing from them in virulence, symptom expression, biochemical and physiological properties, or host range. See **biotype**.

**phytopathogenic** Capable of causing disease in plants. See **pathogenicity**.

**phytopathology** Plant pathology, the science or study of plant diseases, their nature and effects on plant hosts, causes, and control.

**phytosanitation** Any measures to remove or destroy infected plant material likely to be a source of inoculum. See **sanitation**.

**phytotoxic** Poisonous, injurious, or lethal to plants (usually used with reference to chemicals).

**pin nematode** Any member of the genus *Paratylenchus*.

**pl.** Plural.

**plant disease** See **disease, of plants**.

**plant health care** (abbr. **PHC**) A holistic approach to plant care focusing on the health and growth of plants and on proper care to prevent stress in plants.

**plant pathology** The science or study of diseases of plants, their nature and effects on plant hosts, causes, and control; phytopathology.

**plectoid** Resembling nematodes in the genus *Plectus*.

**pleomorphic** See **polymorphic**.

**plica** (pl. **plicae**) A highly convoluted fold in the epidermis in nematodes, over which new cuticle is formed during molting, thus permitting rapid growth after the molt.

**polycytous** Pertaining to nematodes having a large number of hexagonal cells (128–8,224) in the intestinal epithelium.

**polydelphic** Having three or more ovaries (gonads).

**polymorphic** Having different forms.

**polymyarian** Pertaining to the somatic musculature of nematodes with a large number of longitudinal rows of muscle cells between chords, best viewed in transverse body section.

**polynucleate** Having more than one nucleus per cell.

**polyphagous** Feeding on many kinds of food. n. **polyphagy**.

**postemergence** The period after the appearance of a plant (seedling) aboveground.

**posteriad** Directed backward, as opposed to anteriad.

**posterior bulb** See **basal bulb**.

**postuterine sac** A rudimentary or vestigial ovary in nematodes, which may serve to store spermatozoa; also called the postvulvar-uterine sac.

**postvulvar-uterine branch, postvulvar-uterine sac** See **postuterine sac**.

**potato cyst nematode** Either of two species of cyst nematodes, *Globodera pallida* (the white or pale potato cyst nematode) and *G. rostochiensis* (the golden potato cyst nematode).

**potato rot nematode** *Ditylenchus destructor.*

***Pratylenchus*** The genus of lesion (root-lesion) nematodes and meadow nematodes.

**preadult** The last juvenile stage before adulthood in nematodes; the fourth-stage juvenile.

**predator** An animal that feeds on other, usually smaller or less powerful animals (its prey), usually killing the prey quickly and eating most or all of it.

**predispose** To make prone to disease, infection, or stress.

**preemergence** The period before plant emergence.

**prerectum** A differentiated portion of the posterior end of the intestine of some nematodes in the order Dorylaimida, separated from the main intestine by a constriction in the lumen. The granules of its cells are smaller, less numerous, and much lighter in color than those of the intestine.

**preservative** A substance used for killing or stopping the growth of microorganisms in or on a substrate.

**primary inoculum** Propagules or vegetative structures of a pathogen, usually from an overwintering source, that cause initial rather than secondary outbreaks of disease.

**primary symptom** A symptom produced soon after infection, at the site of entry (also called a local symptom), in contrast to a secondary symptom, which follows more complete invasion (colonization) of the host. See **secondary symptom**.

**probola** (pl. **probolae**) A prominent, sometimes ornate, specialized cuticular structure on or near the lips of certain nematodes, often a modification of the labial and cephalic regions.

**procorpus** The cylindrical portion of the corpus, anterior to the metacorpus, in nematodes.

**prodelphic** Having a uterus anteriorly directed at the origin.

**progeny** The young or offspring of an organism.

**prorhabdion** See **rhabdion**.

**prostom** See **protostom**.

**protandry** A form of hermaphroditism (occurring in some nematodes) in which the gonad produces sperms that are stored for subsequent fertilization of eggs produced later by the same gonad. See **syngonic**.

**protodeum** The cuticle-lined invagination incorporating the rectum in female nematodes and the cloaca in males, into which the rectal glands open.

**protoplasm** The living material in cells, on which all vital functions of nutrition, secretion, growth, and reproduction depend; an essential semifluid, viscous, translucent colloid contained in all plant and animal cells.

**protoplast** The organized living unit of a cell, including the cytoplasmic membrane, cytoplasm, nucleus, and cytoplasmic organelles, but exclusive of the cell wall.

**protostom** The middle portion of the stoma in nematodes, between the cheilostom and the telostom. It is sometimes subdivided into the prostom, mesostom, and metastom (the anterior, medial, and posterior sections, respectively).

**protuberant** Bulging out, prominent.

**proximal** Toward the center of the body; toward the end

nearest the center of the body; next to or nearest the point of attachment or origin. See **distal**.

**proximate** Nearest.

**pseudo-** (prefix) False; spurious.

**pseudocoelom** The body cavity behind the nerve ring in nematodes, generally considered to lie between the somatic muscles and the alimentary tract. It contains a fluid that functions as a respiratory and circulatory system and in which internal organs are suspended.

**pseudocoelomic fluid** The watery, lymph-like, nutritive fluid in the pseudocoelom.

**pseudocoelomyctes** Fixed cells in the pseudocoelom.

**punctation** A minute dot, oval, pit, or depression in the deeper layers of the nematode cuticle. Punctations occur in transverse or longitudinal rows or in various patterns.

***Punctodera*** A genus of round cyst nematodes.

**pure culture** A culture in which only one species or biotype of an organism is growing.

**putrefaction** Anaerobic decomposition of organic substances, especially proteins, by microorganisms, producing disagreeable odors.

**pyriform** Pear-shaped.

**quadricolumella** See **crustaformeria**.

**quiescence** A rest period caused by external conditions unfavorable to germination or growth.

**quiescent** Dormant; quiet.

**race** A physiologic race; a genetically and often geographically distinct mating group within a species. See **physiologic race**; **strain**.

**rachis** A central strand or axial chord of nonnucleated tissue, to which oocytes may be attached, in the ovary in nematodes.

**radial muscle fibers** Muscle fibers acting to dilate the lumen of the esophagus in nematodes.

***Radopholus similis*** The burrowing nematode.

**range** The geographical region or regions in which a pathogen is known to occur. See **host range**.

**ray** Any of the muscular genital papillae (tactile organs) in the bursa of certain nematodes, primarily animal parasites.

**rectal glands** The three to six large glands in the region of the rectum of some nematodes in the class Secernentea. In *Meloidogyne* spp., they produce the gelatinous matrix in which eggs are deposited; their function in other nematodes is unknown.

**rectum** The narrow, dorsoventrally flattened tube lined with cuticle at the posterior end of the alimentary tract in female nematodes, separated from the intestine by a sphincter muscle; the hindgut.

**recurved** Curved backward, inward, or downward; curved in an opposite direction.

**red ring nematode** *Rhadinaphelenchus cocophilus,* also called the coconut red ring nematode.

**reflexed** Bent or folded back.

**relax** To inactivate nematodes, generally by gentle heat prior to fixing; to kill or anaesthetize.

**renette** A ventral cell or group of cells emptying through the excretory duct or ducts in nematodes in the class Adenophorea.

**reniform** Kidney-shaped.

**reniform nematode** *Rotylenchulus reniformis.*

**resistance** The ability of an organism to overcome, to some degree, or retard the activity of a pathogen (infection) or some other damaging factor. Plants may be slightly, moderately, or highly resistant to certain pathogens. See **immunity**; **susceptibility**; **tolerance**.

**respiration** A series of enzymatic reactions (oxidation processes) in which a living organism produces energy for cellular activities, usually by oxidizing carbohydrate and releasing carbon dioxide and water; the oxidation of food by plants and animals to yield energy for cellular activities.

**reticular, reticulate** Having netlike ridges or markings. n. **reticulation**.

**reticulum** (pl. **reticula**) A netlike or weblike structure; an anastomosing system of lines, veins, folds, and ridges.

**retroarcuate** Curved backward.

**retrorse** Backward-projecting; in a backward or downward direction, as opposed to antrorse.

**rhabdion** A plate in the cuticular lining of the stoma in nematodes. There are three major types:

- **cheilorhabdions,** forming the wall of the cheilostom (the most anterior part of the stoma, at its entrance)
- **protorhabdions,** forming the wall of the cylindrical stoma; they may be subdivided into pro-, meso-, and metarhabdions
- **telorhabdions,** forming the wall of the telostom (the most posterior portion of the stoma, connecting it with the esophagus)

**rhabditoid** Resembling nematodes in the genus *Rhabditis.* See **esophagus**.

**rhabditoid bursa** A bursa in which the caudal alae meet posteriorly and anteriorly, forming a complete oval; a wide bursa.

**rhizoplane** The surface of roots. See **phylloplane**.

**rhizosphere** The microenvironment of the soil immediately surrounding a living root (within 5 mm of a root). The microflora of the rhizosphere is frequently richer than and different from that of soil away from roots.

**rib** Any of the six rigid supporting structures (four sublateral, one dorsal, and one ventral) in the cephalic framework in nematodes, radiating transversely and then posteriorly to fuse with the anterior edge of the basal ring and six radial bars.

**rice root nematode** Any member of the genus *Hirschmanniella.*

**rice stem nematode** *Ditylenchus angustus.*

**ring nematode** Any member of the genus *Criconemella* or related genera.

**root devitalization** Cessation of root elongation caused by nematode feeding at or near the root tip, resulting in stubby-root, coarse-root, or curly tip symptoms.

**root knot** A plant disease characterized by the formation of galls on roots, caused by *Meloidogyne* species, which attack over 3,000 plant species. The disease is most common in sandy soils and in untreated greenhouse beds.

**root-knot nematode** Any member of the genus *Meloidogyne.*

**root-lesion nematode** Any member of the genus *Pratylenchus.* Species in this genus are also called lesion nematodes or meadow nematodes.

**rot** A state of decomposition and putrefaction; the softening, discoloration, and disintegration of living plant tissue by enzymes, usually produced by fungal or bacterial infection. Rots may be hard, dry, soft, mushy, wet, black, brown, white, etc. See **decay**.

***Rotylenchulus*** The genus of reniform nematodes, including *R. reniformis,* commonly called the reniform nematode.

***Rotylenchus*** A genus of spiral nematodes.

**roundworm** See **nematode**.

**rudimentary** Reduced in size; poorly developed; vestigial.

**ruga** (pl. **rugae**) A fold or wrinkle.

**rugose, rugous** Wrinkled; covered with coarse, netlike lines; roughened.

**S** In de Man's measurement code, the length of the stylet of a nematode divided by the diameter of the body at the base of the stylet.

**saccate, sacciform, saccular** Like a bag, sac, or pouch.

**sanitation** Destruction or removal of infected or infested plants or plant parts; elimination of inoculum and insect vectors; decontamination of tools, equipment, containers, work space, hands, etc.; cultural methods of disease control by reducing levels of inoculum. v. **sanitize**.

**saprobe** A saprophytic or saprozoic organism, feeding on dead organic matter, commonly causing its decay; also called a saprotroph. adj. **saprobic**. See **saprophyte**.

**saprogen** adj. **saprogenic, saprogenous**. See **saprophyte**.

**saprophile** adj. **saprophilous**. See **saprophyte**.

**saprophyte** An organism (e.g., a plant, fungus, or bacterium) that feeds on dead organic matter, commonly causing its decay; a necrotroph feeding on dead material that is not part of a living host; also called a saprogen or saprophile. adj. **saprophytic**. See **necrotroph**; **parasite**.

**saprotroph** See **saprobe**.

***Sarisodera*** A genus of cyst nematodes.

**scaffold** A very heavy, refractive guiding ring in nematodes.

**scanning electron microscope** (abbr. **SEM**) An instrument in which a scanning electron beam is used to produce images (photographs) of a specimen coated with a thin layer of gold, palladium, or aluminum, thus allowing observation of surface features of the specimen at high magnification. See **electron microscope**.

**sclerotized** Hardened (used with reference to hard, refractive structures in nematodes).

**scutellum** A greatly enlarged, shieldlike phasmid in *Scutellonema* species and other nematodes in the subfamily Hoplolaiminae.

**S-E system** The secretory-excretory system of nematodes, which exhibits a high degree of structural and functional diversity in different species.

**Secernentea** A taxonomic class in the phylum Nematoda (Nemata). Most plant-parasitic nematodes are members of this class.

**secondary organism, secondary pathogen** An organism that multiplies in tissue that is already infected by a different pathogen (the primary pathogen).

**secondary symptom** A symptom appearing after the initial symptoms of a disease, usually in virus infections. See **primary symptom**.

**secrete** To form and release a secretion.

**secretion** A substance produced by and released from a cell, gland, or tissue to perform a function in the organism outside the part in which it was produced.

**sedentary** Remaining in a fixed location; sessile; stationary.

**sedentary endoparasite** A parasitic nematode that enters a host, migrates to a feeding site, and becomes nonmotile. See **endoparasite**; **migratory endoparasite**.

**seed gall nematode** Any member of the genus *Anguina* inducing the production of galls that contain the nematode in infectious stages, in place of healthy seeds.

**seed infection** Colonization of seed tissues by a pathogen, with or without conspicuous symptoms.

**seed infestation** Contamination of seed with propagules of pathogens carried on the outer surface of the seed or dispersed in seed lots.

**seedborne pathogen** A pathogen associated with seeds and potentially capable of causing disease in seedlings or older plants. Numerous bacteria, fungi, nematodes, phytoplasmas, and viruses are seedborne.

**Seinhorst extraction technique** A method of extracting nematodes from plant material and soil, utilizing funnels, a constant supply of fresh water, and suitable settling basins, in an apparatus known as a Seinhorst elutriator or Seinhorst chamber

**self-fertile** Capable of sexual reproduction as a result of the fusion of eggs and sperms produced by a single individual. n. **self-fertility**.

**self-incompatible** Self-sterile.

**SEM** See **scanning electron microscope**.

**semi-** (prefix) Half; in part.

**semiendoparasite** A parasitic nematode that only partially embeds itself in host tissue. See **endoparasite**.

**semifenestra** (pl. **semifenestrae**) Either of the pair of openings in the vulval cone of *Heterodera* cysts with a vulval bridge, which separates the openings.

**seminal receptacle** The organ in which sperm cells are stored in the female reproductive system in nematodes.

**seminal vesicle** The enlarged tube- or pouchlike structure functioning as a temporary storage organ for sperm in the male reproductive system in nematodes.

**semipermeable** Differentially permeable, allowing some fluids or solutes to pass through but not others. Cell membranes are semipermeable.

**senesce** To decline or degenerate, as with maturation or physiological aging; to grow old. In plants, the process is often hastened by environmental stress, disease, or insect attack. n. **senescence**. adj. **senescent**.

**sensilla pouch** See **amphid**.

**sensillum** (pl. **sensilla**) Any of the nerve endings contained in the sensilla pouch in nematodes.

**sensitivity** The tendency to develop severe symptoms in response to the attack of a pathogen. adj. **sensitive**.

**seriate** Arranged in a series or connected order. See **biserial**; **uniserial**.

**serology** The study, detection, and identification of antigens, antibodies, and their reactions. Methods based on the specificity of the antigen-antibody reaction can be used to detect and identify antigenic substances (proteins) and the organisms that carry them.

**sessile** Permanently attached; not capable of moving around.

**seta** (pl. **setae**) An elongated cuticular structure articulating with the cuticle and functioning as a tactile sensory organ in nematodes. Setae are usually located around the oral opening and also occur at caudal, labial, and somatic sites.

**setose** Covered with setae or bristles.

**severity** See **disease severity**.

**sexual dimorphism** The state in which the males and females of a species are pronouncedly different in morphology.

**sexual reproduction** Reproduction by the union of two compatible nuclei or gametes, involving nuclear fusion and meiosis.

**shaft** The distal, main body of the spicule, usually curved, and often bearing a velum (a thin flange of cuticle). See **spicule**.

**sheaflike organ** The vaginal remnant in the vulval cone of cysts of *Heterodera*.

**sheath** A retained or extra cuticle in some species of nematodes.

**sheath nematode** Any member of the genus *Hemicycliophora*.

**sign** Visible evidence of the presence of a pathogen or its parts and products on a diseased host. See **symptom**.

**signature** See **disease signature**.

**sinuate, sinuous** Serpentine; crooked; wavy; flexuous.

**slime** A wet, generally sticky substance; mucus.

**soil texture** The relative percentages of sand, silt, and clay in a soil.

**soilborne** Transmitted in soil (used with reference to pathogens); living in soil (used with reference to microorganisms).

**somatic musculature** The single layer of more or less spindle-shaped cells attached to the hypodermis throughout the length of the body in nematodes, functioning in body movements.

**soybean cyst nematode** *Heterodera glycines*.

**sp.** (pl. **spp.**) Species (used only following a genus name). The singular abbreviation signifies an undetermined species; the plural abbreviation signifies two or more species without naming them individually. See **species**.

**spear** The stylet of a nematode.

**species** (pl. **species**) The taxonomic rank subordinate to genus but superior to subspecies, variety, race, and strain; usually the lowest level of classification of organisms; a group of closely related individuals of the same ancestry, resembling one another in certain inherited characteristics of structure and behavior and relative stability in nature. Individual members of a species ordinarily interbreed freely and maintain themselves and their characteristics in nature. The scientific name of a species is a Latin binomial, consisting of the name of its genus and a specific epithet, both of which are italicized. adj. **specific**. See **genus**; **Latin binomial**; **race**.

**species aggregate** A group of closely related species having more in common with each other than with other species in the same genus.

**specimen** An individual taken as a representative of a larger group; a portion or sample of material.

**sperm, sperm cell** A male sex cell or nucleus; a male gamete, typically motile; in nematodes, a spermatozoon.

**spermatheca** (pl. **spermathecae**) The enlarged portion of the female gonad between the oviduct and the uterus, functioning as a sperm storage organ, in some nematode species; also called the seminal receptacle. It is usually present in amphimictic species and absent in automictic species.

**spermatocyte** A cell giving rise to a sperm cell.

**spermatogonium** (pl. **spermatogonia**) A cell derived from the cap cell and giving rise to a spermatocyte in nematodes.

**spermatozoon** (pl. **spermatozoa**) The mature, functional sex cell (gamete) of a male animal.

**spermoplane** The surface of a seed. See **phylloplane**; **rhizoplane**.

**sphincter** A circular muscle that functions to open or close a natural duct or orifice.

**sphincter-Z** A constriction between the uterus and the spermatheca in some *Xiphinema* species.

**spicate, spiciform** Arranged in the form of a spike; having spikes.

**spicula** [1] (pl. **spiculae**) A spicule. [2] Plural of **spiculum**.

**spicular pouch** The pouch containing the spicules in nematodes; often called the spicular sheath, in parasitic nematodes.

**spiculate** Having a small, erect point.

**spicule** The copulatory organ of male nematodes. Spicules are heavily sclerotized structures occurring in pairs.

**spiculum** (pl. **spicula**) A structure formed of fused spicules.

**spikkles** Small, localized swellings caused by *Ditylenchus dipsaci* in leaves of *Narcissus* and several other plants.

**spinate** Having spines on the surface of the body, like nematodes in the genus *Criconema.*

**spinneret** The porelike opening of a caudal gland, situated in the tail, terminally or subterminally, in some nematodes in the class Adenophorea.

**spinulose** Delicately spiny; having little spines.

**spiral** Coil-like (used with reference to a type of amphidial opening in nematodes).

**spiral nematode** Any member of the genera *Helicotylenchus, Rotylenchus,* or *Scutellonema.*

**spp.** Plural abbr. of **species** (used only following a genus name to refer to two or more species without naming them individually). See **species**.

**spring crimp nematode** *Aphelenchoides fragariae,* also called the dwarf nematode.

**squamose, squamous** Covered with or consisting of scales.

**ssp.** See **subspecies**.

**stem and bulb nematode** *Ditylenchus dipsaci.*

**sterile** Infertile; free from living (reproducible) microorganisms; uncontaminated; nonsporing.

**sting nematode** Any member of the genus *Belonolaimus.*

**stock cultures** Cultures of known species of microorganisms maintained in the laboratory for various tests and studies.

**stoma** The cylindrical or subglobular structure connecting the oral opening with the anterior portion of the esophagus in nematodes, and either bearing teeth or armed with a protrusible spear or stylet; also called the buccal capsule.

**stomatostyle, stomatostylet** See **stylet**.

**stomodeum** The cuticle-lined anterior part of the alimentary tract and associated glands in nematodes, formed by an invagination of the body wall, and composed of the mouth, lips, stoma (buccal capsule), and esophagus or pharynx.

**strain** An organism or group of similar organisms differing in minor respects from other members of the same species or variety; the descendants of a single isolate in pure culture. See **biotype**; **form**; **isolate** [2]; **physiologic race**.

**stria** (pl. **striae**) A minute, superficial groove, furrow, projection, ridge, or cleft in the cuticle in nematodes, encircling the lips or body or extending longitudinally.

**striate** Marked with parallel or radiating striae (used with reference to the cuticle of some nematodes).

**striation** See **stria**.

**stricture** A contraction of the lumen in nematodes.

**stubby-root nematode** Any member of the genera *Paratrichodorus* or *Trichodorus.*

**stunt nematode** Any member of the genus *Tylenchorhynchus.* Species in this genus are also called stylet nematodes.

**stunted** Reduced in size and vigor as a result of disease, unfavorable environmental conditions, or abiotic agents.

**stylet** The usually long, pointed, stiff, slender, hollow, protrusible feeding organ in the mouth portion of parasitic nematodes, for piercing and withdrawing nutrients from plant cells or animal prey. There are two types:

**stomatostyle (stomatostylet),** a spear that apparently evolved as a fusion of the sclerotization of the stomatal walls. The anterior aperture is ventrally oblique. This type is common in nematodes in the order Tylenchida.

**odontostyle (odontostylet, onchiostyle, onchiostylet),** a spear evolved from a large tooth and derived prior to each molt from a single cell in the esophageal wall. The anterior aperture is broadly oblique. This type is typical of nematodes in the order Dorylaimida.

**stylet aperture** The anterior opening of the stylet in nematodes, located obliquely dorsally in the superfamily Dorylaimoidea and generally obliquely ventrally in the order Tylenchida.

**stylet extension** The sclerotized structure extending from the base of the stylet to the esophagus musculature in nematodes; also called the odontostyle extension or odontophore.

**stylet knobs and flanges** Protuberances (usually three) at the base of the stylet in nematodes, functioning as posterior points of attachment of the stylet muscles.

**stylet nematode** Any member of the genus *Tylenchorhynchus.* Species in this genus are also called stunt nematodes.

**subcrystalline layer** A superficial, waxy, translucent layer on cysts of some *Heterodera* species.

**subcuticle, subcuticular layer** See **hypodermis**.

**subdorsal** Situated 30° laterally from the dorsomedian axis and perpendicular to the anteroposterior axis of the body; dorsolateral.

**subglobose** Almost spherical but slightly elongated; subspherical.

**sublateral** Situated ventral to the lateral position and perpendicular to the anteroposterior axis of the body.

**submicroscopic** Too small to be seen with a light (compound) microscope.

**subsp.** See **subspecies**.

**subspecies** (abbr. **subsp.** or **ssp.**) A population within a species having more than one sharply defined character (usually a morphological character) and often possessing a distinct habitat.

**subspherical** Almost spherical but slightly elongated.

**substrate** [1] A surface or medium on or in which a microorganism grows or is attached and from which it may get its nourishment. [2] A substance acted upon by a chemical agent, such as an enzyme.

**subventral** Situated 30° laterally from the ventromedian

line and perpendicular to the anteroposterior axis of the body; ventrolateral.

**sugar beet nematode** *Heterodera schachtii.*

**summer dwarf nematode** *Aphelenchoides besseyi.*

**supplements** Genital papillae in some male nematodes, located preanally on the ventral side, derived from cuticle but in some cases provided with glands, and functioning during copulation; also called supplementary organs

**suscept** Any living organism attacked by or susceptible to a given disease, pathogen, or toxin; a susceptible individual or species. See **host**.

**susceptibility** The inability of an organism (e.g., a plant) to resist the effect of a given pathogen or damaging factor. See **resistance**; **tolerance**.

**susceptible** Not immune; lacking resistance or having poor resistance; prone to infection.

**suture** A seam or impressed line formed by the union of two adjacent inarticulate margins of body parts.

**swing-out rotor** A centrifuge rotor with buckets free to swing out horizontally when the rotor is spinning, so that the gravitational force is directed along the long axis of the centrifuge tubes.

**symbiosis** A close, usually obligatory association of two or more different species living together, not necessarily to their mutual benefit (often used exclusively for an association in which the partners benefit, which is more properly called mutualism).

**symmetrical** Of equal morphology about a given axis.

**symptom** Any indication of disease by the reaction of the host; the visible external or internal effects produced in or on a host (e.g., a plant) by the presence of a pathogen or disease agent. See **disease signature**; **sign**.

**syncytium** (pl. **syncytia**) A multinucleate mass of protoplasm surrounded by a common cell wall, resulting from continued nuclear division or cell wall breakdown and subsequent fusion of protoplasts. See **giant cell**; **nurse cell**.

**syndrome** The totality of effects produced in an organism by a given disease, whether all at once or successively; a pattern or sequence of disease development; the symptoms caused by and signs of one or more disease agents. See **disease signature**.

**syngonic** Pertaining to hermaphroditic reproduction (occurring in some nematodes) in which both sperm and eggs are produced by the same gonad. See **protandry**.

**synonym** In biology, an alternative name for an organism (species or group), especially a later or illegitimate name; a rejected scientific name, other than a homonym; a taxon regarded as identical to another. A synonym may be junior or senior, depending on the date of authorship.

**synoptic key** An artificial key to taxa, in which all characters of the taxa are listed, without ranking or sequence, and the taxa positive for each character are listed (usually identified by number); a specimen is identified by eliminating taxa with dissimilar characters until a single taxon is left.

**syntype** Any of the specimens (strains or cultures) of equal nomenclatural rank forming all or part of the material described by the author of the original description of the taxon, when the author did not designate or indicate a holotype.

**systemic** Throughout the whole body (used with reference to the distribution of a chemical or a pathogen in a host).

**T** In de Man's measurement code, the percentage of the body of a male nematode occupied by each gonad (the length of the gonad divided by length of the body and multiplied by 100).

**tail** The portion of the nematode body between the anus and the terminus.

**tardigrade** Any member of the Tardigrada, a subdivision of the phylum Arthropoda; commonly called a water bear.

**taxon** (pl. **taxa**) A taxonomic group of any rank, such as a family, genus, or species; a taxonomic unit in a systematic scheme.

**taxonomy** The science or "scientific art" dealing with the systematic description, naming, and classification of organisms based as far as possible on their natural relationships.

**telamon** A rigid, cuticularized thickening of the ventral cloacal wall of some male nematodes, apparently serving as an accessory guiding structure for the spicules from the spicular pouch into the cloaca.

**telogonic** Producing germ cells only at the distal end of the gonad. See **hologonic**.

**telorhabdion** See **rhabdion**.

**telostom** A short valve connecting the stoma to the esophagus in nematodes; the basal component of the stoma.

**terminus** The tip of the tail in nematodes.

**tessellate** Checkered; blocklike; patterned with longitudinal ridges broken by transverse striations, forming rows of squares (used with reference to describe the cuticle of some nematodes).

**tessellate stylet nematode** *Tylenchorhynchus claytoni.*

**test plant** A plant species, variety, or cultivar used for diagnosis or for evaluating the infectivity of a pathogen. See **indicator plant**.

**testis** (pl. **testes**) The male reproductive organ, producing sperm or spermatozoa. See **ovary**.

**tissue** A group of cells, usually of similar structure, performing the same or related functions.

**titilla** (pl. **titillae**) Any of the small projections on either side of the distal end of the gubernaculum in male nematodes.

**tobacco cyst nematode** Either of two species of cyst nematodes, *Globodera solanacearum* and *G. tabacum.*

**tobacco stunt nematode** *Tylenchorhynchus claytoni.*

**tolerance** The ability of a host (e.g., a plant) to maintain normal or nearly normal vigor, without excessive injury or loss of yield, when a pathogen is established in it. See **resistance**; **susceptibility**.

**topotype** A specimen collected from the same locality as the type of a taxonomic group.

**toxic** Of, composed of, or acting as a poison; capable of causing injury to a living organism.

**toxicant** A poisonous or toxic substance.

**toxicity** The relative capacity of a substance to produce injury or interfere adversely with the vital processes of a living organism; quality, state, or degree of poisonousness.

**toxin** A poisonous secretion produced by a living organism; a nonenzymatic metabolite produced by one organism and injurious to another.

**translucent** Clear enough to transmit light (allow light rays to pass through) but not transparent.

**transmission** The transfer or spread of a pathogen from host to host or from one host generation to another. Transmission may be horizontal (occurring by physical contact, vector, or pollen) or vertical (from mother to progeny, as in seed transmission).

**transverse** Crosswise; at right angles to the longitudinal axis.

**transverse striae** Superficial lines in the cuticle in nematodes, extending around the body and supporting the annules.

***Trichodorus*** The genus of stubby-root nematodes.

**trilaminar** Consisting of three thin, parallel layers.

**tripartite** Three-part; divided into three.

**triploblastic** Possessing three germ layers (in nematodes, the ectoderm, mesoderm, and endoderm).

**triradiate** Having three radiating arms or branches (used with reference to the lumen of the esophagus in nematodes).

**triturate** To grind or crush into very small particles. See **comminute**; **macerate**; **mince**.

**tropism** Movement, growth, or curvature due to an external, one-sided stimulus that determines the direction of movement.

**truncate** Ending abruptly as though the end were cut off squarely.

**tubercle** A small, wart-like process.

**tulip root** A disease of oats caused by *Ditylenchus dipsaci* and the characteristic symptom of the disease.

**turbid** Cloudy; not clear.

**tylench** A common name for any nematode in the order Tylenchida or the superfamily Tylenchoidea.

**tylenchoid** Having the anatomical characteristics of members of the major group of plant-parasitic nematodes with a stomatostylet (straight with three basal knobs), an esophagus with a median bulb, two ovaries, and an annulated cuticle. See **esophagus**.

***Tylenchorhynchus*** The genus of stunt or stylet nematodes.

***Tylenchulus semipenetrans*** The citrus nematode.

**type** The specimen or specimens on which the name of a taxonomic group (taxon) is based, serving as the criterion of the characteristics of the taxon.

**type species** The species designated as the type of its genus. It does not need to be the species most typical of the genus.

**type specimen, type strain** The single specimen designated as the type of its species or of a subspecies group.

**typotype** The specimen used to prepare an illustration of the type of a taxon.

**underbridge** A structure extending across the vulval cone, below and parallel to the vulval bridge in cysts of some *Heterodera* species.

**undulate** Having a wavy (not flat) surface or margin; wavy up and down, not in and out.

**unilateral** Of, on, or affecting one side only.

**uninucleate** Having one nucleus per cell.

**uniserial, uniseriate** Arranged in single series or row. See **seriate**.

**uterus** (pl. **uteri**) The portion of the female reproductive system between the oviduct and the vagina in nematodes; a structurally modified portion of the oviduct in which eggs may develop.

**V** In de Man's measurement code, the position of the vulva expressed as a percentage of the length of the body of a nematode (the distance from the lips to the vulva divided by the length of the body and multiplied by 100).

**v.** Verb.

**vacuole** A relatively clear, bubble-like, membrane-lined space in the cytoplasm of a plant cell, filled with a watery solution of sugars and other plant products and byproducts.

**vagina** The cuticle-lined canal that connects the uterus or uteri with the gonopore or vulva in female nematodes. See **vulva**.

**vaginal remnant** The remainder of the vagina in the vulval cone of *Heterodera* cysts.

**valve** In nematodes, a sclerotized structure that regulates the rate or direction of intake of materials (e.g., the esophago-intestinal valve, or cardia).

**var.** Variety.

**variability** The tendency of an organism to change its characteristics from one generation to the next as the result of mutation or adaptation to environmental forces.

**variety, botanical** A group of closely related plants of common origin in a species, differing from other varieties in distinct morphological characters, such as form, color, flower, or fruit; a subgroup of a species given a Latin name according to the rules of the International Code of Botanical Nomenclature, the name being written in italics. A taxonomic variety is known by the first validly published name applied to it, so that nomenclature tends to be stable. See **cultivar**.

**variety, cultivated** See **cultivar**.

**vas deferens** (pl. **vasa deferentia**) The slender, tubelike gonoduct in male nematodes, united posteriorly with the rectum to form the cloaca, and leading from the vas efferens to the seminal vesicle and the ejaculatory duct. The vas deferens may be muscular or glandular or both.

**vas efferens** (pl. **vasa efferentia**) The distal, specialized tubular part of the seminal vesicle in male nematodes, connecting the growth zone of the testis and the seminal vesicle proper.

**vector** A living organism (e.g., an insect, mite, nematode, parasitic plant, or human) able to acquire a pathogen (e.g., a bacterium, fungus, nematode, phytoplasma, or virus) and transmit it from one host to another.

**velum** (pl. **vela**) A delicate, flange-like extension of the cuticle (cuticular membrane) on the ventroanterior side of the spicules of male nematodes.

**ventral** On, near, or toward the front or lower surface of the body of an organism or one of its parts. See **dorsal**.

**ventral supplements** Secretory structures located ventrally in the posterior portion of male nematodes.

**ventriculus** A glandular region of the esophagus in nematodes.

**ventrolateral** Situated 30° laterally from the ventromedian line and perpendicular to the anteroposterior axis of the body; subventral.

**ventromedian line** The true middle line on the lower side of the nematode body.

**vermiform** Long and slender; worm-shaped; having a wormlike motion.

**vernier** A graduated scale used to measure fractional parts of the smallest divisions of an adjacent larger scale. Verniers are used on mechanical stages of microscopes to measure relatively large microorganisms.

**vestibule** The narrow anterior portion of the stoma in nematodes, acting as the stylet guide in stylet-bearing nematodes.

**vestige** A trace or visible sign left by something vanished or lost; a minute remaining amount; a small and degenerate or imperfectly developed body part or organ, the remains of one that was more fully developed in an earlier stage of the individual, in a past generation, or in closely related forms. adj. **vestigial**.

**virulence** The degree or measure of pathogenicity of a pathogen; the relative ability of pathogens to cause disease. See **pathogenicity**.

**virulent** Highly pathogenic; able to cause severe disease; able to produce all the typical symptoms of a disease in a susceptible host.

**viruliferous** Containing or carrying a virus (used particularly with reference to virus-laden insect and nematode vectors capable of transmitting viruses).

**viscid, viscose** Sticky, gluey, slimy, mucilaginous, glutinous, lubricious, viscous, clammy, half-liquid. See **gelatinous**.

**viviparous** Bearing living young.

**vulva** The external opening (gonopore) of the female reproductive system in nematodes, in the form of a transverse slit or circular pore on the ventral side.

**vulval bridge, vulvular bridge** A narrow connection (persisting vulvar slit) across the fenestra of the vulval cone, forming two semifenestrae, in cysts of some *Heterodera* species.

**vulval cone, vulvular cone** The conspicuous elevation at the posterior end of *Heterodera* cysts.

**warts** Cuticular crests arranged in single or double longitudinal rows on one side of the body of some free-living nematodes.

**water soaking** A plant disease symptom by which plants or plant parts appear wet, dark, and usually sunken and translucent.

**waterlogged** Lacking soil aeration as a result of excess water and poor drainage.

**wheat nematode** *Anguina tritici.*

**white tip** A disease of rice, caused by *Aphelenchoides besseyi.*

**wilt** [1] Lack of rigidity and turgor or drooping of leaves, stems, and flowers due to lack of water (inadequate water supply or excessive transpiration). [2] A vascular disease of plants, caused especially by certain species of fungi and bacteria, interrupting the uptake of water by the plant and disturbing the normal distribution of water in the plant.

**wing** A longitudinal projection of the cuticle in certain nematodes. Wings are usually lateral, but some nematodes have numerous wings evenly spaced around the body.

**wound** In plants, an injury in which the surface is cut, scraped, torn, punctured, or otherwise broken.

**wound parasite** A pathogen that can enter a host only through a wound.

**xer-, xero-** (prefix) Dry; drought.

***Xiphinema*** The genus of dagger nematodes.

**yellowing** As distinct from chlorosis, a process in which green tissue turns yellow as a result of the underdevelopment or partial destruction of chlorophyll. See **chlorosis**.

**Z-organ** See **organ Z**.

**zygote** A diploid cell (containing a double set of chromosomes, denoted 2*n*) or fertilized egg resulting from the union of two haploid gametes (isogametes or heterogametes); a cell in which two nuclei of opposite sex have undergone fusion; the beginning of a new organism as a result of sexual reproduction.

# Index